BUILDING CONSTRUCTION FOR THE Second Edition FIRE SERVICE

BUILDING CONSTRUCTION FOR THE Second Edition FIRE SERVICE

Francis L. Brannigan, SFPE
Professor (retired)
Fire Science Curriculum
Montgomery College
Rockville, Maryland

**National Fire Protection
Association**
Batterymarch Park, Quincy, Massachusetts

First Edition, 1971

NFPA No.: FSP-33A
Standard Book Number: 87765-227-9
Library of Congress Card Number: 78-178805
Printed in USA

Production Editor: Debra Matson
Art Coordinator: Carmen Johnson

Seventh Printing, December 1990

TABLE OF CONTENTS

ABOUT THE AUTHOR

Francis L. Brannigan has had a lifetime of varied professional fire protection experience. During World War II he directed a naval fire fighting school, commanded a seagoing fire fighting unit, and served as a district chief in the unique Army-Navy-Pancanal fire protection organization. He remained with the Navy after the war to help develop a competent career fire service for the Naval Shore Establishment.

He served for many years as the Public Safety Liaison Officer of the federal Atomic Energy Commission. He developed the Chain Reaction Training Program for fire officers in the correct handling of radiation accidents, and the Fire Loss Management program for the protection of life and property from fire.

At Montgomery College, Rockville, MD, he developed a model Fire Science Program, assembling an outstanding adjunct staff, each a nationally recognized expert in his field. He is currently a member of the adjunct staff of the National Fire Academy, Emmitsburg, MD and The Fire and Rescue Institute, University of Maryland, College Park, MD. In association with his wife, Maurine, he has assembled extensive collections of slides on all aspects of building construction and fire loss management. Jointly they conduct seminars across the country.

He was honored by the Society of Fire Protection Engineers with full membership despite the fact that his degree was not taken in engineering. He served for many years on National Fire Protection Association technical committees. He received the Fire Angel Award from the Cleveland fire fighters and the Training Officers Conference Award.

The Chesapeake Chapter of the International Society of Fire Service Instructors founded the Francis L. Brannigan Instructor of the Year Award in his honor.

FOREWORD

Francis L. "Frank" Brannigan is a man with a mission. He believes the terrible phrase "structural collapse" can be stricken from the records as a cause of fire service deaths. That is not to say that buildings will not collapse under attack from fire or that design errors of omission and commission will not cause failures. Rather, Frank believes that even if those things are "givens," something can be done about it. And that "something" is the reason for this book, presented now in its second edition after more than a decade of instructing the fire service in the dangers of building collapse, through the pages of the first edition.

Through years of fire ground experience and observation of the effect of fire on structures, Frank has concluded, and rightfully so, that there is no mystery surrounding what appears to be sudden and inexplicable building failures that often, on first glance, seem unrelated to the fire itself. If you just look close enough *before* fire occurs and from that knowledge apply the right tactics during the fire, the risk to fire fighters from the building itself can virtually be eliminated is Frank's credo.

In this book, Frank draws upon his long practical experience to explain in laymen's terms the hazards of various types of building construction encountered. This is the fire fighter's view — not the designer's or construction engineer's. He discusses in detail the many obvious and not so obvious dangers inherent in fighting building fires. He traces the paths fire can take in spreading throughout buildings of various designs and materials. He discusses at length the things a good observer should notice in anticipating what will happen to structural stability under fire conditions. "Know your buildings!" is the message, and Frank is convinced that if every fire officer and fire fighter took the time to study the peculiarities of the constructions in their response areas, their chances of not becoming fatal statistics are immeasurably improved.

This is Frank Brannigan's book and the opinions and recommendations expressed in the text are, of course, the sole responsibility of the author and do not necessarily reflect the views and recommendations of the National Fire Protection Association. The text has not been acted upon or reviewed by any of the Technical Committees of the NFPA. It is being published by the NFPA as a valued addition to the literature of fire suppression in buildings and is dedicated to preservation of life in the most hazardous of occupations.

Gordon P. McKinnon
Editor-in-Chief, NFPA

AUTHOR'S PREFACE

The first edition of this book was published a little over ten years ago. It filled a serious gap in fire protection literature and went through ten printings. It is widely used as a text in fire science courses and in fire service training programs. It also has been found useful by persons in a variety of related fields, such as architecture, building management, and building inspection.

Since publication of the first edition, we have had the opportunity to crisscross the country from Florida to Alaska, from Maine to California, conducting seminars on the dangers to fire fighters from weakness and faults in different types of building construction. And in the course of our travels we have had occasion to photograph hundreds of buildings under construction, repair, and demolition; in use; and, occasionally, afire. A view from that activity is that there have been many significant changes in building design and construction in the last few years and most of them are detrimental to the fire service.

The book, in large measure, has been rewritten to reflect the changes that have occurred. What has not been entirely rewritten has been updated and rearranged to assist the reader's comprehension of the text material.

Chapter 1, Introduction, has been entirely rewritten. It introduces the reader to the concept of fire loss management and stresses the need for an organized information recording and retrieval system on the construction features, both good and bad, of the buildings in a response area.

Chapter 2, Construction Principles, was strengthened and clarified throughout. A distinction is drawn between fire load and rate of heat release. Additional emphasis was placed on trusses and hazards of connections between building components.

Chapter 3, Wood Construction, takes into account the explosive development of the use of wood trusses and I beams as substitutes for sawn beams.

Chapter 4, Ordinary Construction, has been reorganized to group exterior and interior collapse together, and to include the most important material from a chapter in the first edition on mill construction.

Chapter 5 is a new chapter on garden apartments based on the author's ten years as a consultant on this subject to the National Bureau of Standards.

Chapter 6, Principles of Fire Resistance, has been updated and now includes a discussion of the path by which fire resistance required by law actually takes effect.

Chapter 7, Steel Construction, has been rearranged to make more clear the difference between unprotected steel construction and construction where the steel is protected. It also presents the author's opinion that many of the 165 deaths in the Beverly Hills Supper Club Fire in 1977 were due to fumes from a typical metal roof deck fire.

Chapter 8, Concrete Construction, has been rewritten to include a number of significant building collapses which have occurred in recent years.

Chapter 9 is a new chapter covering flame spread and fire growth.

Chapter 10 is a new chapter covering smoke movement in buildings.

Chapter 11 is a new chapter covering the structural conditions which can make one high-rise fire very different from another. In addition, it focuses attention on multi-floored fire resistive buildings, many of them built as "fireproof" a century or more ago.

Finally, the author would appreciate correspondence from those who wish to offer corrections and comments or suggestions for improvements so that he may benefit from their constructive interests.

Scientists Cliffs
Port Republic, MD
August 1982

ACKNOWLEDGMENTS

The author faces the blank page in the typewriter quite alone. But there are many who help; some like those among the thousands who attended the author's slide seminars on building construction, and who raised questions and suggested illustrations but who cannot be identified, should not go unmentioned.

The author's associates in the fire service, particularly Battalion Chief William Grimes of the New York City Fire Department, and Chief Charles Monzillo of the Willimantic (CT) Fire Department; and the staff of the Operational Risk Analysis Program of the National Fire Academy all made valuable contributions. Wayne Powell and Anne Currier of the Academy staff were especially helpful.

Harold Nelson and Daniel Gross of the National Bureau of Standards were the usual mine of accurate technical information. Richard Gewain of the American Iron and Steel Institute and Henry Collins of Underwriters Laboratories Inc. provided needed assistance.

During the preparation of this text, the author was under contract to the National Fire Academy to develop educational field programs to instruct the fire service in recognizing hazards of building constructions. His associate was Ms. Susan Hills, senior associate of IMR, who was also on the project. Her constant incisive probing questions were of immeasurable assistance in sharpening this text.

A note of thanks goes to Stanley T. Dingman, consulting editor to the NFPA, who demanded tightness and organization where it was badly needed and provided the pressure which was needed to keep the author to the task.

It is customary for the author to thank his family, if they haven't left him, and in this case it is special. Assistant Professor Vincent Brannigan, JD, made significant legal contributions. John Brannigan of the Library of Congress and Eileen Brannigan Woodworth of the Enoch Pratt Library, in Baltimore, contributed research. Anne Brannigan Kelly provided architectural information, Christopher Brannigan did sketches, and Mary Ellen Brannigan Schattman provided information on fire resistive buildings in Texas.

My wife, Maurine Brannigan, is my lecture program collaborator. She participates in all our seminars, censors my ramblings for clarity, and, in general, acts as ombudsman for the students.

Many of the illustrations are from the author's personal collection.

Cover design: Miriam Recio   Technical illustration: Redhouse Associates

CHAPTER 1

Introduction

The purpose of this book is to provide the fire ground commander and fire suppression forces with as much practical information as possible about the various types and methods of building construction and their influence on fire tactics and fire fighter safety. The comments and observations have been distilled from the author's experience and from that of others who have first-hand knowledge of the hazards to be found and the tactics to be used in buildings attacked by fire.

The book can also be of benefit to fire inspectors, building officials, fire protection engineers, architects, and builders. It can bring them insights into fire prevention and fire protection which they would not obtain in their usual work. However, as it has been said, the primary focus of this book is on fire ground problems.

In this day of collective management, committee decision, shared responsibilities, and buck-passing, it is refreshing to reflect that there is one post which embodies the sentiment expressed by the Spanish poet Ortega:

> Bullfight critics ranked in rows
> Crowd the enormous plaza full
> But there is only one who *knows*,
> He's the man who *fights* the bull.

That post is held by the fire ground commander. He's the person on the spot when the fire occurs in the middle of the night while those who had a share in creating the problem (possibly even negligently or criminally), the owner, the tenant, the code writer, the architect, the building official, are resting for the night or eternally.

The fire ground commander is well aware that the fire recognizes no authority. It is the ultimate anarchist. It can be overcome only by superior force properly directed. It is unforgiving of mistakes. *

---

*The author spent a good part of his professional life in authoritarian military and civilian organizations. It comes as a shock to powerful executives to be told, "You can disregard this recommendation. You may even fire me, but you are not in charge of the fire. It will do its own thing."

The Fire Problems

The fire problems, as seen by the fire suppression forces, stem from the relationships between buildings and fires. These are not easy to define because they vary from building to building and from fire to fire. They vary, to some extent, because of the multiplicity of viewpoints about buildings and fire.

Aesop tells us of the blind men who went out to "see" an elephant by feel. One thought it was a tree trunk, another a snake, a third, a big stone wall, and so on. No one was able to comprehend the whole elephant.

When it comes to fire protection, there are not only those who can't see, but also those who won't see. The latter are probably those who contribute most to the fire problem. Among their widely held points of view are these:

"The building is fireproof, so what's the problem?"

"It meets the code."

"The insurance company approved it. It must be OK."

"You are depriving the citizens of economical construction."

"We can't do anything about the errors already in existence."

"It's not your job to tell us how to build buildings. Your job is to fight fires."

The people who build buildings are not primarily concerned with the likelihood of fire or the problems that will be met by fire suppression forces. Buildings are designed, built, altered, enlarged, and rearranged, not so much with thought to their fire resistance and life safety, as with thought to their appearance, usefulness, or economy, and certainly with almost no thought at all to the convenience and safety of those who must fight fires on or within them.

The fire ground commander must be aware of these attitudes and the problems they bring if he is to make a successful attack without injury to his fire fighters.

Gravity and Fire

To understand the problems of buildings attacked by fire, consider both gravity and fire and what they do to buildings.

In 2000 B.C., Hammurabi, the King of Babylon, promulgated his famous edict which, in effect, said that if a building collapsed and killed the owner, the architect (builder) would be put to death. This put the fear of collapse into the entire building profession, architects, designers, builders, and officials. As a result, there have been relatively few building collapses over the centuries and not much has been written about those that have happened. The serious student should read Feld's *Construction Failure*[1] and McKaig's *Building Failures*[2] for a real understanding of building collapses apart from fire.

We in the fire service are often puzzled as to why serious fire hazard situations, which seem self-evident to us, are ignored by others. The problem lies in the difference between fire and gravity as assaults on buildings.

Gravity is the eternal enemy of any structure. Twenty-four hours a day, every day, the force of gravity is trying to pull the building down. By some system or other, the builder puts together an aggregate of components which defeats the law of gravity and stands up. The instant the system fails, the law of gravity acts immediately.

The probability of gravity is "one," that is to say, a certainty. If a building is not designed to withstand gravity, it will fall.

The probability of fire has no certainty. It is whatever a person judges it to be. Buildings erected without any thought to fire prevention or fire resistance have stood for years because they have not been attacked by fire. On the other hand, some so-called fire resistive buildings have been severely damaged if not destroyed by fire. Lacking our detailed knowledge of fire behavior, the designer, the owner, and the building official judge the probability of fire differently than we in the fire service.

The difference in the ways people look at the results of fire and gravity is apparent in the aftermath of the collapse of the tension-supported walkways in the Kansas City Hyatt Regency Hotel in 1981. Immediately, there was a cry for sounder structural design. Had the collapse been the result of fire, the attitude probably would have been, "Let's not get too hasty with this fire protection stuff. We must consider the economics of building a high-rise building," or "If we make the code so tough that collapse like this is impossible, investors won't build in our city."[*]

In 1970, The Smithsonian Museum of American History (then History and Technology), suffered a serious fire. The Star-Spangled Banner and the National Stamp and Coin Collection were saved from destruction by a fire door. This door had been bitterly opposed by a museum executive. The author needled him about this during a post-fire investigation. His reply was most interesting: "What I have learned from these sessions is that if you build a building to last 75 years, it will have a serious fire. Therefore, limiting a fire is a design criterion just like a snow load on the roof."

If the management of enterprises who hire architects and builders can understand this, the fire problem is on the road to being solved. That happy day is far in the future, but in the meantime, understanding the nature of the problem may help to develop a solution.

Most of us learned that Isaac Newton discovered gravity when an apple fell on his head. Perhaps this story is the reason for the inordinate preoccupation with gravity as a force that causes unsecured objects to be drawn to the center of the earth.

But gravity has a consequence other than possible collapse of a building structurally weakened by fire. It is almost completely ignored as the force which makes the heated toxic products of combustion rise as it pulls down the heavier, colder air.

[*] In an article in the *Wall Street Journal* of Feb. 12, 1982, H. Klein and H. Lancaster draw attention to serious and even fatal flaws in recent construction. They lay much of the blame to the soaring costs of construction and high interest rates which pressure architects to use the most economical materials. The pressure of "fast track" construction breeds mistakes. They point out that lighter construction reduces safety margins so that if a mistake is made, it will have more serious consequences than heretofore.

All around the world, even in the most primitive locales, high-rise buildings are successfully erected, resisting the law of gravity as expressed in collapse, even when attacked by fire. But, almost universally, it is only where a strong code, strictly enforced, is in effect, that the fact that gravity will force the lighter toxic products of combustion up through the building is taken into consideration.

If the architectural profession took the problem of the distribution of toxic combustion products to humans who can't breath them successfully, as seriously as it does collapse, many of our fire problems would disappear. Designers are acutely conscious of the effects of wind and gravity on a structure and consider adequate control of these forces as the essence of their art. As for fire, they do little except what is required by outside interests, such as the state or an insurer. These requirements may not consider all aspects of the problem.

Scope and Organization of This Text

Basic to the discussion of how structures react to assault by fire and gravity is an understanding of the principles that govern structures. These principles are presented in Chapter 2, Principles of Construction.

Chapters 3, 4, and 5 are interrelated. Chapter 3, Wood Construction, discusses buildings in which all the structural loads are carried on wood. Chapter 4, Ordinary Construction, discusses combustible buildings in which exterior loads are carried on masonry and interior loads mostly on wood. This is the so-called ordinary construction. Buildings of these two types represent the majority of buildings in which fires occur.

Chapter 5, Garden Apartments, looks at the problems of combustible low-rise multiple dwellings, usually called garden apartments, which house millions of people. It is the author's considered opinion that there is probably no area in which intelligent fire fighting can make so great a difference as in this type of structure. (In some areas, these are incorrectly called condominiums. Buying a condominium is a way to buy a building, not to build it. There are condominium high-rise apartments, condominium office buildings, and condominium marinas.)

The buildings mentioned so far share one vital characteristic. They are not designed, or required, to be resistant to collapse in a fire. It is a fact that some types of combustible buildings, as built, happen to achieve some degree of resistance to collapse sufficient to permit fire fighters to operate on or under burning structures. Newer construction methods, alterations, or features desired by the owner, may eliminate this accidental fire resistance. We in the fire service must protect ourselves with our knowledge of the ways in which different types of buildings can fail and with specific information about the particular building on fire.

Chapter 6, Fire Resistance, discusses the methods of testing building elements designed to resist fire and the administrative procedures by which these elements are incorporated into codes and practices.

Chapter 7, Steel Construction, deals with the fire characteristics of steel and discusses buildings in which steel carries the principal structural loads.

Concrete buildings present unique problems for fire fighters, both when the buildings are under construction and after they have been completed. The characteristics of such buildings and their fire problems are discussed in Chapter 8, Concrete Construction.

The spread of flame and fire growth are not specific to any structural type of building. They are almost universal and the material in Chapter 9, Flame Spread, is applicable to any building.

Smoke movement also is a phenomenon independent of the structural nature of the building and is treated as a unit in Chapter 10, Smoke and Fire Containment.

Chapter 11, High-Rise Construction, deals with the fire differences of various types of high-rise buildings, pointing out that "fire resistive" and "high-rise" are not synonymous and that many "fireproof" buildings built a century or more ago are still in use.

This whole subject of building construction and its relationship to fire problems is not as straightforward as arithmetic. Rather, it is much like a merry-go-round. The same elements and the same problems may be found at any point. This means there is some duplication of material in the various chapters. This is unavoidable, but not entirely without merit because the repetition may help to fix the material in the reader's mind.

Fire Loss Management

In an attempt to organize the thinking of industrial executives and scientists about the fire problem, and to focus their attention on cost-effective solutions,

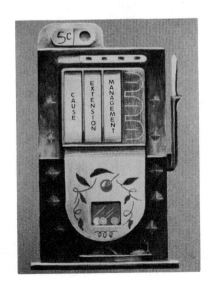

The author's "Fire Slot Machine" designed to focus attention on fire loss management problems and solutions.

the author developed the "fire slot machine" concept of fire loss management.

The casino slot machine has three wheels which revolve independently, each carrying various symbols. When three like symbols line up on jackpot, the money payoff occurs.

The three wheels of the "fire slot machine" are the "cause" of the fire, the reason for "fire extension," and what "management" has done or failed to do. When all three line up at the same time, the fire jackpot is DISASTER.

The cause of the fire can be friendly (necessary and useful) sources or unfriendly (arson, lightning, spontaneous ignition, and similar) sources. However, the cause of a fire is of no concern here as this text is concerned with what happens after the fire has started.

Extension, which does concern us, is influenced by:

- Fire load and rate of heat release.

- Surface spread and fire growth.

- Structural resistance to fire.

- Containment of fire.

Management is concerned with two elements: who manages the fire problem (owner, designer, tenant, insurer, government: federal, state, or local?), and how a particular fire is controlled, from discovery through alarm, evacuation, suppression and recovery.

Just as the casino slot machine can be "gimmicked" down to the point of no return at all, so can the fire slot machine, and this is the key to fire loss management — find the least expensive "gimmick" which prevents the fire slot machine from showing "disaster."

The clay flue liner is an example. It permits us to have a "cause" (a raging fire) and the means of "extension" (a combustible building) in proximity. The "gimmick" (the liner) is a function of "management" which prevents the cause and extension wheels from showing up at the same time.

The factors influencing the extension of fire will be discussed at appropriate places in the various chapters of this book. Here, attention will be focused on one of the elements involved in the "who" of fire loss management, the local government.

Local Government Functions

Local governments have four primary fire loss management functions:

- Building code.

- Fire prevention code.

- Water supply.

- Fire suppression.

All but water supply will be discussed. Not all local governments exercise all of these functions fully, but the thesis holds generally.

Building Codes

Building codes are concerned principally with buildings that are to be built. Once a building is approved and a certificate of occupancy is issued, the building may never again come to the attention of the building department. In some few cases, routine inspections to study building stability are made where there has been a history of failures, as in southern Florida reinforced concrete construction.

Building codes are concerned with a number of features of a building. The principal ones of interest here are structural stability and fire containment.

Structural Stability

All buildings, under normal conditions, are required to have structural stability. With respect to fire, buildings of more than a certain size are required, by most codes, to be "fire resistive," that is, within certain limits, to resist collapse in case of fire. If a building is not required by law to be fire resistive, the designer is not required to give any consideration to its potential collapse in a fire. Some non-fire resistive elements, such as sawn wooden beams, have historically demonstrated certain resistance to collapse. If the builder chooses to substitute elements of literally no fire resistance, such as wooden I beams, which will carry the normal load, the codes do not object.

Fire Containment

To insure containment of the fire to the building of origin, typical code provisions include:

- Regulation of wood shingles.

- Requirements for masonry exterior walls (fire limits).

- Distance between buildings, or alternates such as fire resistive exterior walls, rolling shutters, exterior sprinklers, etc.

For limitation of the fire to specified areas within the building, typical provisions include:

- Gypsum board sheathing in combustible buildings.

- Fire resistive floors.

- Fire walls to subdivide floor areas.

- Closure devices on openings in barriers, such as fire doors, enclosed stairways with self-closing doors, etc.

To insure safe egress of the occupants, typical provisions include:

- Outside fire escapes.

- Fire tower stairways.

- Enclosed interior stairways.

- Exitways calculated in accordance with the number of permitted occupants.

- Limited flame spread in corridors.

Making it safer to fight the fire is rarely a primary consideration. Some code writers have told fire suppression officers, "That's your job, not ours."*

It is apparent that the basic thrust of building codes is to get the occupants out, confine the fire to some area supposedly manageable by the fire department, and let the fire department extinguish the fire manually.

Building codes essentially provided "static" fire protection — protection which functioned by being in place.** More recently, there has been a shift to requirements for "dynamic" fire protection — for systems which react to the fire and perform some fire control or fire suppression function.

A typical example of dynamic fire protection is automatic sprinkler protection. Such protection may be required for life safety, as is exemplified by the requirement for sprinklers in high-rise factory buildings in New York City.

---

*A few minor requirements come to mind. In New York City, radio aerials must be 8 feet above the roof. Years ago, a fire fighter was cut severely in the neck by the 100 feet of copper wire the old crystal set required. Windows on a shaftway must be marked "shaftway," and where basement stairs have been removed, a sign "chute — no stairs" is posted.

**The author always has had trouble with classifying a fire wall as "static" fire protection when it requires the "dynamic" closing of a fire door to make the wall effective, but let's not be picky.

Sprinkler protection may be encouraged by granting "trade-offs" of other code provisions. Typically, NFPA *101®* , *Life Safety Code®* allows fewer exits and greater travel distance to exits if the building is sprinklered. *

Recently, there has been a tendency to require sprinklers on the very logical basis that local fire suppression forces cannot muster the resources demonstrated as being necessary to control a fire in an unsprinklered high-rise building.

The U.S. Fire Administration may be the victim of budget cutting but it has made a lasting contribution to the control of fire in the United States by sponsoring the development of special sprinklers for residential occupancies. As these are perfected, it will be possible for codes to require them and eventually the loss of life from fire which occurs in dwellings will be controlled.

Deficiencies of Building Codes

Even the best building codes can have technical deficiencies. Provisions are handed down from code to code, often without any valid basis. Others are based on the influence exerted on code-making authorities by proponents of one material or another. Building code provisions are voted on in meetings at various levels and then by the appropriate political body. All these votes are subject to the political process in all its ramifications. It is not inaccurate to state that building codes are political rather than technical documents.

It is disturbing to hear a fire officer say, "This building can't have any problems. It was built to the latest code."

Retrospective Code Provisions

The building code as written usually covers only buildings yet to be built. Public outcry after a disaster often provides the opportunity for fire departments to push for retrospective improvements to existing buildings.

This often is met with loud cries of unconstitutionality. Some municipal attorneys, not up on the law, may agree and the proposed legislation is killed.

The U.S. Fire Administration asked Assistant Professor Vincent Brannigan, J.D., of the University of Maryland, to research this matter. He found no appellate case where a retrospective firesafety code was found to be unconstitutional. The full report was published in "Record of Appellate Courts on Retrospective Fire Codes," *Fire Journal*, November 1981.

The following excerpt from the article neatly sums up the entire problem:

> In 1917, Judge Gavegan of the New York Supreme Court (a trial court) was faced with the issue of the constitutionality of the

---

* If the building is sprinklered for life safety, then it is the author's contention that the building should be evacuated if the sprinklers are out of service. This vital point will be repeated several times in this text.

retrospective laws adopted by New York City after the Triangle Shirt-waist Company fire. In this case, he was confronted with every legal and economic issue available both at that time and today. The building owners complained that the statute violated due process of law and equal protection, and was retroactive... The Court dismissed the constitutional challenges to the retropective law...

Judge Gavegan showed a special sensitivity to fire-safety problems.

"The state can enact remedial laws only after it has been demonstrated that there are evils to be remedied. Actual experience, even human sacrifice, is often required, in order to arrest the attention of the public and shock it into a realization of the necessity for remedial laws. Experience in the matter of fire danger shows that so-called fireproof structures sometimes burn; that automatic fire alarms sometimes fail to alarm; that inside fire apparatus sometimes proves ineffective; that smoke from inflammable* contents of fireproof buildings causes suffocation and death; that elevators sometimes refuse to work; that outside fire escapes are more serviceable as a means of entrance for firemen than as a means of escape for occupants; that the protection of the best fire department in the world is sometimes inadequate, and that persons of the highest intelligence often lose their presence of mind and become helpless victims of unreasoning panic. All the aforementioned precautions are good, but they do not go far enough. The people of the state of New York are no longer satisfied with measures that are merely good in matters involving life and death. They insist, within reasonable limits, upon the best measures of precaution that human and legislative foresight can devise, to the end that a recurrence of such tragedies as the Triangle Shirtwaist Company fire may in the future be prevented."

This language was adopted by the appellate court and the judgment was unanimously affirmed by the New York Court of Appeals.[3] Due to a citation error, this passage has never appeared in another case.

Fire Prevention Code

Without being too precise, a fire prevention code can be described as being concerned with how a building is operated as distinct from the building code which is concerned with how it is built.

A fire prevention code is enforced by an organization which may be a part of the fire department, or entirely separate from it. In any event, in many locations there is a distinct and tragic split between fire prevention and fire suppression.

---

*Note that today the word "flammable" is preferred because of the generally negative meaning of the prefix "in."

It may be that part of the problem lies in the very name. Only a small part of what a fire prevention division does is actually fire prevention. Most of its work is concerned with minimizing the consequences of a fire that occurs. Exits don't prevent fires. They make it possible for people to flee. Sprinkler systems do not prevent fires. They are simply a technically superior way to remove the heat from a fire, the same function as is performed by the hand hose stream, and so on. More important than the possibly erroneous name is the fact that many in the fire suppression force interpret that, "Fire prevention is their business, and thus, not ours."

One of the serious and pervasive fire hazards that faces all of us today is that exits, constructed at great cost to meet codes laboriously developed, are often not available because they are locked or blocked by the management of the premises. The author once proposed to a chief of fire prevention that all hands be enlisted in a project to correct this condition. The author further proposed that fire fighters be equipped with card forms to report blocked exits that they noticed whenever in any public building, off duty. The idea was rejected with horror. "That would generate so much work we wouldn't get anything else done." In another case, a federal official remonstrated with the management of a store in which the exits were locked. The local fire marshal was incensed and threatened to have him arrested for impersonating a fire inspector.

The fire suppression forces should be more actively involved in many things that are usually handled by fire prevention forces. In more and more buildings, built-in fire protection is shifting from static to dynamic systems that react to the fire. All these systems perform one or another fire suppression function, with the expectation that the systems will perform faster, better, and more safely than fire fighters. These systems are just as important to the fire ground commander as are the pumps, ladders, hose, and SCBA that the force brings to the scene, and the efficiency of these systems is surely within the fire ground commander's purview.

Too often the fire suppression forces are found in happy ignorance, while whatever the fire department knows about the special systems is locked up in "plan review" in the fire marshal's office.

To the author's mind, the ideal organization would place the district commander in charge of all the fire protection problems in the district. To back up the commander technically, a force of specialists out of headquarters would provide special expertise. It is ridiculous for an inspector to drive by several fire stations to investigate "a fire prevention matter" that could easily be handled by the local station.

What Codes Do Not Cover

It is important to understand what the building code or the fire code does not cover, particularly to help clear up misconceptions held by people we are encouraging to take a course of action.

It is not part of the municipality's legal function to assure that organizations take the necessary steps to stay in business. That is the problem of the management of the business. The building code may produce a fire resistive building but that is no guarantee that a particular business will survive the fire or the fire fighting effort.

The author's college employer installed a complete computer system. A cursory look disclosed several fire deficiencies. An offer to provide a competent consultant without charge was rejected, "The fire marshal and the building inspector okayed it." The computer manager was incredulous when told that the fire marshal's only interest was that the staff get out alive, but this produced no change of attitude.

The United States Government lost, by fire, the service records of thousands, if not millions, of service men and women in a fire resistive records repository. Press reports told of one ex-soldier being dragged off to jail by MPs for being AWOL (which he wasn't) because his records had been destroyed.

Fire Suppression

The thrust of this text, however, is not code enforcement, but safe and efficient fire suppression. Most fires occur in buildings. Fighting an advanced fire in a building surely ranks with the world's most hazardous occupations. It would seem to be obvious that the fire combat forces should know as much as possible about the building before the fire. Unfortunately, in many cases, the building might as well have just landed from outer space.

Prefire planning is only rudimentary in most fire departments. The ultimate objective should be to have available to the fire ground commander every scrap of useful information at the time of the fire. We cannot depend on the memory of Captain Truehart. Information meticulously filed in the fire prevention office might as well not exist.

The information should be systematically organized in a form which is usable on the fire ground. It may be in plastic-covered pages in a loose-leaf book or recorded in a record and retrieval system using tape or microfiche. The unit should be mounted in a command vehicle that responds to alarms.

As computers develop, it may be possible to store the vital information in a central computer and transmit it by radio to a display unit at the scene, with the ability to produce hard copy. A back-up system, independent of the computer, should be available because of the many problems computers can develop.

The specific record and retrieval system is much less important than that the job be started. It can be a monumental job, but the work can be spread over the available time and the recording be started now. There are many items of information that should be in such a system. Only some of them are within the purview of this text.

Suggested Guide to Building Prefire Plan Information

The following material is intended to provide a guide to some of the informa-

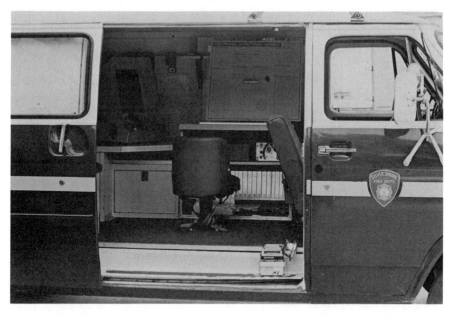

View inside the command unit of the Silver Spring Fire Department. All the information the fire department has on target buildings is on the tape reels. The fire ground commander can call up information on a particular building at the flick of a switch. Photo courtesy Silver Spring, MD Fire Department.

tion on the structural aspects of a building that fire departments should have at hand. A typical downtown "Main Street, USA" building, discussed in Chapter 4, Ordinary Construction, is being considered. Some suggestions for fire resistive buildings are found at the end of Chapter 11, High-Rise Construction.

In no way is this offered as an all-inclusive check list. It suggests items which are discussed in the text but which are not usually found in such fire department preplans and building surveys as have come to hand.

The items that are important to a particular fire department are those that exist in buildings in which that fire department might fight a fire. There are undoubtedly many hazardous conditions that are not described in this text, though a real effort has been made to at least mention every hazardous condition noted in looking at literally thousands of buildings. The attitude of the surveyor must not be that of filling out a check list. Imagination is vital. The system should be designed for updating and inserting additional material. You will learn that when you find a previously unrecognized condition in one structure, it will be necessary to return to other similar buildings and check them.

The information must be specific. In some plans, a building is described as a brick building. There are many types of buildings, each radically different, that can present a brick exterior. It is necessary to know more than what such a building "looks" like. One must determine what the actual construction is. What is behind the brick? What keeps the brick in place? The building may be of wood, steel, concrete, or composite construction. These are not distinctions without differences. The reaction to fire and the potential for injury vary greatly from type to type. A study of the following chapters will reveal the many ways brick may be used in building construction.

Similarly, there is no such thing as a "stucco" building. Stucco is a finish that can be applied over wood, concrete block, brick, terra cotta tile, or reinforced concrete construction.

The same may be said for any building. Its outside appearance may give no clue to its actual construction and may lead to erroneous, if not fatal, assumptions as to how it will behave under attack by fire. Get to know individual buildings. This book is designed to show you how.

Defects and potential hazards that may come into being during the fire should be noted. Where appropriate, specific warning instructions should be entered in the plan. This is particularly true for buildings that are so hazardous that they should be handled as defensive operations initially.

Some hazards and defects to be aware of are the types and condition of walls; roofs, roof supports, and coverings; interior supports; floor supports, flooring, and finishes; void spaces and vertical openings; ceilings; alterations; fire load; and sprinkler protection.

Different lists could be developed for each principal type of building. Note again that only items pertinent to the subject of this text are mentioned here. There are many others that should be instantly available to the fire ground commander.

The author would be pleased to hear from any fire department that develops an information system along the lines suggested and would be happy to assist in getting it made available to the fire service generally.

References

[1]McKaig, T. H., *Building Failures*, McGraw-Hill, New York, 1962.
[2]Feld, J., *Construction Failure*, Wiley, New York, 1968.
[3]Cockroft vs Mitchell 130 NE 921 (1921).

CHAPTER 2

Principles of Construction

This is the hardest chapter in this book for the student to learn. For many instructors it is the toughest to teach. As a result, it is sometimes passed over lightly. This is a great mistake. There is a whole new vocabulary to learn, and in a few cases, some definitions and concepts to "unlearn." This takes work, real mental effort, and understanding, not just reading. The author seeks diligently the way to make complex concepts understandable, but he cannot do the student's or the instructor's jobs for them. The learning process is often difficult, but the alternative, staying ignorant, is terribly costly. There are those who dismiss this chapter's subject as "theory," nice to know, but not necessary. The material is "theory," — not in the sense of "speculation" which is one meaning, but rather in the true sense of being "the general principles of a body of facts."

Why Study Building Construction?

There are two good reasons for learning the principles and language of building construction. This text is directed without apology to the fire combat officer. Fortunately, the "iron curtain," which has separated the misnamed "fire prevention" division from the fire suppression forces, has disintegrated in at least some fire departments. All officers, therefore, need an accurate knowledge of building terms. If a fire officer makes glaring errors in terminology, such as referring to "I beam columns," it will be very difficult for a professional in the building field to give credence to recommendations, no matter how excellent. The most important reason for knowing building terms is safety. There are too many instances of fatalities which have occurred because fire officers had not been trained to recognize construction elements and the conditions under which they are likely to fail. By far, most fires are fought in buildings, and the study of fire tactics should properly start with the study of buildings.

As an example, the fire chief of a major city told the author of his experience. There had been a fatal collapse of a *bowstring* trussed garage roof. All units were then told to locate and note trussed roofs in their area. The question, "What's a truss?" was asked. "You know — a big hump up in the center."* At a fire in a

---

* Some fire service writing shows the common error which philosophers call "arguing from the particular to the general." Get a logic text from the library and study this error to avoid it.

restaurant, the chief inside, seeing a large clear area asked the officer on the roof, "Do we have a truss?" Seeing no hump, the officer replied, "Negative." The roof collapsed "without warning" and there were several fatalities. Now they know about *parallel chord* trusses.

Definitions of Loads

Specific terms are used to describe the different loads placed on a building. It is important to understand and use them correctly. The various terms defined, however, are not mutually exclusive. For example, a load may be a live load and an impact load at the same time.

Dead Loads

Dead load is the weight of the building itself and any equipment permanently attached or built in. It does not, as is often supposed, have anything to do with whether the load is alive, as a person, or inanimate. *

In years gone by, deadweight was often piled on the building without regard for the consequences, or even in the mistaken belief that strength was being added. Of the fires which led to the development of present standards of fire resistance, none has been more important than the Parker Building in New York City, early in the 20th century. One of the many defects that were discovered in the investigation of the fire which destroyed that building was that as much as 30 inches of cinders had been added to the top floor to level it, thus making the dead load exceed the design live load.

The modern designer understands that deadweight breeds deadweight. Every effort is made to lighten all parts of a building. Removal of a pound of deadweight at the top of a building enables the builder to shave ounces at many points in the supporting structure. Buildings can be considered as being bought "by the pound."

Unfortunately, fire resistance is closely related to mass. All other things being equal, a heavier steel beam will take longer than a lighter beam to reach its yield point due to heating. A 4-inch by 10-inch wooden beam will support its load while burning longer than will a 2-inch by 10-inch beam. From this can be derived a valid general principle: *Any structural element which is lighter than the element previously used to carry an equivalent load is inherently less fire resistant.*

Added Dead Loads

Dead load is often added during alteration. An example is the addition of air

---

*The British use the term "self-weight," a more accurate term. Older buildings particularly should be surveyed for substantial added dead loads, and this should be taken into account in fire operations.

conditioners to the roof of a building previously without air conditioning. These are often added without any strengthening of the structure. At a fire in a large city, fire fighters were overhauling a fire in a restaurant, seeking out hidden pockets of fire in the overhead. They attacked the fire-weakened supports of the added air-conditioning units and the units fell, causing several fatalities.

A grocery supermarket in Maryland was converted to a Japanese restaurant, the type with many grills. Each grill required a heavy fume hood. These hoods were hung from the roof. When the building was a supermarket, the roof needed only to keep out the rain. Now the same roof must carry the hoods. The factor of safety has been greatly reduced.

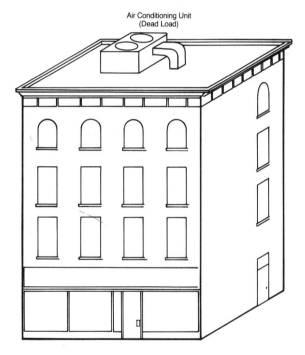

Air Conditioning Unit
(Dead Load)

Modernization of a building by adding such equipment as an air-conditioning unit to the roof, often adds dead load, thus, absorbing some of the designed factor of safety and the building loses some of its resistance to collapse in a fire.

Live Loads

Live loads are any loads other than dead loads. Dead loads are determinate, that is, they can be calculated as accurately as is necessary. Live loads are indeterminate. The live load must be estimated based on the projected use of the building, or, as in the case of snow, wind, or rain, on meteorological records.

The designer designs the building with a given use in mind. If there is a building code it specifies the floor loads for specific types of buildings. Typical building code minimum requirements as shown in the Building Officials Conference of America (BOCA) Code are:

	pounds per square foot
Classroom (fixed seats)	40
Classroom (movable seats)	100
Classroom corridors	100
Libraries (reading room)	60
Libraries (stacks)	150
Dwelling units	40
Stores (upper floors)	125

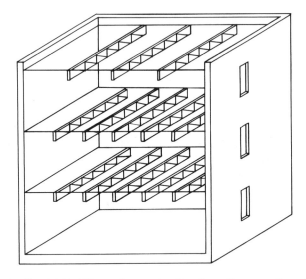

Examine a steel bar-joist building under construction. Very often you will see that the roof is obviously designed to carry a lesser load than the floors.

When the use of a building changes, the design of the building should be reviewed to determine whether the structure is adequate for the new use. This is often not done. A typical case is the conversion of a residence into a commercial structure. Often nothing is done to strengthen the structure for the heavier loads.

The fire and the fire fighting efforts can change or increase the loading of a structure. At the same time, the ability of the structure to resist gravity is being attacked by the fire. Unless a building is designed and built to be fire resistive, there is no provision in the design of the structure to prevent failure due to fire. The vast majority of structures built are not fire resistive by design. Whatever resistance the structure has to fire-caused collapse is purely accidental.

We must constantly bear in mind, therefore, the fact that the building's tendency to collapse is being increased by the fire, while the fire suppression forces place additional demands on the structure. When the structure, or part of

The lumber pile to the left is less of a load on the balcony than the one on the right, because it is spread over the length of the beam. Both locations should be avoided in fire fighting operations.

Changes in occupancy or use often create excess live loads without any change being made in the structure making it much more susceptible to collapse in a fire.

it, is no longer able to resist the loads, collapse occurs *instantly*. Gravity always acts.*

---

*Joshua blew down the walls of Jericho with bugles, but no number of bugles can delay the force of gravity one second. (Non-fire service readers who need this explained should ask the nearest fire chief.)

Excess Live Loads

Buildings which appear to be overloaded, or in which the present use provides loads in excess of the original design, or structures which are internally braced such as with adjustable posts, should all be noted on prefire plans.

Added Live Loads

A most important live load can be the water trapped in a building or on a roof. Flat roof buildings often are built with little reserve strength. Water from rain or snow may pond on such a roof and the fire department may be asked to assist. The flooded roof may be near collapse. Don't add to the weight. A possible solution is to build a salvage chute in a suitable location and puncture the roof from below. The problem is that the mere puncturing of the roof may precipitate collapse, and in these litigious times, you may get sued.

Be aware of the loading and collapse possibilities of such roofs from overspray when using master streams on a fire in an adjacent building.

The author, wintering in Florida to escape snow, received a call from a fire department up north. "We have this 5-&-10 which had a big snow load. We got the snow off with hose streams, but we can see some cracked trusses. How should we brace them?" The advice was emphatic: "Get out of there and tell the owner to hire somebody competent, like a wrecking company." Even staying outside was not enough. A collapse of the roof would set up an air ram which would blow the windows out. We agreed that the best fire department service was to set up barriers well away from the building.

An elevated water tank is designed for a certain live load — the water in the tank. If the tank overflows during freezing weather, the water will become an ice coating on the structure and overload it to the point of collapse.

Where heavy streams are used, all must be conscious of the tremendous added weight. A 600-gpm master stream will add 25 tons to a structure in 10 minutes. Some of this water will be absorbed into contents. Paper and fabrics of any type, particularly, will absorb huge quantities of water. Five fire fighters died when piles of scrap from the manufacture of tissues, soaked by water from sprinklers, collapsed on them. Some fire departments have set specific "drain times" to permit water to drain from structures, subjected to heavy master streams, before personnel are permitted to enter the building.

Heavy water loads used to be a problem only to large municipal departments, but today, with large hose, and well-organized tanker and dam operations, heavy water loads can be delivered in many suburban and rural locations. Water thrown on a theater marquee, in Hagerstown, MD, in freezing weather, turned to ice and collapsed the marquee, trapping several men. Fortunately, a fire fighter had access to a construction crane and the marquee was lifted.

At times, the load situation in a structure is such that even the weight of hose lines, ladders and fire fighters might be too much additional load, at least at a specific location.

Water retained in a building increases the live load at the same time the building is being destroyed by fire.

Impact Loads

Impact loads are loads which are delivered in a short time. A load which the structure might resist, if delivered as a static load, may cause collapse if delivered as an impact load.

Easy does it! The impact load of a fire fighter jumping onto a roof may cause it to collapse.

It cannot be repeated too often that buildings have a habit of standing up as long as they are left alone. A fire is a violent insult to a building. All sorts of inter-relationships among various building elements, which have mutually supported one another and provided strength not even calculated by the designer, are disturbed. An explosion, the overturning of heavy live loads, such as a big safe, the collapse of heavy, nonstructural, ornamental masonry, or even the weight of a fire fighter jumping onto a roof may be enough of an impact load to cause a collapse.

There is a formula for tensile impact in the fourth edition of the *Civil Engineering Handbook*.[1] If a tension bar is supported at the upper end and a weight is allowed to drop from a height, the stress produced in the bar caused by the drop of the weight is as follows:

$$s_1 = s + s \sqrt{1 + \frac{2h}{e}}$$

In this formula, s_1 is the unit stress in pounds per square inch of the impact; s represents the unit stress in the bar if the weight were quiescent (just sitting there); e denotes the total deformation in the bar in inches when the weight is quiescent; and h is the distance in inches that the weight falls. The test bar is a piece of steel rod hanging freely. The weight dropped is like a doughnut around the rod. The rod is enlarged at the bottom so the "doughnut" can't fall off.

Any structural member under load must change shape. Compressive loads shorten a member; tensile (pulling) loads lengthen a member. If there is no change of shape, there is no load. The dropping weight exerts a tensile (pulling) load on the bar. The same weight quietly in place on the bottom of the bar also exerts a tensile load on the bar. In both cases, the bar is elongated. The value "e" in the formula represents the elongation when the weight is quiescent. The elongation is greater at the moment the weight hits.

The whole point of looking at this equation is to solve it using the value of zero for "h" (the height through which the weight drops), thus simulating what we might think of as "no impact."

Solution. Impact stress s_1 = quiescent stress s plus s times the square root of 1 plus *zero* (since h is *zero* the whole term is *zero*) or s plus 1s which equals 2s.

Thus, impact stress equals twice the quiescent stress even when the height through which the object drops is zero, which most of us would believe would cause no impact. The point, of course, is that there is no such thing as "no impact." No matter how gingerly personnel and equipment are placed on a structure, there is a substantial increase in the stress, at least momentarily. In any event, violent assaults on a structure already under attack by fire are to be avoided.

Impact loads can produce disastrously high stresses. This can be particularly significant when it is realized that the impact load, as in an explosion, can be delivered from a direction from which the designer expected little or no stress. A cooking-gas explosion caused the multistory collapse of the living rooms of a 24-story apartment house in England. The undesigned lateral impact load blew out one bearing wall. The floor it was supporting fell and became an impact load on the floor below, and the successive collapse of the floors followed. Backdraft explosions can be detonations when the CO-O_2 (carbon monoxide-oxygen) mixture is exactly right, and blow the building apart. Building codes typically require that explosives be handled in buildings of "4-hour fire resistive construction," thus, almost guaranteeing the creation of a bomb.

In industry where codes do not apply, an opposite tack is often taken. Buildings housing hazardous processes are isolated and built of friable (easily disintegrated) construction. Typically, a steel frame is covered with cheap, easy to replace cement-asbestos board. If an explosion occurs, the structure vents, the board becomes dust-like particles, not missiles, and much of the equipment survives the explosion, though pictures appear to show total devastation.

Special purpose buildings may be designed so as to channel the force of an internal explosion in the desired direction. Transformer stations are often built with heavy walls to protect one transformer from an explosion in the adjacent transformer and to direct the blast upward or outward in a preferred direction.

A national laboratory, which is proud of its achievements in the physical sciences, decided to build a room of three sturdy walls and one "blowout wall" for handling hydrogen. The blowout wall was built of wood studs, which would make terrible, jagged missiles, and faced out on the lawn where many of the staff ate lunch on pleasant days!

The safety factors built into ordinary buildings are rarely large enough to assure that there will not be progressive collapse in the wake of the first excessive impact load. The force of gravity always acts.

Static and Repeated Loads

Static loads are loads which are applied slowly and remain constant. A heavy safe is an example of a live, static load. It is not a dead load.

Repeated loads are loads which are applied intermittently. A rolling bridge crane in an industrial plant applies repeated loads to the columns as it passes over them successively.

Wind Loads

Wind load is the force applied to a building by the wind, which is constantly trying to overturn the building. The designer must counter this force. He does so in a variety of ways which need not be of concern here except to note that the framing of lightweight, unprotected-steel buildings is tied together to resist wind forces in such a way that fire-caused failure of one part may cause the collapse of other sections. Wind shear walls in framed buildings may be important to fire suppression forces and are discussed under Walls.

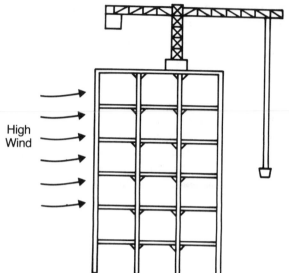

High
Wind

*The wind is constantly try-
ing to shear the building from
the earth. The designer must
counter this force (wind load).*

The wind load can be very significant to the fire fighter when operating in or near buildings under construction, when the full benefit of the interconnection of all the parts of the completed building has not yet been realized.

In addition to its effect on structures, the wind is a force to be reckoned with when the movement of smoke in buildings is considered. This force appears to be ignored by some whose solution to the high-rise fire problem rests first on sophisticated* smoke-handling equipment. Fire forces faced with such a system might well ask, "How will this system function when faced with a 30-mile per hour wind and all the windows in a substantial area are out?" Demand specific proven answers.

The push to reduce dead load (and thus cost) in buildings produced an interesting result in the case of some high-rise buildings. The Empire State Building (built in the 1930s) is said to weigh 23 pounds per cubic foot.** A modern high-rise weighs about 8 pounds per cubic foot. The result is that the wind can have a profound effect on a modern high-rise, producing noticeable oscillations on the upper floors. To counteract this, some buildings have been provided with dampers. These are huge weights of concrete or lead sliding on films of oil. A computer senses the wind situation and adjusts the damper accordingly, much as a heavy weight on the end would take the whip out of your fly rod.

Concentrated Loads

Concentrated loads are heavy loads located at one point. A steel beam resting on a masonry wall is an example of a concentrated dead load. In a typical con-

---

*The author's cynical definition of sophisticated: "You can't get parts when it breaks down."
**One reference gives the weight as 15 pounds per cubic foot.

crete block wall, solid block, brick, reinforced concrete, or a wall column may be inserted in the wall to carry the weight of the concentrated load. If a wall is being breached and the structure is found to be stronger than normal, choose another location. You are right under a concentrated load.

A safe is a concentrated live load. Early "fireproof" floors were brick segmental arches. The wooden floor was leveled in some cases by supporting it on little piers. At some locations there was a gap of several inches between the floor and the arch. When the floor burned, a safe fell the few inches. The impact was too much for the arch which had carried the safe for years, and the safe wound up in the basement.

The author observed a fire at a YMCA which was in a converted dwelling. An agitated employee arrived. "Tell the firemen there is a big safe on the top floor." "Don't worry lady. It's in the basement." Fortunately, no fire fighters were between the safe and the basement as gravity acted. In preplanning, make note of concentrated loads, particularly in non-fire resistive buildings. The more heavily a structural unit is loaded, the sooner it will collapse.

Imposition of Loads

Loads are also classified as to the manner in which they are imposed on the structure.

Axial Loads

An axial load is a force whose resultant passes through the centroid (or center of mass) of the section under consideration and is perpendicular to the plane of the section. In simpler words, an axial load is straight and true; the load is evenly applied to the bearing structure. All other conditions being equal, a structure will sustain its greatest load when the load is axial.

Structural elements are not always loaded in the most efficient manner. Other considerations may be more important, for instance: we are aware that a ladder is strongest when absolutely vertical, so that a person represents an axial load. But the nature of the human mechanism forces us to forego axial loading and slant the ladder so that the person can climb it.

Eccentric Loads

An eccentric load is a force whose resultant is perpendicular to the plane of the section but does not pass through the centroid of the section, thus bending the supporting members. A key factor in the famous Washington, DC, Knickerbocker Theater collapse in 1922, which caused the deaths of 97 persons, was the fact that the main roof trusses bore eccentrically, rather than axially, on the hollow tile walls. The ends of the trusses did not rest on the full width of the walls

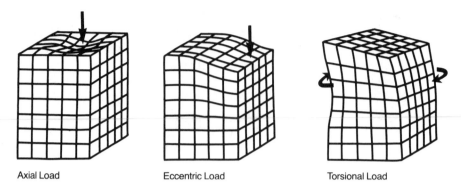

Axial Load   Eccentric Load   Torsional Load

Loads are applied in one of three manners. A serious fire can cause a load to shift from one manner to another, thus causing collapse.

but on only a portion. The wall was not reinforced to carry this eccentric load. The roof survived for several years until an unusually heavy snowstorm deposited 2 feet of snow on it.

Place a stack of books on the floor. Sit on the stack squarely. Note how your evenly applied weight stabilizes the stack. Shift your weight till it bears on only part of the stack. Note the tendency of the stack to bend outward.

Torsional Loads

Torsional loads are forces that are offset from the shear center of the section under consideration and are inclined toward or lie in the plane of the section,

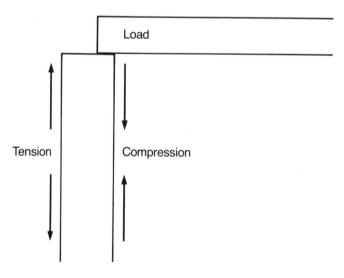

One side of a column under eccentric load is in compression while the other side is in tension. Unless the column is designed for these two opposing forces, failure will occur.

thus twisting the supporting member. The failure of one part of a steel frame building will often place undesigned torsional (twisting) stresses on other parts of the building, thus extending the area of collapse. As will be seen later, the maintenance of the structural integrity of unprotected steel may be the most important operation of a fire department at certain fires.

Axial loads are more easily dealt with by the designer than are eccentric and torsional loads. A structure which carries an axial load successfully may be incapable of supporting the same load when it is applied as an undesigned eccentric load. If the loading shifts from axial to eccentric or torsional during a fire, collapse may result. An eccentric load shifting to torsional may have the same effect.

Undesigned changes in loading can cause disaster. A serious fire upsets the delicate balance of forces in a building. Prefire plans should examine possible changes in loadings which may occur during a fire.

Fire Load and Rate of Heat Release

The terms defined in this chapter are standard engineering terms recognizable by any engineer, with one exception. Fire load is not a standard engineering term; it is a fire protection term.*

The fire load is the potential fuel available to the fire. To the extent that the building is combustible, the building is part of the fire load, as are all combustible contents. In the previous edition of this book, fire load was defined as also including the element of the rate at which heat would be evolved. It is now better practice to confine the term fire load to simply the amount of fuel present and use the term rate of heat release (RHR) to describe the intensity of the fire, the rate at which the fuel burns.

The basic measurement of the fire load is the British thermal unit (Btu), or the amount of heat it takes to raise one pound of water one degree fahrenheit (the metric equivalent is the kilojoule: for rule of thumb, one Btu = approximately one kilojoule).

Each combustible material has its own caloric (or calorific) value in terms of Btu per pound. For estimates, two rules of thumb are used. For wood and paper and similar materials, 8,000 Btu per pound is the caloric value; for plastics and combustible liquids, 16,000 Btu per pound is the caloric value.

The weight of the fuel is multiplied by the appropriate Btu value and divided by the area to arrive at Btu per square foot, the measure of the fire load. You may find fire load expressed in pounds per square foot, a practice which dates back to when there was only one kind of fuel, i.e., wood and paper, estimated at 8,000 Btu per pound. In such cases, plastics are converted into "equivalent

---

*When *Engineering News-Record* carried the story of the American Iron and Steel Institute report of the McCormick Place Fire (see Chapter 7, Steel Construction) in its March 16, 1967 issue, the term "fire load" was explained parenthetically, indicating that the editor doubted that the definition of the term was known to his engineer readers.

pounds" on the basis of 1 pound of plastic equals 2 pounds of wood.

A fire load of 80,000 Btu per square foot or 10 pounds per square foot (psf), is taken to be the equivalent of the 1-hour exposure to the standard fire of the ASTM E 119 Fire Endurance Test, 160,000 Btu per square foot is equivalent to 2 hours, etc. (See Chapter 6, Principles of Fire Resistance, for an explanation of fire resistance.)

If these figures are used cautiously and broadly, rather than precisely, it is possible to indicate that a given fire resistive building is "overloaded" from the fire load point of view, in a specific area.

Rate of heat release (intensity) refers to the rate at which the fuel burns. A 5-pound solid chunk of wood will burn more slowly than 5 pounds of wood chips, and 5 pounds of excelsior will burn extremely rapidly. The number of Btu remains the same, but they are delivered more rapidly.

There is no real data available on which to base estimates of rate of heat release (RHR). It is necessary that some estimates be made because this is an important criterion in the development of prefire plans.*

It is only in recent years that the concept of calculating the fire load in advance has gained acceptance even in fire protection circles. Yet, the absorption of heat by water is the essence of fire suppression by fire departments.

Many prefire plans note that a building is sprinklered. How many contain any analysis of whether the sprinkler system is capable of delivering the amount of water needed to absorb the heat that must be absorbed if the fire is to be stopped from spreading?

Building Classification Fallacies

Building codes make a crude and unsophisticated effort to get at the fire load problem by classifying buildings by their intended use. A warehouse will usually contain more potential heat per square foot than a department store. A warehouse may be required to have a 4-hour fire resistance while a store may be required to have only a 2-hour fire resistance. The warehouse may store sheet metal and the store may merchandise foam rubber. But this is rarely taken into account in the code.

Some codes require sprinklers for certain mercantile basements as a result of the serious hazards faced by fire fighters in fighting basement store fires. The code related to the type of occupancy, not the fire load. Consider two adjacent identical buildings. In one, the ABC Company sells office supplies. It is *mercantile* and the building is sprinklered. Next door, the XYZ Company operates an *office*. There, paper is typed on, and the paper is stored in the basement. No sprinklers are required in this case; the fire load is identical.

We can hardly expect building codes to be structured to provide graduated fire

---

*Our engine companies possibly should be called "heat removal units." If we did so, then perhaps we would ask, "How much heat is it likely we will have to remove?" and from this develop water flow requirements.

protection requirements based on fire load until we understand and use the term properly ourselves.

The Characteristics of Materials

Buildings are built of various materials, natural and man-made. Each material, and different forms of the same material, has certain physical and chemical characteristics which make it more or less desirable for its intended function.

Some materials are good in compression, others in tension. Some are heat insulators, others absorb sound. Some resist weather, others lack structural strength but provide a good smooth surface which takes a "nice coat of paint." Some are merely cheap.

The fire characteristics are often misunderstood or ignored unless the law directs attention to this vital facet of the suitability of a material for its intended use.

As each material is covered in the text, its fire characteristics will be discussed. A brief summary of some important characteristics of common materials that can influence their behavior in fire is provided here.

- *Wood* burns and loses structural strength. The vast majority of structural fires are fought by fire fighters standing on or under wooden structures.

- *Brick* is quite fire resistant but brick or mortar may be deteriorated.

- *Stone* of many types may spall, lose part of its surface when heated.

- *Marble* can go to chalk and lose all strength.

- *Cast iron* has basically good fire characteristics. The casting method determines whether it is good or bad, and this cannot be determined by examination. Poor connections are probably the chief failure cause.

- *Steel* elongates substantially at about 1,000°F. If restrained, it will buckle. It fails about 1,300°F. Water doesn't cause failure, it simply cools steel. Cold drawn steel, such as cables, fails at 800°F.

- *Reinforced concrete* is a composite material. Failure of the bond between concrete, which provides compressive strength, and steel, which provides tensile strength, causes failures of reinforced members. Inherently, concrete is only noncombustible. It may be formulated to be fire resistive, i.e., designed to perform its intended function under fire conditions within certain limits (see Chapter 6, Principles of Fire Resistance).

- *Gypsum* is an excellent material. It absorbs heat from the fire as it breaks down under fire exposure.

- *Plastics* have a wide variety of fire properties beyond the scope of this book. It is important not to generalize about the fire hazards of plastics. Some ignite and burn readily. Others do not. Some burn with heavy smoke production, others do not. Until recently, plastics have been sold and used without adequate attention to the hazards they represent.

In many fires, the most important heat to absorb with water is that heat which is being absorbed by structural materials. Heat absorption by a material changes the structural characteristics of that material which the designer relied on to carry the loads.

For our own safety, we must be aware of the effect of fire on building materials and thus on the delicate interrelationship of the forces by which a building remains standing.

The Meanings of "Combustible"

The word "combustible" and its antonym "noncombustible" (or "incombustible") are used in fire protection in a number of different contexts. In the regulation of flammable liquids, for example, the word "flammable" is used to describe liquids which have a flash point below 100°F and the word "combustible" for those with a flash point above 100°F.

With respect to structures, in past years, combustible was defined indirectly by defining noncombustible. The 11th edition of the NFPA's *Fire Protection Handbook*, published in 1954, said: "Noncombustible as applied to a building material or combination of materials means that which will not burn and ignite when subject to fire, such as steel, iron, brick, tile, concrete, slate, asbestos, glass, and plaster." This was a nice comfortable definition of noncombustible and by antithesis, of combustible; however, such a definition was found not to be precise enough nor does it take into account that some materials are "more combustible than others."

Distinguishing degrees of combustibility requires numerical values. NFPA 220, *Standard on Types of Building Construction*, 1975,[2] defines noncombustible material as: "a material which, in the form in which it is used and under the conditions anticipated, will not ignite, burn, support combustion, or release flammable vapors, when subjected to fire or heat..." The distinction between true noncombustible materials and materials that are to some degree combustible is provided in the definition of "limited combustibility," also from NFPA 220: "a material, *not complying with the definition of noncombustible material*, which, in the form in which it is used, has a potential heat value not exceeding 3500 Btu per pound (8141 Kj/Kg),[3] *and* complies with one of the following paragraphs (a) or (b)."

(a) Materials having a structural base of noncombustible material, with a surfacing not exceeding a thickness of ⅛ of an inch (3.2 mm) which has a flame spread rating not greater than 50.

(b) Materials, in the form and thickness used, other than as described in (a), having neither a flame spread rating greater than 25 nor evidence of continued progressive combustion *and* of such composition that surfaces that would be exposed by cutting through the material on any plane would have neither a flame spread rating greater than 25 nor evidence of continued progressive combustion.

Paragraph "a" above enables building codes to more properly classify gypsum board. At one time, the definition of "noncombustible" was tortured to include gypsum board, so that it would not be classified as combustible because of the paper facing.

Typically, building codes contain a classification called noncombustible construction. Certain specified elements of the building are required to be noncombustible, but there may be substantial combustible elements in the form of such unmentioned (therefore permitted) elements as fiberboard sheathing, wooden mansard and balconies, cornices and other decorations and combustible roofs, specifically the often unrecognized insulated metal deck roof.

Many confusing terms also are used to describe materials and to set requirements in building codes. One authority lists them as: flameproof nonburning, self-extinguishing, slow-burning, flammable,* nonflammable, fire retardant, burning, flame retardant, fire resistant and noncombustible.

The terms are definable but only when linked to a given material and a definite mode of fire testing. A sample examined by one fire test may be "self-extinguishing" while it may rate as "combustible" if another test is used.

Effect of Energy Conservation

The loss of energy in heating and cooling buildings has received great attention recently due to the high cost of fuel.

Insulation added to buildings changes the fire characteristics. In one test, a wood frame and steel building was subjected to a fire in a 140-pound wooden crib. The exposed corner of the structure was damaged and the fire went out. Various types of insulation were then applied to the building. In each case, the fire characteristics were altered. In some cases, the building was heavily damaged.[3]

Insulation can cause heat to be retained in light fixtures designed to disperse

---

*The term "inflammable" may still be found. For many years, the NFPA has striven to have it removed from fire protection literature as confusing, since many understand the prefix "in" to be negative, i.e., *not* flammable.

heat. Paper vapor seals on insulation placed directly on fixtures can become ignited.

Fixed sashes of double-paned glass were a factor in a motel fire in which several persons died. Such windows cannot be readily broken by what is available in a motel room. They should be equipped with emergency opening devices.

Insulation added to fire resistive floor and ceiling assembly (see Chapter 7, Steel Construction) may cause heat to be retained in the grid and bring about premature failure.

Ways of Applying Forces to Materials

Stress and Strain

Materials are tested to determine their ability to resist compression, tension, and shear, and values for common materials are published.

The external force which acts on a structure is load. The internal forces which resist the load are called stress and strain.

Stress is usually measured in pounds per square inch (psi). Occasional references to pounds per square foot (psf) are also found. The unit area — square inch, foot, yard, meter, etc. — is at the discretion of the person making the calculations. Always be careful to note the unit area used when examining any calculations. Load and stress are often used synonymously. KIP, a term meaning 1,000 psi, is used in engineering calculations where the number of pounds per square inch would be so large as to be unwieldy.

All materials deform (change shape) when stressed. Strain is the measure of this deformation.* In compression, the deformation takes the form of shortening. In tension, the deformation takes the form of elongation. Strain is measured in decimal fractions of an inch of deformation per inch of original length.

Forces are applied in various ways:

- *Compression* Compressive forces are crushing, pushing the mass of the material together.

- *Tension* Tensile forces tend to pull the material apart.

- *Shear* Shear forces tend to cause the molecules of the material to slide past one another.

Roget's Thesaurus indicates stress as a synonym for strain, and in the vernacular, they are so considered. This indicates the importance of understanding technical terms precisely. Engineering texts of an earlier generation used the term deformation for what is now called strain. We might understand more easily had its use been continued.

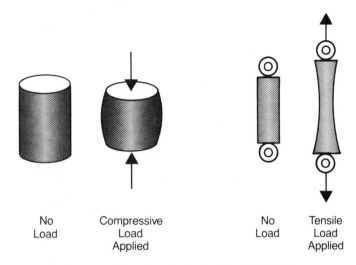

Compressive and tensile loads deform materials.

Observe a suspension bridge. The cables are in tension. The cables deliver the load to the towers which carry it to the earth in compression. The ends of the cables are buried in concrete. The shear resistance of the cable anchors in the concrete keep the cables from pulling out.

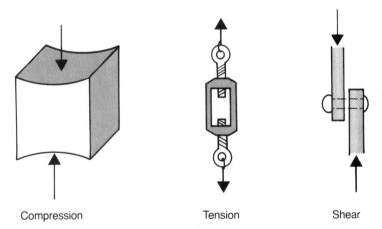

The three basic forces that act on building materials.

Elastic Range and Plastic Range

If the material returns to its original shape after the load is removed, the load was within the elastic range of the material. If the material remains permanently deformed after the load is removed, the load was within the plastic range. The residual strain is called the permanent set.

Yield Point

Deformation (strain) is proportional to the load (stress) in the elastic range. The point at which the deformation increases suddenly (although the increase in stress is at the same rate) is the yield point, sometimes called the elastic limit.

Modulus of Elasticity

For each material used in construction, the stress (in pounds per square inch) is divided by the strain (in fractions of an inch per inch of length). The dividend is a very large number. In the case of steel, it is about 29,000,000. This is called the "modulus of elasticity." In formulas, it is designated by the letter E. It measures the ability of the material to yield and then return to its original condition.

The modulus of elasticity is used in calculations of the strength of structural members. It provides a numerical value for the particular material being used. The modulus of elasticity of concrete is about 3,000,000. The moduli of elasticity of various woods range around 1,000,000 plus.

Compressive and Tensile Strength

In compression tests, crushing strains are exerted on the material until it buckles or crumbles. Concrete is tested in compression since it has little tensile strength.

In tensile tests, the material is drawn out until failure occurs. This is called the ultimate strength. Failure is not the important measure, however. The point at which the elongation rate suddenly increases, even though the pull (tensile strength) is being increased at a steady rate, i.e., the yield point, is the important measure of the suitability of a material for a particular use.

Some materials are strong in tension and weak in compression. Manila hemp rope is a good example. For years, it has been debated whether the Indian fakir really can make the rope rigid enough so the boy can climb it. To do so would require that the rope have compressive strength which it does not have. There appears to be no point in recruiting fakirs for high-rise rescues.

Other materials are strong in compression and weak in tension, such as natural stone and concrete. The Romans perfected the stone arch because all members of the arch are under compression, and little tensile load is placed on a material weak in tensile strength.

Steel is almost equally strong in compression and in tension. Usually steel is tested only in tension. If the sample passes, it is assumed the steel also has the required compressive strength.

Wood varies greatly in its compressive strength, depending on the direction in which the load is applied. Wet wood has less strength than dry wood. Plywood is manufactured with the grain of alternate pliers laid at right angles to develop approximately equal strength in either direction.

The compressive strength of soil and rock varies by a factor of over one hundred. Since all structural loads must be transmitted ultimately to the earth, accurate knowledge of the ground on which the building is to be built is vital to its stability.

Effects of Shape

The shape in which a given amount of material is used affects its ability to resist a compressive load or deflection (bending combines both compression and tension). However, shape is not a consideration in tensile loads.

Set up two stacks of books of the same height about 8 inches apart. Lay two sheets of typewriter paper across the gap. The paper, in the form of thin sheets, is barely able to support itself, and in fact, it may sag so badly that it falls.

Roll the two sheets of paper lengthwise into tubes about an inch in diameter. Secure the tubes from unrolling with rubber bands or clips. Place the tubes across the gap. The tubes are beams which will carry a sizable book without collapsing. Another form of beam can be created out of the sheet of paper by folding it into pleats, giving us a folded plate, used at times for concrete roofs. Corrugations give steel a greater ability to span a gap without unacceptable deflection (bending) than the same steel would have as a flat plate.

Take two more sheets of paper and roll them into tubes as before. Stand all four on end as columns. Carefully set a large volume, such as the NFPA's *Fire Protection Handbook*, in place on the columns. You will find that it is possible to stack at least two of these books on the columns. A heavy weight is supported on less than a half ounce of paper arranged in the ideal shape for a column. Note in this exercise that the load must be perfectly axial and that any eccentricity in the loading will cause the structure to fail.

Four heavy volumes of the NFPA's Fire Protection Handbook, 15th ed. are supported on four columns, each made from a single sheet of paper of the size held by the woman. Less than half an ounce of paper in the efficient shape of columns supports nearly 25 pounds of books.

Resistance to tensile loads, on the other hand, is directly related to the cross-sectional area of the material. Regardless of the shape, the paper would tear at the same pull if tested for tensile strength. T. H. McKaig in his book, *Building Failures*,[4] says: "A standard 3-inch pipe (outside diameter 3½ inches) weighs 7.58 pounds per foot; a 1¹¹⁄₁₆-inch round bar weighs 7.6 pounds per foot, practically the same. However, according to standard design theory, the pipe, used as a column 7 feet high, will safely carry a superimposed load of 36,050 pounds, whereas the rod will only support a load of 13,900 pounds (when used as a column)." Conversely, the pipe and the rod, made of the same quality steel and of equal cross-sectional area (because they are of equal weight per foot of length), will have the same tensile strength (when the load is suspended).

Suspended Loads

The fact that relatively slender tensile members can carry a load which would require a compressive member of much greater cross section is being used currently in the design of many buildings.

Designers are eliminating interior columns at selected locations by hanging the ends or beams from the overhead structure. A slender rod can replace a 4-inch by 4-inch wooden column. Several fire problems develop. The tension member, having less mass, is inherently less fire resistive. In addition, the load cannot be delivered to the earth in tension; it must be converted into a compressive load. This requires one or more connections up in the overhead someplace where fire temperatures are often greater than on the floor. The weakest (to fire) connection is the most significant connection to the fire fighter. Suspended structures should be noted and studied as part of the prefire plan.

This concept of suspended beams has been applied in the construction of some high-rise buildings. Columns are replaced with cables suspended from beams cantilevered out from the top of the central reinforced concrete core. This provides some economy of construction, unobstructed floor space, and an open space at the entry level to the building. This type of structure is discussed further in Chapter 11, High-Rise Construction.

Safety Factor

The safety factor represents the ratio of the strength of the material just before failure to the safe working stress.

The term "safety factor" is sometimes misapplied. It is recognized that it is not practical to use a material in a structure so that it will be loaded to its ultimate strength as shown in tests. The material used may not be as good as the material tested; the workmanship may be inferior; the material may deteriorate over the years. For these reasons the design load is only a fraction of the tested strength of the material.

If the design load is only a tenth of the tested strength, the factor of safety is said to be ten. If the design load is half the tested strength, the factor of safety is said to be two.

Steel made under controlled conditions has a safety factor of two; masonry constructed in place might have a safety factor of ten. The less that is known about the characteristics of a material and its role in a building assembly, the greater the factor of safety. The factor of safety is truly a factor of ignorance. Reductions in the factor of safety made on the basis of knowledge are not reductions in true safety. In other cases, reductions in the factor of safety may be significant when a fire occurs.

Consider a brick masonry wall. It was designed for a certain load. During the lifetime of the building, a substantial additional load is added to the wall. The addition has absorbed some of the factor of safety. It is not known how much, or even if it is significant. All that is really known is that the wall is more heavily loaded that the design called for, and thus bears watching. Steel used in excavation bracing is loaded to twice what would be permitted in a structure, because it is temporary. If a fire occurs, the steel will fail much faster because the failure temperature of heated steel decreases as the load increases.

A building built as a residence will usually be designed for a floor load of about 40 pounds per square foot. This includes a factor of safety which has been found adequate for residences. If the building is converted to mercantile or office use, the load will probably be much greater than the residential load. Often nothing is done to improve the building. The factor of safety has been infringed upon, possibly to the point where some slight increase in the load might cause collapse.

Composites

Composite Materials

At times, two materials are combined to take advantage of the best characteristics of each. For instance, concrete is a relatively inexpensive material which is strong in compression but weak in tension. Steel is strong either way but is more costly than concrete. By providing steel at the locations where tensile stresses develop, a composite material, called reinforced concrete, is developed.

All elements of a composite material must react together if there is to be no failure. If the materials separate, the composite no longer exists, and the two materials separately may be unable to carry the load successfully. We will return to this when reinforced concrete is discussed.

Composite Structural Elements

Two different materials may be combined in a structural element. A "flitch plate girder" is made by sandwiching a piece of steel between two wooden beams. The girder is much stronger than a piece of wood of the same dimensions yet it can be installed by carpentry techniques. Brick and block composite walls are another common example. Both elements must react together under the load.

Composite Construction

This term is sometimes used to describe buildings in which two different materials carry structural loads. Some concrete buildings, for instance, have a top floor or penthouse of lightweight steel.

Structural Elements

Having examined the loads imposed on buildings, and the effect of these loads on construction materials, attention is turned to the components of structures.

All loads generated within a structure or received from any source must be transmitted from the point received to the earth without any discontinuity in the transfer of the load from structural element to structural element. If a connection fails or if the earth yields to compression or shear, the structure will fail.

Structural members are differentiated by the type of load each carries and the manner in which each transfers its load to the next element of the structure. Regardless of the material used, certain principles apply to each structural member. Each of the structural members will be examined in turn.

Beams

The beam is probably the oldest structural member. It is not hard to imagine primitive man, intrigued by the fallen tree which provided a path across a stream, felling a tree, lopping off branches (thus removing unnecessary deadweight), and placing the tree across a stream to form the first bridge.

A beam transmits forces in a direction perpendicular to such forces to a point of support. (Consider a load placed on a floor beam; the beam receives the load, turns it at right angles, divides it, and delivers it to the points of support.) Note well that the definition takes no account of the "attitude" (i.e., vertical or horizontal). While beams are ordinarily thought of as horizontal members, this is not necessarily so. A vertical or diagonal member that performs the functions of a beam, although it may have another name, such as a strut, is actually a beam.

When a beam is loaded, it deflects (bends downward). The load may be its own deadweight, or it may be a superimposed load in addition to the deadweight. Some beams are built with a slight camber (upward rise) so that when the design load is superimposed, the beam will be more nearly horizontal.

Deflection causes the top of a beam to shorten so that the top is in compression. The bottom of the beam elongates and thus is in tension. The line along which the length of the beam does not change is the neutral axis. It is along this line that the material in the beam is doing the least work and can be most safely removed.

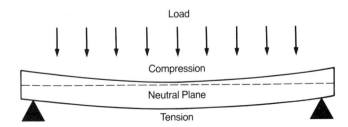

A beam deflects under load. The top is in compression and becomes shorter; the bottom is in tension and is elongated. Along the neutral plane, the length is unchanged.

Along the byways of West Virginia, the simple suspension bridge, spanning a creek to the home on the other side, is a common sight. This simple, sturdy, but shaky, bridge illustrates the designer's dilemma. The cable is the ideal beam. Fully in tension, it makes the most economical use of the available material.* For a beam to be safe and economical is not enough in most circumstances. It must deflect so little that the deflection will not be noticed. This is accomplished by using additional material, or by rearranging the shape of the material. In other words, stiffnesss can be achieved by mass of material or by geometry. In recent years, the economics of using geometry over mass has had a tremendous effect on structures.

A sawn wooden beam contains more wood than is necessary to carry its designed superimposed load. The additional wood assures that the deflection under the designed load will be so slight as to be unnoticeable. When a fire is consuming the beam, it will continue to carry its load as long as only this extra strength is being consumed. The first sign of noticeable deflection, therefore, is a sign that the beam is dangerously close to collapse. While steel and masonry are not consumed by fire, they can distort or disintegrate. Failure may be sudden or gradual.

The extra material and deadweight that are required for rigidity represent a challenge to the designer. This challenge is being met with new construction systems in which arrangement of the material of the beam is used to accomplish its purpose while reducing mass. Trusses and space frames are typical examples.

When a steel beam is rolled, in effect a plastic is being extruded. The mill can give the beam any cross-sectional shape desired. It has been determined, though, that the best shape is a beam formed in the letter I (I beam). The top and bottom flanges are made wide to resist compression and tension. The web connecting the two flanges is quite thin. The purpose of the web is to keep the top

---

*Cable-supported roofs are being used for large open areas such as the Madison Square Garden in New York City.

and bottom of the beam apart, as the load-carrying capacity of a beam is determined by its depth.

Wooden beams (usually called joists) are rectangular in cross section because wood is a fibrous material in which the fibers are only approximately longitudinal. Examine a 2-inch by 6-inch beam and estimate what proportion of the fibers actually extend from one end of the beam to the other.

It is not practical to mill out the excess wood to make the joist the ideal I shape. This excess wood makes it possible for fire departments to operate on burning wooden structures. As long as only the excess wood is burning, the structure is relatively safe.

Wooden I beams are now in use. They have 2 by 4s for top and bottom flanges and a piece of plywood for the web which holds the flanges apart. Such beams have no reserve. As soon as burning starts, vital strength is lost.

These and other substitutes for sawn beams demand adequate preplanning and a change in traditional tactics.

Carrying Capacity and Depth of Beams

All other considerations being equal, the load-carrying capacity of a beam increases by the square of the depth of the beam.

Consider a 2-inch by 4-inch (full size) wood beam carrying a certain load. If another 2-inch by 4-inch beam is laid alongside the first beam, the carrying capacity is doubled. However, if the same amount of wood is sawn into a 2-inch by 8-inch beam, thus multiplying the depth by two, the carrying capacity is increased by the square of 2, or 4. A standard manual of carpentry provides a table of "Safe Long Time Loads, Uniformly Distributed, for 1,800-pound Structural Grade Lumber." For convenience in determining loads for beams of different thicknesses, the table is arranged by sizes from 1 inch by 4 inch to 1 inch

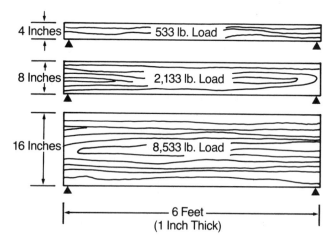

Increasing the depth of a beam four times increases the load capacity sixteen times.

by 16 inch, even though no beams are as thin as 1 inch. To find the strength of a 3-inch thick beam, the table figure is multiplied by 3.*

The table shows that a 1-inch by 4-inch beam spanning 6 feet can carry 533 pounds, a 1-inch by 8-inch beam spanning 6 feet can carry 2,133 pounds (4 times as much), and a 1-inch by 16-inch beam spanning 6 feet can carry 8,533 pounds. This latter beam is 4 times the depth of a 1-inch by 4-inch beam. All other conditions being equal, the load-carrying capacity (strength) of a beam increases as the square of the depth, but increases only in direct proportion to increases in width.

To look at this in another way, from a standard table a 1-inch by 8-inch beam spanning 10 feet can carry 1,280 pounds. A 4-inch by 8-inch beam also spanning 10 feet, therefore, can carry 5,120 pounds. In both cases, the beam is described according to standard terminology, that is, width first.

Suppose the 4-inch by 8-inch beam is used as an 8-inch by 4-inch beam. It now represents eight 1-inch by 4-inch beams side by side. A 1-inch by 4-inch beam across a 10-foot span can carry 320 pounds. Eight times 320 is 2,560 pounds. Thus, the 4-inch by 8-inch beam can carry only half the load when used as an 8-inch by 4-inch beam.

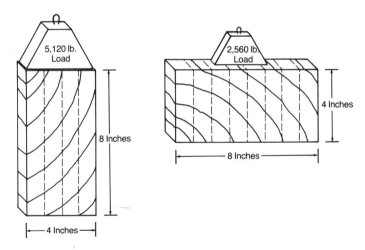

Any beam can carry a greater load when set so that its depth is greater than its width.

In building construction, the greater efficiency of a deeper beam must be balanced against other considerations, such as the desired thickness of the floor and the deflection of flooring between widely spaced beams (floor boards are themselves beams). For a variety of reasons, therefore, the almost universal spacing for wooden beams in ordinary construction is 16 inches and the depth of the beam is determined by the design load and the span.

It is noticed by now that the length of the span, that is the distance between

---

*Note that in all cases the computing is done on the basis of full thickness. If, as is usually the case, the lumber is dressed, a proportional deduction is made for the wood that has been removed.

supports, is a determinant of the safe load of a beam. As the length of the span increases, the safe load decreases in direct proportion.

Note the permitted loads for a 1-inch by 10-inch beam at various spans.

Feet	Pounds
6	3,333
12	1,666
18	1,111
24	833
30	667
36	555

These figures assume a uniformly distributed load. If the load is concentrated at the center of a beam, the permitted load is half of the distributed load. The figures also include the weight of the beam, which must be deducted to get the usable load. For convenience, the figures for wooden beams have been used. However, the principles involved are the same for all beams.

Types of Beams

Different terms are used for various beams:

- A *simple beam* is supported at two points near its ends, while a *continuous beam* is supported at three or more points. Continuous construction is structurally advantageous because if the span between two supports is overloaded, the rest of the beam assists in carrying the load. In simple beam construction, the load is delivered to the two support points and the rest of the structure renders no assistance in an overload. Continuous construction can provide economies in construction.

- A *fixed beam* is supported at two points and is rigidly held in position at both points. This rigidity may cause collapse of the wall if the beam collapses and the rigid connection does not yield properly.

- A *cantilever beam* is supported at only one end, but it is rigidly held in position at that end. In cantilevered beams, the tension is in the top, the compression is in the bottom. Cantilever structures are very likely to be unstable in a serious fire because the fire may be destroying the method by which the beam is held in place. The cantilever beam resembles a playground seesaw. We can all recall being lifted up in the air by a nasty playmate who then ran off and dropped us to the ground. Cantilevers are being used widely for both architectural and design economy considerations. In a fire the question is, "What's happening to the other end of the seesaw?"

- A *suspended beam* is a simple beam, with one end or both ends suspended on a tension member such as a chain, cable or rod. The typical theater marquee is a suspended beam. The fire may destroy the anchoring connection. The beam becomes an *undesigned* cantilever. An old theater was converted into a furniture store. Late in the fire the marquee collapsed killing six persons. In a fire, the question is, "What is happening to the connection of the tension member inside the building?" In short, "Who is holding the rope?"

- A *girder* is a beam which supports other beams. A girder used as a beam may be built of steel plates and angles riveted together, as distinguished from a beam rolled from one piece of steel.

- A *spandrel girder* is a beam which carries the load on the exterior of a framed building between the top of one window and the bottom of the window above.

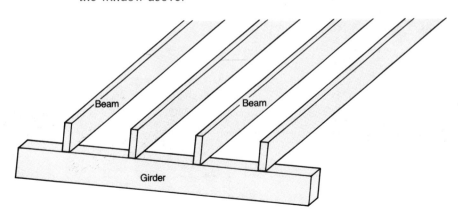

A girder is a beam that supports other beams.

- A *lintel* is a beam which spans an opening in a masonry wall.

- *Grillage* is a series of closely spaced beams designed to carry a particularly heavy load.

- A *transfer beam* has a function in delivering loads to the ground. Ideally columns should be aligned, one above the other to deliver their loads in compression to the foundations. If it is necessary to change the vertical alignment, a transfer beam must be designed to receive the concentrated load of the column and deliver it laterally to supports. Improper alterations of buildings may produce undesigned transfer beams which are points of weakness.

Beam Loading.   Refers to the distribution of loads along a beam. Consider a given beam as a simple beam. It can carry eight units of distributed load. If the load is concentrated at the center, it can carry only four units. If the beam is cantilevered and the load is distributed, two units can be carried. If the load is at the unsupported end, only one unit is safe.

Reaction and Bending.   The reaction of a beam is the pressure in pounds exerted by a beam on a support. The total of the reactions of all the supports of a beam must equal the weight of the beam and its load.

The bending moment, or moment of a beam, is that load which will bend or break the beam. The amount of bending moment depends not only on the weight of the load, but also on its position. The farther a load is removed from the point(s) of support, the greater the moment.

The Truss

Consider a structure in which three columns support two simple beams as shown.

Each beam is 10 feet long and can support 1,000 pounds. The center column is in the way. If it is removed, the beam is 20 feet long.

The two beams above could carry a total of 2,000 pounds (1,000 pounds on each beam). The 20-foot beam can carry only 500 pounds (double the length = half the capacity). What can be done to clear the space and yet carry the load?

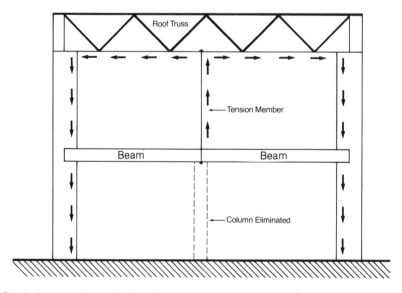

To eliminate a column, the load is taken up on a tension member to the roof truss, then delivered to the wall. The wall then delivers the load to the ground in compression.

Suppose only part of the column is cut off and the load is picked up on two tension members. The new structure can carry 2,000 pounds. A truss has been designed. Specifically, this is an inverted king post truss.

The truss is defined as a framed structure consisting of a group of triangles arranged in a single plane in such a manner that loads applied at the points of intersections of the members will cause only direct stresses (tension or compression) in the members. Loads applied between these points cause flexural (bending) stresses.

The Triangle Principle

The rigidity of the truss rests in the geometric principle that only one triangle can be formed from any three lines. Thus, the triangle is inherently stable. An infinite number of quadrilaterals can be formed from four lines, so the rectangle is inherently unstable. The economy of a truss derives from the separation of compressive and tensile stresses so that a minimum of material can be used. There are many designs of trusses, each with its own name. Sketches of many types can be found in any good construction book, such as Huntington's *Building Construction.*[5]

The top chord is the top member of a truss. The bottom chord is the bottom

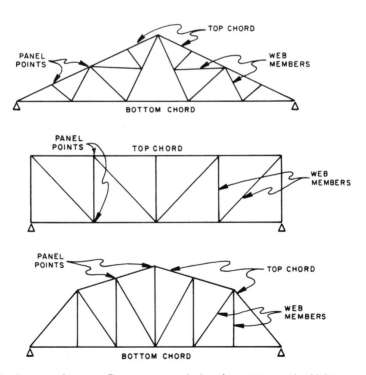

Three basic types of trusses. There are many designs for trusses, each with its own name.

member of a truss. The web is the term applied to members connecting the top and bottom chords. Struts are web members carrying compressive stresses. Ties are web members which carry tensile stresses.

The bowstring truss gets its name from the curved shape of the top chord.

The parallel chord truss gets its name from the fact that the bottom and top chords are parallel.

The vierendeel truss consists of rectangles with rigid joints. In effect, the triangles are found in the joints. The exterior of the World Trade Center in New York is formed of a series of vierendeel trusses. These were carefully fabricated and tested in place as they provide the principal wind resistance for the giant structure.

The bar joist is a truss. *

A space frame is a three-dimensional truss, thus contradicting the phrase "in a single plane" in the standard truss definition given above.

A serious problem in truss design has always been the connectors. In recent years, there have been tremendous strides in the development of connectors capable of transmitting heavy loads. Trusses can provide huge clear spans at a deadweight considerably less than that of a corresponding ordinary beam. In some cases, no ordinary beam can do the job.

Hazards of the Truss

In examining a building for prefire planning, the presence of trusses should serve as a red flag of danger. Trusses are vulnerable to collapse under fire conditions because the truss is composed of relatively lightweight members compared to the weight of material which would be required if ordinary non-truss construction were used. We have already seen that, to a large measure, the inherent fire resistance of any structure is a function of the mass of material.

Trusses are built of wood or steel, or wood and steel combined. In old construction, cast-iron struts may be found. There is some use of reinforced concrete. Steel construction will be discussed later, but for the moment consider the following:

Steel can be considered to be a thermoplastic, but a heavier piece of steel can absorb more heat without failure than a lighter piece. Because it contains less mass of material, the truss is inherently less fire resistant than an ordinary steel beam designed to carry the same load.

The failure of any element of the truss can cause the failure of the entire truss. Connectors, adequate for their design load, may fail rapidly in a fire, thus causing failure of the entire truss. The tying of adjacent trusses together to resist wind load may cause successive failure. In fact, successive failure of trusses appears to be the rule rather than the exception.

In recent years, lightweight wood, or wood and metal parallel chord trusses have become popular for floor or roof construction in non-fire resistive buildings.

---

*Metal aerial ladders are of truss construction.

A whole new situation with respect to structural integrity and void spaces, which is barely recognized by fire suppression forces has been created. Trusses are entirely different structural elements from the sawn joists or rafters previously used. Such joists provided unnecessary wood ("fat") which could burn away before structural strength was impaired. Trusses have no "fat." The failure of any element entitles the truss to fail. If fire involves the floor area, early sudden collapse should be anticipated. In addition, carbon monoxide may accumulate in the voids and cause a backdraft explosion.

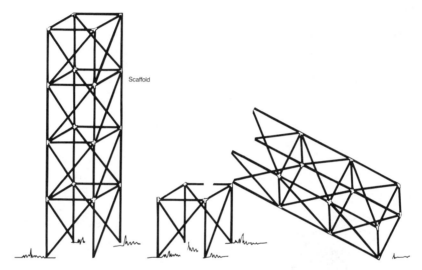

If one of the connections to the legs fails, the trussed tubular scaffold will buckle and collapse.

Reason for Truss Collapse

The following is better understood after columns have been studied, but it is included here for completeness.

The top chord of a truss is under compression, as is a column. Therefore, it is subject to the same laws of the safe relationship of length to width (the slenderness ratio) as any column.

The connections to the web members serve to reduce the length of the "column." Should they fail, the length of the column is suddenly increased. If the load is then excessive, as is most likely, collapse will result.

The bottom chord of a truss is under tension, like a rope. For a rope to fail it need be cut in only one place. A failure at only one point in the bottom chord may cause a truss to fail. Trusses of wood top and bottom chords, and tubular steel web (seen most often by this author in roofs), are particularly dangerous in this regard. At each panel point (connection) wood is drilled out to accommodate the steel pin. Not only is there less wood but also the heated steel causes pyrolytic decomposition of the wood at the weakest point; thus, at the point where there is the least wood, the wood is being attacked both from inside and outside.

There are many roofs which cannot be ventilated with safety using present techniques. Take a good look at the roofs of fast-food outlets and commercial buildings to name just two types of buildings in which truss roofs of the types described are almost standard.

Trusses will be studied as each type of construction is studied, but, in the spirit of a succinct highway hazard warning noted in Ireland that says "you have been warned": BEWARE THE TRUSS!

Columns

A column is a structural member which transmits a compressive force along a straight path in the direction of the member. We usually think of columns as vertical, but any structural member which is compressively loaded is governed by the laws of columns, no matter its attitude (i.e., vertical, horizontal or diagonal). Nonvertical columns are often called by other names, such as struts or rakers (diagonal columns bracing foundation piling).*

A *bent* is a line of columns in any direction. If a line of columns is specially braced to resist wind, it is called a wind bent. A bay is the floor area between any two bents.

The walls of the two buildings are the "beams" and the horizontal bracing members are the "columns." This mutual support arrangement is reminiscent of two revelers after a party — "who is holding up whom?"

---

*Columns by themselves are often used for monuments.

Shapes of Columns

The most efficient shape for a column is one which distributes the material equally around the axis as far as possible from the center, i.e., the cylinder. The thin-wall tubing used for the legs (columns) of a child's swing set is an example of an efficient use of steel in forming a column. In theory, column material could be paper thin, but then it would be subject to local damage. Most codes provide for minimum wall thickness for columns to prevent local damage.

It is more difficult to attach beams to round columns than to rectangular columns so rectangular columns are most often used. In cast iron buildings, interior columns are usually circular while wall columns (columns which carry the weight of exterior curtain walls) are rectangular. Rectangular columns often fit better into floor plans. Structural design is always a compromise between competing needs.

When beams were discussed, it was learned that a 2-inch by 8-inch joist set on the narrow edge would be a much more efficient use of wood than the same amount of wood in the 4-inch by 4-inch shape. The opposite is true of columns. The 4-inch by 4-inch shape would be the most efficient use of materials in a column, because it is most nearly circular in cross section. Large wooden columns, almost always ornamental as well as structural, are hollow. The column consists of curved, usually tongue-and-grooved, sections glued together. While steel beams are I-shaped, steel columns are H-shaped, box-shaped, or cylindrical. Beams are shaped like the letter I because the depth is the determining strength factor. Columns are shaped like the letter H and of a dimension that permits a circle to be inscribed through the four points of the H. A builder may use an I-shaped steel section as a column but it is contradictory to speak of "I-beamed columns."

Masonry walls under construction are often braced against an expected high wind. Available scaffold planks, usually 2 inches by 8 inches or 2 inches by 10 inches, are most often used. The load on the braces is a compressive load. One builder described the 2-inch by 8-inch planks as "snapping like matchsticks," at a construction site before the walls fell in a windstorm. If four 2-inch by 8-inch planks had been spiked together to make a hollow column and set diagonally against the wall, with 2-inch by 8-inch planks laid flat against the walls as beams, the wind load would have been much better resisted with the available material.

The capacity of a column per square inch of total area depends upon its slenderness ratio. This is indicated by l/r, or the length of the column section l divided by the radius of gyration, r. The distribution of the column material determines the radius of gyration. The farther from the center of the column the material is located, the greater is its load-carrying capacity.

Types of Columns

There are three types of columns differentiated by the manner in which each fails:

- Short, squat columns fail by crushing.

- Long, narrow columns fail by buckling. In buckling, the column assumes an S shape.

- Intermediate columns can fail either way.

Long, slender columns will be studied, not to learn to design them, but to learn the principles applicable to columns and other structures under compressive loads, notably the top chord of trusses.

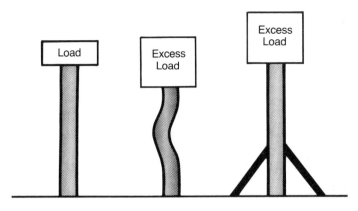

When a column fails, it assumes an S shape. Bracing a column increases the load-carrying capacity.

Euler's Law Columns

Very long, thin columns are known as Euler's law columns. A straight column, axially loaded, may suddenly collapse. Euler discovered that there is a critical load for a column and that the addition of even a single atom over the critical load can cause sudden buckling and collapse. The critical load, however, is not the governing factor in the design of column loadings because of unavoidable crookedness of the column and eccentricity of loadings. This crookedness and eccentricity will cause flexural stresses, which combined with the basic compressive load will cause failure before the critical load is reached.

Once yield stresses are reached in column action, there is generally very little reserve strength left and the column is on the verge of total collapse. By contrast, in beam action there is generally considerable reserve strength beyond initial yielding. The slightest indication of column failure therefore should cause the building to be cleared.

The failure of a column is likely to be much more sudden than the failure of a beam. The failure of a column may precipitate the collapse of the entire portion of the building dependent upon that column, or it may not, depending on whether the building is pinned or rigid-framed. (These terms are explained farther along in this chapter.)

Euler's Law Formula: Examine the formula for Euler's law for the understanding it will give of the three vital variables in column stability. The formula is:

$$P_c = \frac{\pi^2 EI}{L^2} \quad \text{in which}$$

P_c is the critical load, the absolute maximum load.

π^2 is 3.1413 squared.

E is the modulus of elasticity of the material in the column.

I is the moment of inertia, a mathematical value for the geometric cross-sectional shape of the column.

L^2 is the length of the column squared.

Euler's law applies only to long, thin columns, and a designer uses other formulas for intermediate length columns, but that is of no concern here. It is important to know the critical elements in determining the stability of a column. These are: the material of which the column is constructed; the cross-sectional shape of the column; and the length of the column, squared.

The value for the material from which the column is made is expressed in the equation by the modulus of elasticity, which has been discussed already.

The shape of the column is expressed in the equation by the moment of inertia. It has already been seen that a cylinder is the most efficient column. Any column which is wider in one direction than another will tend to deflect in the most limber direction. Stand an ice cream stick on end and push down on it. It will bow out on the flat side. It is impossible to make it bow out in the opposite direction.

The fact that the square of the length is the divisor in the formula for safe loads of columns is a most important item to bear in mind when considering any problem involving the stability of a column.

Consider a long, slender column. The load on the column will cause it to tend to buckle. If the column is braced rigidly at the mid-point, the effective length of the column (L in the formula) is cut in half. Since the square of the length is the divisor in the formula, cutting the length of the column in two reduces the value of L^2 to one quarter of what it was initially. This means that P_c (the critical load) is four times higher than it was initially. Shortening the effective length of a column by intermediate bracing pays dividends in load-carrying capacity.

We are not interested in designing columns, but we are vitally interested in possible causes of column failure. Consider one such possibility.

Effect of Loss of Bracing

A long, narrow column is securely braced at one or more points along its length. The effective length of the column, for this purpose, can be considered to be that of one of the segments. During a fire, the bracing collapses and falls away from the column. Instantly, the effective length of the column increases. The increase is by squares and thus extremely significant. The L^2 in the formula is the square of the now effective length, and thus the critical load, P_c, is drastically reduced. If the critical load value is less than the actual load, the column will collapse.

Truss Failures

Examine the scaffolding erected to reach high locations at buildings under construction or repair. The columns are slender tubes. How can a slender tube 100 feet high support the platform and workers? It is braced at regular intervals. In fact, it is not a column 100 feet high but a series of columns 8 feet high or so, one on top of the other. It is easy to visualize that if the bracing were cut off the column, the structure would fail.

Consider the top chord of a truss (for some reason a bowstring truss is easier to understand but the type makes no difference). It is under compression and resembles a long slender column, which is braced at the panel points and thus is in effect a series of smaller columns. If a rivet pulls out at the panel point, the bracing is lost. Two 10-foot "column" sections now become one 20-foot section. The load-carrying capacity of the section immediately is greatly reduced. The truss will likely fail by buckling since trusses have little reserve to cope with undesigned changes in loading.

Emergency Shoring and Bracing

At a collapse or cave-in, it might be necessary to provide shoring from available material. Given a choice, materials most nearly circular or square in cross section make the best struts. Always bear in mind that it is the nature of the loading which is significant. Struts used horizontally to span between planks shoring up an excavation are in fact columns and should have a column-type cross section. Thus, a 4-inch by 4-inch piece of lumber is much superior to a 2-inch by 8-inch piece for a cross strut (column), while the 2-inch by 8-inch piece is superior for shoring because it is loaded as a beam.

Temporary Bracing

When a building is under construction, there are many elements not in position at any given time. It is possible that columns might not have the full benefit of the bracing which will be provided by the completed building. A guy or temporary bracing is common. This bracing may not be adequate to resist high winds or other unexpected loads, or it may be vulnerable to fire.

Excavation Bracing

When a basement is excavated, the sides of the excavation must be braced unless the excavation is in solid rock. The load of the adjacent buildings is delivered to the earth. The earth under the building is trying to shear sideways into the opening of the excavation. The shoring prevents this. Excavation bracing is always done at the lowest possible cost. Excessive cost cannot be recovered by charging rents higher than the market.

The development of underground parking garages has provided a new hazard dimension. As is traditional in fire protection, the problem will receive no attention until a catastrophe occurs. In the past, excavations were carried down only a few stories as the space was undesirable. Now, 10-story-deep excavations are not uncommon.

A typical procedure is to drive steel piles called soldier beams (all lined up next to one another). As the excavation deepens, wood planks are placed between the soldier beams. The beams are connected laterally by a girder called a waler. Walers are braced by diagonal columns called rakers. These "columns" are more heavily loaded than would be permitted for a permanent structure. The more heavily loaded a steel structure, the lower the temperature at which it will fail.

Typically, excavations contain huge fire loads of plywood, plastic insulation, LPG and other fuels. It is not at all self-evident that fire fighters responding to a

Diagonal bracing members, often found in excavations, are called rakers. Failure of the rakers may cause adjacent buildings to fall into the excavation.

fire in an excavation would recognize that the most important task is to cool the rakers. Even if the problem is recognized, the cooling might be impossible due to obstructions. It is not necessary to heat up the whole column. Heating a short section to 1,200°F would cause buckling. The consequences are beyond our competence, but it is not incredible that adjacent buildings might collapse into the excavation.

The massive assembly of rakers and cross braces interferes with the construction of the basement floors. As an alternate, where ground conditions permit, holes are drilled into the rock and high-tensile-strength cables are anchored at the base of the hole. The cables are then drawn through holes in bearing plates which are installed on the surface to be supported. Tension is applied to the cables in a manner similar to that described for tensioned concrete.

The ends of the cables stick out for several feet. They would act as heat collectors. The cold-drawn steel cables used totally lose their prestress at about 800°F. If the heated end of the cable delivers 800°F to the part of the cable under tension, the cable will fail and the excavation is in danger of collapse. *

If such a collapse occurs, an august body will meet and deliberate and announce that they have discovered the importance of "fireproofing" for steel supporting significant loads *at all times*, not just in finished structures.

By contrast, steel building columns exposed to fire during the early stages of construction are less vulnerable to collapse than they will be when fully loaded, since the extent to which a column is bearing its designed load determines the temperature at which it will fail.

Walls

Walls transmit to the ground the compressive forces applied along the top or received at any point on the wall. A wall resembles a wide slender column. The wall may also be required to resist flexural forces, as does a beam. Wind load is an example of flexural force when the wind is received on the flat surface of a wall.

Load-bearing vs Non-load-bearing Walls

All walls are classified in two main divisions: load-bearing and non-load-bearing.

Load-bearing walls carry a load of some part of the structure in addition to the weight of the wall itself.

Non-load-bearing walls support only their own weight. Veneered walls, curtain walls, panel walls, and partition walls are some examples of non-load-bearing walls.

---

*The author posed this problem to a prominent structural engineer who lectures, writes, and consults on structural failures. His answer was, "Don't even think about that."

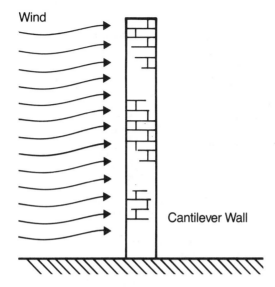

Wind

Cantilever Wall

Turn this illustration sideways to appreciate that a freestanding wall is a cantilever beam with respect to wind or any other lateral load, e.g., an aerial ladder leaning against it.

A load-bearing wall is inherently more stable than a non-load-bearing wall of identical construction because the weight of the superimposed load (the load carried by any building component in addition to its own deadweight) provides stability. Stack up a "wall" of books or blocks. The wall is easily pushed over. Superimpose your weight on the "wall"; it is much more difficult to push the "wall" over. However, the consequences structurally of the collapse of a load-bearing wall may be much more serious than those following the collapse of a non-load-bearing wall, because of the collapse of building elements supported by the bearing wall. It makes little difference, though, if a wall is load-bearing or non-load-bearing to a fire fighter caught in a collapse.

Within these two broad classes, other descriptors can be applied to walls. Some which are important to us are:

> **Cross Wall** A cross wall is any wall at right angles to any other wall. The walls should brace each other. Sometimes the bond is so poor that there is no benefit from the cross wall.

> **Veneer Wall** A veneer wall is a single thickness of masonry (one wythe) designed to improve the exterior appearance of the building. Decorative masonry such as stone, brick, or marble may be veneered over common stone masonry, concrete block, reinforced concrete, or steel. By far the greatest use of veneered walls is brick veneer on wood frame.* The veneered wall totally depends for stability on the underlying wooden wall. It is catastrophically unsafe if left to itself.

*A steel-framed brick veneer building is described in Chapter 7, Steel Construction.

Composite Wall In order to save costly brick in masonry walls, brick and concrete block (in earlier years, terra cotta tile) are used to form a composite wall. All elements of such a wall must react together to carry the load. Formerly, some bricks were turned on end (headers) to cross the wall and tie it together. This was unsatisfactory. A steel wire masonry truss is now laid in the wall. As a result, the bricks are all laid as stretchers (lengthwise). Thus, it is no longer possible to differentiate a brick veneer wall from a brick masonry wall at a glance, since now both walls can be laid as all stretchers.

Panel Walls and Curtain Walls These terms are used interchangeably to describe non-load-bearing enclosing walls on framed buildings. Technically, panel walls are one story in height and curtain walls more than one story.

Party Walls A party wall is a bearing wall common to two structures. If structural members of both buildings are located in common sockets, fire can extend through the opening.

Fire Walls Fire walls should be able to stop the fire with little or no assistance from the fire fighting forces. All penetrations should be equal in fire resistance to the fire wall. Openings should be protected with operating fire doors of the proper rating. *

Partition Walls Partition walls are non-load-bearing walls which subdivide areas of a floor. They may be required to have some degree of fire or smoke resistance. Variously, they may be required to extend to the under-side of the floor above or only to the ceiling line, leaving a void above.

Shear Wall A shear wall is a wall in a framed building designed to assist in resisting the force of the wind. It is usually incorporated in some required enclosure such as an elevator shaft or stair shaft. If you are breaching such an enclosure and find reinforced concrete, examine the other walls, they may be required only to be fire resistive and may be of block, or even gypsum on studs.

---

*Fire walls in steel-framed buildings are probably unstable if there is fire on both sides of the wall. See Chapter 7, Steel Construction, for additional information.

Cantilever Wall

When a masonry wall is under construction, it acts as a cantilever beam with respect to wind loads received on the face of the wall. When the roof is in place, the wall, with respect to wind loads, becomes a simple beam supported at both ends. High, freestanding walls are common in the construction of large "one-high-story" buildings, such as shopping centers, churches and industrial buildings. Severe winds may topple a freestanding wall.

Bracing should be adequate to prevent damage, but it is often skimpy or inadequate in design. The use of planks for struts (these are really columns) and stakes so short that they move when the rain turns the ground to mud are two common bracing errors.

Eccentric loading of the wall by hanging work platforms on one side is another cause of failure. If the wall is braced at one end by another wall at right angles (a cross wall), the unbraced end will be more vulnerable to collapse. During high winds, there are often gusts which greatly exceed the already high wind speed. Failure of what had been regarded as well-braced walls has been laid to gusts. Avoid the collapse and missile zone of all such walls during high winds.

Precast concrete "tilt slab" walls are vertical cantilevers when being erected and are braced by tormentors (temporary bracing poles). The roof of the building provides the permanent bracing. If the roof fails in a fire, the walls revert to undesigned vertical cantilevers and may collapse.

Wall Bracing

A wall can be compared to a column extended along a line. Most walls have a high height to thickness ratio which is comparable to a tall slender column. They are therefore similarly unstable. President Thomas Jefferson tried to build a thin garden wall. It failed. He reasoned that it was buckling like a column and built the buckle into it to brace it internally. The result is the famous "serpentine wall" at the University of Virginia.

Walls are braced or stiffened by:

- *Buttresses* Masonry structures built outside the wall. The flying buttresses of Gothic cathedrals are built away from the wall.

- *Pilasters* Masonry structures built inside the structure.

- *Wall columns* Columns of steel, reinforced concrete, or solid masonry (such as brick or solid block) in a block wall. Concentrated loads such as main girders are applied to the wall at the point where the wall is braced. If you are breaching a wall and find any evidence that the wall is strengthened at that point, stop and start elsewhere. You are likely right under a concentrated load.

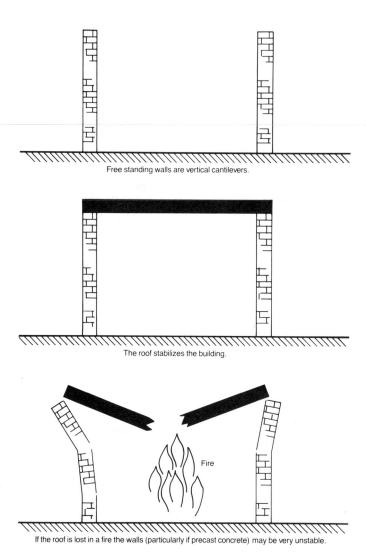

Free standing walls are vertical cantilevers.

The roof stabilizes the building.

Fire

If the roof is lost in a fire the walls (particularly if precast concrete) may be very unstable.

In many buildings, the roof is necessary to the stability of the structure.

- *Cavity walls* Walls are also stiffened by spacing out the material. A wall of 4 inches of brick, 4 inches of space, and 4 inches of brick (12 inches in all) is more stable than an 8-inch brick wall just as a column is more stable the further the material is from the center. In the past, such hollow (or cavity) walls were used to control rain. Rain penetrating the outer wythe drains down the cavity to weep holes at the bottom. Currently, the void is often filled with foamed plastic, either rigid board or foamed in place. Some of what the author has examined has not been fire inhibited. The exact fire problem is not known, but one can postulate degradation, possibly toxic fumes, and if oxygen becomes available, possibly violent ignition.

Wall Breaching

A homogeneous wall seems to sense openings in the wall. The load coming down the wall splits above an opening and comes down either side. Draw a line vertically above an opening, its height half the width of the opening. Draw a triangle connecting the point at the top of the line and the ends of the opening. Only the masonry in the triangle is resting on the lintel.

The proper way to breach a wall is to open a triangular hole. Breaching should be stopped if a wall stiffener or bracing as described above is found, or if wall elements fall apart easily indicating that the wall is in poor condition.

Roofs

Roofs are very important to the fire service. They are often our working platform. For the designer, however, the roof is usually only a cover to keep the weather out and to support necessary mechanical equipment. Where snow load is not a factor, a roof may be of very flimsy construction. A published report of a Southern California fire, in which a roof collapse cost the life of a fire fighter, recounted that the roof was supported on shingles driven into the brick work.

Generally, the roof is of much lighter construction than the floors. Many buildings of "noncombustible" construction have combustible roofs. These roofs are usually of very light truss construction, either steel or wood. Such roofs throw into serious question some of our operating procedures, particularly ventilation from the roof and inside operations when there is fire in the roof void.

The roof is vital to the stability of precast concrete tilt slab buildings. If the roof fails, collapse of the side walls is an imminent possibility.

Arches

The arch combines the function of the beam and the column. The arch is under compression along its entire length. For this reason, the Romans used the arch to span distances that could not be spanned by stone beams because of the low tensile strength of stone. If the arch is properly built, connections are not important to its strength.

Arches tend to push outward at the base and therefore must be either braced or tied. Arches usually are braced by buttresses. Failing masonry arches often are tied with steel rods. Others, such as steel arches, are tied together under the floor by tension rods or cables. Cutting of these floor ties during alterations can cause collapse. In some 19th century buildings, the floor arch ties are visible across the ceiling.

The word "arch" brings to mind the graceful, segmental, curved Roman arches. A segmental arch is one which, as the name implies, describes a portion of the circumference of a circle, but this is not the only arch form. Arches can have one, two, or three hinges, points at which an arch changes direction. Gothic and trefoil arches are found in churches and similar buildings. By tapering the masonry units, flat arches are made. Hollow tile flat arches were common in 19th-century "fireproof" buildings.

The removal of any part of an arch can cause the collapse of the entire arch. Cutting through tile arch floors, such as for a distributor or cellar pipe, is dangerous because the removal of one tile can cause the entire arch to fail.

Many masonry arches are completely or partially false. What is made to appear to be an arch is actually masonry carried on a lintel. Wood lintels can burn through and cause collapse. Steel lintels can elongate and push masonry out. A severely unbalanced load or a load outside the center third of an arch may cause it to collapse.

The *rigid frame* is a development of the arch. Steel rigid frames are widely used for industrial and commercial buildings where clear space is required. Wooden rigid frames are often used for churches. Precast concrete rigid frames are also used.

Such frames must be tied together at the bottom to resist the characteristic outward thrust of the arch. The ties are made of steel reinforcing rod. The rods are usually embedded in the concrete floor. Therefore, the floor is structurally necessary to the building. It is possible that in future years, a person cutting into the floor may consider the tie just another rod and cut it with surprising results. In one huge hanger, the ties were cast into concrete beams to prevent this. In one church it was observed that the ties apparently passed through an ordinary combustible floor-ceiling void. A fire in the basement could cause the ties to fail and perhaps collapse the frame.

Shells and Domes

A shell is a thin plate that is curved. Shells are built of concrete and wide areas can be spanned with extraordinarily thin shells. Shells can be less than 2 inches thick. The shell transmits loads along the curved surface to the supports. An eggshell provides a good example of the strength of a shell in relationship to the weight of material.

The dome is a shell. It can also be considered a three-dimensional arch. The hyperbolic paraboloid (often abbreviated HP) is another form of shell. These produce roofs of varying architectural design.

Geodesic domes are domes formed of a large number of triangles of equal size. They provide structures with very high volume to weight ratios. The sheath may be any desired material. The fate of the geodesic dome sheathed in plastic at Expo '67 in Montreal is described in Chapter 9, Flame Spread.

Transmission of Loads

All loads must be transmitted continuously to the ground from the point at which they are applied. Any failure of continuity will lead to partial or total collapse. There is a tendency among those who are concerned about building stability to make light of partial collapse and consider it not very important.

A partial collapse is very important to two people, the person under it, and the person on top of it. If you are caught in a collapse, it is of little importance to you whether the building collapsed entirely or partially.

The collapse of the walkways in the Kansas City Hyatt Regency Hotel, which claimed more than a hundred lives, was only a "partial collapse."

From Floor to Ground

Consider how a load is transmitted to the earth. Loads are usually placed on floors. Almost all floor elements are really beams. In some older "fireproof" floor designs, brick or tile arches are sprung between beams. (An arch is said to "spring" from the point of support.)

Consider a wood floor on joists, which are in turn supported at one end by a masonry wall and at the other end by a girder. The ends of the girder rest on masonry walls, but it is supported at two intermediate points on columns. When a load is placed on the floor board between the joists, the floor board deflects and transmits the load to the joists on either side of the load point. The amount of the load transmitted to each joist depends upon the distance of the load point from each of the joists, the nearer joist receiving the greater load.

The load transmitted by the floor boards to the joists is transmitted by the joists to the wall on one end and to the girder on the other. The proportion of load delivered to each support point again depends on the relative distances from the point at which the load is applied to the ends. The load received by the wall is delivered to the foundation, and thus to the ground. The load delivered to the girder is divided among both walls and both columns if it is a continuous beam. If (as is more likely) it is a simple beam (separated at each support point), the load is delivered to the nearest two supports.

Our example makes use of a simple masonry and wood-joisted building, but regardless of the size of the building or the construction materials, the principles are the same. The weight of a sparrow landing on the roof of the giant World Trade Center in New York City is transmitted to the bedrock of Manhattan by the structure of the building.

Foundations

Ultimately, all loads are delivered to the earth, through the foundation. The nature of the earth and the weight of the structure determines the foundation. Foundations can range from simple footings, to grade beams (foundation under the entire wall) to foundations which literally float the building on poor soil. In some cases, piles are driven either to bedrock or until the accumulated friction stops the pile. Almost all foundations today are of concrete. In some locations, decay-treated wood is used for small houses; our boast, "well, we saved the foundations," may no longer be valid.

Foundation problems can affect fire suppression in several ways. Masonry walls may develop severe cracks which make the wall vulnerable to collapse. Fire doors may not close properly. Openings may develop in fire walls, or in floor-wall connections permitting passage of fire. Dry-pipe sprinkler systems may not drain properly after "going wet."

Connections

Once a structure more complicated than a Druid's megalith or the caveman's tree-trunk bridge is required, the method by which the building is connected together to resist the forces brought to bear against it becomes most important.

There are two general types of connections. A building is said to be pinned when the elements are connected by simple connectors, such as bolts or rivets. These are usually not strong enough to reroute forces if a member is removed. In a rigid-framed building, the connections are strong enough to reroute forces if a member is removed.

A monolithic (Gr. *mono*-one, *lithos*-stone) concrete building is rigid-framed. If a column is removed, it is quite possible for the building to redistribute the load to the remaining members. If no member is thereby overloaded, the building will not collapse. Buildings built of precast concrete may be pinned, i.e., held together by nuts and bolts or welded joints, or, by the use of "wet joints," may be made monolithic. A particular building may be mixed. In an ordinary riveted steel-framed building, the collapse of a column will cause the collapse of all elements supported by that column. Some steel buildings are built with connections which will redirect overloads to other sections of the building. This is often called "plastic design."

Failure of Connections

The great majority of buildings, once built, tend to stand up. Up to a point the

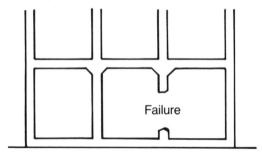

In a monolithic (rigid frame) building, the loss of a column may not result in collapse.

building can cope with undesigned loads, either vertical or lateral. Occasionally, a building not under construction or demolition (it is good to remember that a building on fire is a building under demolition) collapses. Often the failure is due to a failure of the connections. The ways in which connections can fail are endless. Masonry walls shift outward, dropping joists. Temporary field bolting of steel gives way in a high wind. Steel connectors rust. Concrete disintegrates. Of particular interest are connection systems which are of lower inherent fire resistance than the structural elements which they are connecting.

In many buildings, particularly older buildings or buildings built without adequate knowledge of engineering principles, the connections may be quite adequate as long as the building is axially loaded, but an eccentric load, a lateral load, or a shifting wall, floor, or column alignment may cause collapse even after the building has been standing for many years. Be especially careful of buildings built by "practical men" in areas where there was, or is, no building code supervision, such as rural areas that suddenly become urban.

The connections are concealed in finished areas of buildings. Do not pass up the opportunity to examine the basement and attic. What you see there is probably typical of the building. Buildings of the same age were probably built by the same methods.

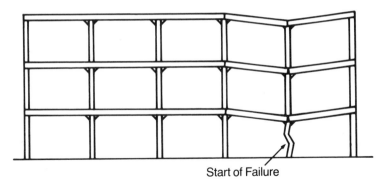

Start of Failure

In a pinned building, the failure of a column will cause collapse of all parts of the structure depending upon it.

Problems of connections are discussed in the appropriate place in other chapters. These are examples of some important connection defects.

- *Sand-lime mortar* Sand-lime mortar was used exclusively until about 1880 when portland cement mortar was introduced and over a period of years supplanted sand-lime mortar. In some cases, the two were mixed for ease of use. Preservationists prefer sand-lime mortar mixtures for restoring old buildings. Sand-lime mortar is water soluble. A fire fighter student, operating with his unit in the basement of a building, noticed that the hose stream had washed the mortar out from the bricks. He alerted the officer. The building was evacuated and shortly thereafter collapsed.

- *Gravity connections*   Many connections in buildings simply depend on the weight of the building element to hold them in place. This is especially true for cast-iron columns. A cast-iron column, gravity fit, was one of the causes of the collapse of the Vendome Hotel in Boston in which nine fire fighters died.

- *Unprotected steel*   Steel and its fire characteristics are the subject of an entire chapter. Steel connections enter into the construction of every building. (Very special efforts were required to produce a building without steel for experiments where the magnetic qualities of steel are undesirable.)
  Wood trusses are held together with steel "gusset plates" or "gang nails." Pyrolytic destruction of the fibers holding the gusset plate releases the plate and the truss fails.
  Steel is often intermixed with combustible construction. Since the building is not required to be "fire resistive," the steel is unprotected. Steel heated to 1,000°F must elongate 9 inches per 100 feet of length. If the girder is restrained and cannot elongate, it will overturn and drop its load of wood joists. Unprotected steel connections can often be found in precast reinforced concrete buildings.

- *Suspended loads*   It was noted earlier that a load can be suspended on a thin rod as contrasted with the bulky column required to support it in compression. This advantage has not been lost on designers. The load cannot stay in tension, however. It must be changed to a compressive load and delivered to the earth. This requires a series of connections. The vulnerable point is the connection most susceptible to fire. This may be floors away from the suspended structure. The steel tension rod may extend into the cockloft where it is connected to a wooden beam. The burning of the beam will cause the connection to fail, dropping the rod and its load. Thus, a cockloft fire might cause an interior collapse.

- *Self-releasing floors*   Many codes require that wood joists in masonry walls be fire-cut. The end of the joist is cut off at an angle to permit the joist to fall out of the wall without damaging the wall. The removal of wood lessens the inherent resistance of the joist to fire and can precipitate collapse.
  Heavy timber buildings are often built with self-releasing floors. Floor girders are set on brackets attached to columns. A wood cleat or steel "dog iron" (a big staple) is used to provide minimal stability. Such a floor can be expected to release sooner than if it were tightly connected. Some designers, recognizing this, require tight connections.

A building with fire-cut or self-releasing floors can be truly said to be "designed to collapse." It is the duty of a fire officer to see to it that his people are not under the structure when the designed collapse occurs.

In this chapter, an attempt has been made to provide an understanding of the most important terms in structural engineering, and to relate structural principles to the behavior of buildings on fire. The serious student should read further. The publications cited as references for this and other chapters, and similar books on construction, all contain a wealth of information. Having acquired a vocabulary and some understanding of structural principles, visit construction jobs and ask questions. There are no stupid questions, but there are stupid actions taken on the basis of ignorance.

References

[1]Urquhart, L. C., *Civil Engineering Handbook*, 4th ed., McGraw-Hill, New York, 1959, pp. 3-23.

[2]NFPA 220, *Standard on Types of Building Construction*, National Fire Protection Association, Quincy, MA.

[3]Degenkolb, John, "Will Energy Conservation Have an Effect on the Fire Protection of Buildings?" *Building Standards*, September-October 1976.

[4]McKaig, T. H., *Building Failures*, McGraw-Hill, New York, 1962, p. 11. (All of Chapter 1.4, Elements of Structural Design, is useful for additional study.)

[5]Huntington, W. C., *Building Construction*, 3rd ed., Wiley, New York, 1963, p. 306. (Also see Chapter 1, Article 2. This very readable book is intended for those who require a knowledge of the construction of buildings. It is recommended in its entirety to the serious student fire officer.)

Additional Readings

Zuk, W., *Concepts of Structures*, Litton Educational Publishing Inc. (by permission of Reinhold Publishing Co.), 1963, p. 37. (This is another excellent and quite readable book, recommended to the serious student.)

Allen E., *How Buildings Work*, Oxford University Press, New York, 1980. Chapter 17 of this book in its entirety should be studied by the serious student. In addition, the rest of the book is a mine of good information on buildings generally, very useful to the many fire fighters who build or alter their own homes.

CHAPTER 3

Wood Construction

At the outset, some basic definitions are in order for a better understanding of the elements of wood construction. Building terms, in particular, are local in origin, and very often the construction trades use a different word for a component than do engineers and laymen. For example, the term "wood frame" is typical of the imprecise terminology common in the building field. A "framed" building is usually understood to be a building with a "skeleton" system of beams and columns. The opposite term is "wall bearing." The wooden-walled building which is usually referred to as a "frame" building is actually a "wall-bearing" building. It can be very confusing.

So that we are all talking in the same terms, here are some definitions pertinent to an intelligent appraisal of the different types of wood construction that are described later in this chapter.

Definitions

> ***Wood*** A hard fibrous material forming the major part of the stems and branches of trees. It is usually milled and otherwise processed for use in buildings.

> ***Lumber*** Wood that has been cut, sawn, and planed.

> ***Timber*** Lumber 5 inches or more in the smallest dimension. (Note: stated dimensions of wooden structural members are nominal or sawn size. In planing, ⅜ inch to ½ inch is removed from each dimension of a sawn timber.)

> ***Heavy timber*** Lumber 8 inches or more in smallest dimension. (Note: some building code definitions of "heavy timber" specify smaller dimensions.)

> ***Chamfer*** To cut off the corners of a timber to retard ignition.

Boards   Lumber less than 2 inches thick.

Dimension lumber   Lumber in sizes between boards and timber.

Rough lumber   Lumber which is left as sawn on all four sides.

Matched lumber   Tongue and grooved lumber (usually lengthwise).

End matched   Lumber with tongue and grooves at the ends.

Splines   Wooden strips which fit into grooves in two adjacent planks to make a tight floor.

Joists   Wooden beams.

Studs   Wooden columns in frame buildings, usually 2 inches by 4 inches or 2 inches by 6 inches.

Wood lath   Narrow rough strips of wood nailed to studs. Plaster is spread on laths. (Practically no longer used, but present in many existing buildings.)

Engineered wood   In the trade, this usually refers to laminated timbers. In this text, it refers to wood modified from its natural state.

Types of Building Construction

The types of buildings in which wood carries the major structural loads are: log cabin, post and frame, balloon frame, platform frame, and plank and beam (western framing).

In addition, wood often carries major interior loads in masonry wall-bearing buildings. These are discussed in the next chapter.

Log Cabin

The log cabin was not invented by Daniel Boone or Davy Crockett. It came to the east coast from Sweden and to the west coast from Russia. (Prefabricated log cabins, ready for delivery, were sold in Moscow in the 17th century.) In many cases, it was easier for early settlers to use whole or slightly dressed trees rather

This building is actually a log cabin that has been sheathed with wood siding which conceals the heavy load of the logs.

than saw the trees to boards. As economic conditions improved, the humble log cabin would be concealed under siding or even brick veneer. Many buildings in the older sections of the country, even multistory buildings, are in fact log cabins. The walls of such a building contain an unexpectedly heavy load and therefore have potential for serious collapse. Fire departments in areas where such buildings exist should be aware of this hazard.

Post and Frame

This came to us from England and Germany, and in Louisiana, from France. Such a building has an identifiable frame or skeleton of timber fitted together by a skilled craftsman. Joints are of mortise and tenon (tongue) construction, skillfully fitted and pinned with wood pegs called "trunnels" (New England dialect for tree nail). Post and frame buildings are the ancestor of the high-rise,

Typical of post and frame structures is this English building. The decorative infill serves no structural purpose.

for they are framed, not wall-bearing buildings. The walls are not structural. They serve only to keep the occupants in and the weather out. Original post and frame buildings are part of our heritage and every effort should be made to preserve them.

In the 1920s, balloon frame houses were built with the exterior finished to resemble post and frame. They were called "English Tudor." Recently, the style has had a renaissance. Buildings of several types of construction have been decorated to appear to be post and frame. Some ordinary brick wall-bearing buildings have been dressed up to give the appearance of post and frame. At Disneyworld and other amusement parks, buildings of steel construction are decorated to resemble post and frame.

Balloon Frame

In 1833, the architect of St. Mary's Church in Chicago hit upon the idea of fabricating a wall of ordinary studs, nailing it together (machine-made nails were becoming available) and throwing up the entire wall at once without the skilled labor necessary to do the cutting and framing required for post and frame construction. The type of construction was derided by others who said it was so light it was like a balloon. It can be compared to a balloon in another way which will be seen. In any event, it became almost the only way to construct wooden buildings until the middle of the 20th century.

In a balloon frame building, the studs run two or more stories high from the foundation to the eave line. At the floor line, a "ribbon board" is nailed to the studs. The joists rest on the ribbon board. The channels between the studs may be open from the cellar to the attic and the joist channels (space between the joists) are open to the stud channels. Thus, fire can spread through the interconnected spaces from cellar to attic and across the ceiling. The author saw this con-

In a balloon frame building, the fire "sees" a combustible chimney extending upward between studs as shown here, and across the building in joist channels.

firmed at a fire in an old frame house. There was a heavy fire in the porch ceiling, and every time it was hit with a fog stream, the fire "ballooned" out at the rear of the building.

A chief officer, a student in a fire science course, described a fire which the first-due company reported as "outside rubbish." He ordered some of the siding removed and found fire in the walls. Shortly thereafter, an officer reported that an upstairs bedroom on the opposite side of the house was untenable, though no fire was showing. The fire had crossed over through the joist channels and had literally surrounded the bedroom. It took a good fight to save the building; damage was extensive.

When a fire has entered the inner structure of a balloon frame building, it can spread to every part of the building in all directions. Cover all parts of the building immediately. Be aware of the potential for intense fire buildup in void spaces. Don't wait for the smoke, heat, and fire to make conditions untenable before you decide to examine the attic. Fire moving up or *down* through exterior stud channels can often best be stopped by removing siding from the outside at the second floor line, the eave line, and the foundation line.

The author was lecturing in Pennsylvania. Fire officers described a fire in a balloon frame wall. They had followed the advice given above and stopped the fire. They were of the opinion that if they had made a conventional inside attack they would have lost the building. The problem with an inside attack is that the fire is moving up through many joist channels. As soon as the first channel is opened, the situation becomes obscured with smoke and steam. It is easy to miss a channel, and the fire may get away. Many fire departments today lack the bent swivel tip which was placed on the old controlling nozzle. With this tip, the nozzleman could quickly give it a dash up and down; this is not so easily done with today's equipment.*

It is safe to assume that any older two-story frame house is of balloon construction; it was the style. Balloon frame buildings are inherently susceptible to extensive fire spread. Today, the scarcity of the necessary long studs makes it more difficult to build a balloon frame house, but the ingenious mechanic finds a way; he uses spliced studs, and perpetuates the problem. Two six-story brick veneer apartment houses built in an Ohio suburb were of balloon frame construction. The fire chief fought successfully to have them torn down before they were ever occupied.

In *Building Construction*,[1] Huntington recommends balloon frame construction as superior to platform frame for masonry veneered houses (see Brick Veneer later in this chapter). The smaller shrinkage in the overall height of the balloon-framed wooden structure, from foundation to eave, due to using studs in the lengthwise attitude, contrasts with greater shrinkage of platform frame construction in which a sizable proportion of the height is made up of 2-inch by 4-inch lumber laid flat.

---

*In an excellent series of articles starting in the February 1980 issue of *Fire Chief*, Vol. 24, No. 2, Chief Donald Loeb discusses the difficulties of fighting fires in the most common type of balloon frame building, the 2½-story dwelling.

Masonry buildings with spans greater than 25 feet must have interior bearing walls. Unless of masonry, these walls are of balloon frame in all older buildings. Interior masonry walls can be distinguished from wooden interior walls by their relative thickness. The masonry walls are much thicker. Large wooden buildings must also have interior bearing walls, which will be balloon frame in older buildings.

Fire burning in balloon frame walls is consuming the structural integrity of the building. Collapse is a serious threat. Fires in balloon frame buildings should be observed by a staff officer positioned far enough away to see the entire building. This officer should watch for signs of fire spread — heavy volumes of smoke pushing out from voids in the building, intense heat buildup and structural failure, which may not be apparent close up.

Platform Frame

In platform frame construction, the first floor is built as a platform; that is, the subflooring is laid on the joists, and the frame for the first-floor walls is erected on the first floor. The second-story joists are then placed, the second-story subfloor is laid, and the second-floor walls are erected on the second-story subfloor. The third floor is built the same way. Three stories is about the limit for platform frame construction. Unlike in a balloon frame building, there is no continuity from top to bottom; thus, the structure is vulnerable to the wind. Because of the method of construction, there are inherent barriers to the spread of fire through the walls. There is a good possibility of confining a fire in the contents to one portion of the structure, even if the fire gains entrance to the concealed space.

Three stories is about the limit for platform frame construction. The structure is vulnerable to wind load as there is no continuous structural member from the foundation to the eaves.

One common construction feature, however, provides a bypass for fire. Soffits (referred to here as the false spaces above built-in cabinets, usually in a kitchen, but also a word to describe the undersides of stairways and projecting eaves) provide a connection without a firestop between wall and joist spaces. Futhermore, in kitchen cabinets only a thin sheet of wood or composition board makes up the top of the cabinet. A kitchen fire that extends into such a cabinet will enter the soffit space quickly. Fire can then extend to the joist spaces in a multistory building or to the attic spaces in a one-story building. Such soffits are often built "back to back" in multiple dwellings so that fire extends immediately to the adjacent occupancy. Interconnection of the soffit void with pipe openings may cause fire to spread vertically throughout the building. This is discussed more fully in Chapter 5, Garden Apartments.

The split-level house is a variation of platform frame construction. There are so many variations that it is impossible to categorize them, but a thorough study of locally built housing is indicated. Many designs incorporate interconnections between void spaces, and the degree to which firestopping is provided varies widely. While the concern here is only with the structure, note the wide open interior of this type of building. A fire can deliver combustion products to any part of the building. The inherent barrier provided by the basement door to the extension of a basement fire to the upstairs area is absent.

Plank and Beam (western framing)

Most people seem to find the 16-inch standard spacing of studs and floor joists to be aesthetically unacceptable, except for storage areas, work shops, etc. Otherwise, we insist that the eye-offending wood structure be concealed behind interior finish. In recent years, architects and interior designers, looking for more economical, or at least different, construction, have adapted industrial plank and beam construction to residential and commercial structures.

Instead of using multiple beams 16 inches apart, heavier beams are used, spaced farther apart. Instead of thin, rough subflooring or plywood, thick finished tongue and groove planks are used; the planks are thick enough to span the wider gap between the beams without deflecting. From the fire fighting point of view, this type of construction has the positive benefit of reducing the volume of concealed space in which fire can burn out of range of hose streams. The finishes used, however, often have high flame spread and smoke-developing characteristics, and the damage may be excessive, even in an easily controlled fire.

Many churches and similar high, open buildings are built of laminated wood rigid frames or arches and heavy plank roofs. Any fire that reaches the surface of the wood will spread rapidly and develop into a huge volume. Such structures would be much less likely to suffer a devastating fire if intermediate structures, such as robing rooms or sacristies in churches were sprinklered or completely noncombustible. Setting the roof on fire and destroying the building requires an intermediate fire. If this possibility is eliminated, the structure is unlikely to be heavily involved.

Noncombustible and Fire Resistive Buildings

Many "noncombustible" buildings have substantial wooden components, particularly roofs and roof adaptations such as mansards which really provide a combustible top floor. An apartment house was of fire resistive construction, but portions of the fifth floor and the entire roof were of wood. An undivided cockloft extended the length of the building. A mansard roof was covered with wood shingles. A loss of over a million dollars was incurred when an electrical fire involved the entire cockloft.

The owners of a fire resistive Maryland motel decided to use the "waste space" at the top of the stairway. A wooden storage room was built into the space. Heavily involved in fire, it came close to collapsing on unsuspecting fire fighters who were coming up the stairway. A post-fire critique developed two points. The attack should have been made via the aerial ladder to the roof. The pile of lumber used for building the storage room had lain on the sidewalk for several days, in the path of fire fighters reporting for duty at the nearby station.

This noncombustible concrete building has a dangerous plywood roof on lightweight wood parallel chord trusses. Ventilation "over the fire" would be disastrous. Ventilation from a safer point will draw the fire laterally.

Fire fighters should be eternally curious. We can take it as a truism that any alteration of a building, except the installation of sprinklers, is detrimental to the fire suppression effort and may create hazardous surprises.

Firestopping

Firestopping consists of pieces of wood or masonry fitted into stud and joist channels to slow down the passage of fire. All firestopping must be in place to be effective. The lack of firestopping in one stud channel (space formed by the sides of two studs and the exterior and interior walls) is sufficient to transmit fire from the cellar to the attic. In older houses, both the exterior sheathing and the lathing on the interior walls are invariably of wood, so all four sides of the chimney-like stud channels are combustible.

Lack of firestopping is particularly critical in balloon frame buildings. With respect to such buildings, it is a fact that firestopping was unheard of by many builders when the buildings were being built 40 to 140 years ago. *Radford's Cyclopedia of Construction*, Volume III[2] published in 1923, treats balloon framing extensively but mentions firestopping only at the sill. The NFPA's *Fire Pro-*

Providing firestopping in this type of roof structure would be defeated by the spacers (arrow) above the joists which raise the roof for the necessary pitch, leaving open spaces between the roof deck and rafters.

tection Handbook, 15th edition,[3] Figures 5-5M, 5-5N and 5-5O, illustrates where firestopping should be installed in wooden buildings.

The author considers firestopping to be of two types. One is "inherent" firestopping which comes as a result of the normal way in which a building is built. This kind of firestopping is incidental to some necessary structural purpose. It is reasonable to assume that inherent firestopping is in place.

On the other hand, there is what the author calls "legal" firestopping, or firestopping which is installed only in accordance with some legal or design requirement. Such firestopping can provide a perfect barrier to the spread of fire in the interior of the building, but there are always the questions, "Was the firestopping properly and completely installed?" "How vigilant was the building inspector?"

There are few building inspectors or mechanics familiar with the basic gas law: If the temperature rises and the volume remains the same, the pressure rises. They have no conception that leaving an "inconsequential" opening in the firestopping will, in effect, create a nozzle. Nor do they realize that if firestopping isn't perfect, it might as well be omitted, and it is rarely perfect. Wood firestopping is often made from the cut-off ends of joists — if they are a little short: "Well, that's all we had." The same practice was observed using the cut-off ends of wooden I beams. There was, of course, no seal because of the reentrant space (the space cut out to make an I) of the I beam. "We always did it that way. It's just a little opening."

Firestopping, once installed, may be removed for the installation of such items as heat ducts, electrical cables, and central vacuum cleaners. In one case, it was removed to install the fire main requested by the fire department and was not replaced. Since the firestopping is not "necessary" in the eyes of the mechanic making the improvement, it is unlikely that it will be replaced. The author has read many articles on the restoration of old houses, and has yet to see a recommendation that firestopping be installed as part of the renovation.

In 1970, the United States Department of Agriculture published *Wood-Frame House Construction*.[4] It says: "In most areas building codes require that firestops be used in balloon framing to prevent the spread of fire through the open wall passages." The structure of this sentence implies, no doubt unintentionally, that firestopping should be installed only where required by law. Contrariwise, the *Manual for House Framing* says that all concealed spaces are firestopped, without reference to legal requirements.[5]

In a fire in a combustible structure, it would be best to assume that no firestopping was installed. Any surprise will be pleasant rather than unpleasant.

It is no doubt a fact that, if draft stopping or firestopping is installed by persons who understand that the effort is to stop a gas under pressure, not simply to satisfy an uncritical inspector, there would be a temporary limitation to the spread of fire. Note, however, that even if the draft stopping is successful, this will not prevent the collapse of the affected area — at least 500 square feet in a single-family residential building and 1,000 square feet in other buildings. This is a large enough area to develop a severe backdraft explosion or to provide a significant collapse. Note that in multiple dwellings, the recommended location for firestopping is along the tenant separations so that the entire floor-ceiling area, above and/or below the unit on fire, is a collapse area.

Trusses are discussed in detail later in this chapter. Trusses are dangerous. Firestopping will not make such construction materially safer. The solution is a change in tactics. Progressive fire departments will make the changes as a result of analysis of the problem. Others will learn as a result of disasters and lawsuits.

Wood as a Building Material

As a building material, wood has many evident advantages; however, it is combustible, yielding about 8,000 British thermal units (Btu) per pound. It varies greatly in strength and modulus of elasticity from variety to variety and within varieties.

Wood construction is subject to insect infestation, wet and dry rot, and other forms of decay, which may cause serious weakening without being apparent to the casual observer. Deterioration of this nature in structural members should be carefully noted on prefire plans.

Following a fire in New York City, a heavy timber building collapsed. Investigation disclosed that dry rot had turned the timbers to punk, which continued to smolder after the visible fire had been extinguished. In a San Antonio building, the author recalls seeing wood colums that had been "fireproofed" with sheet metal cladding. Serious decay could occur unobserved.

Combustibility

Wood's basic defect is its combustibility. It is sobering to realize that most fires are fought by fire fighters standing on or under a combustible structure. Man has long sought to eliminate this undesirable characteristic of wood as a building material. Early fire prevention books extolled the virtues of whitewash in fireproofing wood. In the early 19th century, ads for products like "Blake's Patent Fireproof Paint" were common. Mr. Blake explained that all that his com-

Although the wood in these trusses is fire retardant treated, movement of the steel girder on which they rest could drop the roof. In this case, the fire department required sprinklers.

petitors offered was "worthless counterfeit stuff." His product "turns to stone" in a few months.[6]

Attempts to protect wood by encasing it in cement-like products were found to be dangerous, since the wood, not being exposed to the air, tended to decay.

Wood cannot be fireproofed or made noncombustible. It can, however, be made fire retardant by impregnation with mineral salts to slow its rate of burning. This is accomplished by placing the wood in a vacuum chamber, drawing out moisture and sap from the cells, and forcing mineral salts into the wood. Sometimes the wood is pricked to increase penetration. Such wood is also called pressure-treated.*

Pressure treatment can reduce significantly wood's flame spread, fuel contribution, and smoke developed, as measured in the tunnel test, ASTM E 84.** Of particular interest is that Underwriters Laboratories Inc. subjects the material to the test for 30 minutes rather than the usual 10 minutes. Pressure treatment can significantly reduce the hazard of wood construction where there is insufficient other fuel to provide a strong exposure fire. Given enough exposure to fire, the treated wood will burn, although at a slower rate, yielding the full 8,000 Btu per pound of wood. Any substantially lower number of Btu per pound of finished wood simply reflects the proportion of weight of mineral salts in the treated wood.

When the treated wood is listed by UL, listing marks "A," "B," or "C" are imprinted or applied on paper labels. "A" means that the lumber is listed only as to low (less than 25) flame spread, "B" and "C" include "fuel contributed" and "smoke developed" (less than 25).

---

*The term pressure-treated is also applied to wood which has been treated to resist insects and rot but not fire.

**This test is explained in Chapter 9, Flame Spread.

The phrase, "In a test of 30 minutes duration, flame spread not over equivalent of 25 and no other evidence of progressive combustion," or more simply just "30 minutes" is added to the listing mark for those materials that have passed the 30-minute test. Note that the mineral salts do not penetrate deeply, and the removal of surface wood may destroy the fire retardant treatment.

The fire hazard (flame spread, fuel contributed, and smoke developed) of wood can also be reduced by the application of intumescent coatings. As a substitute for pressure impregnation, one manufacturer suggests that all surfaces be painted on the job with the proper coating (applied according to directions) and touched up as necessary where cut. By this means, an objection of carpenters to the use of treated wood (it is said to dull cutting tools) may be overcome.

One of the major problems with surface coatings is to be sure that they are applied in the necessary thickness. One fire inspection office requires that sufficient can labels be delivered to show that the necessary amount of coating was purchased. The author always recommended that the owner purchase the material and hire a painter just to put it on in an attempt to get the requisite quantity per square foot.

Engineered Wood

For many years, man used wood as he found it and accepted its limitations. The California mission churches, for instance, were long and narrow because their width was limited by the biggest tree trunks which could be found to use for roof beams. Wood splits along the grain. Sawn wooden beams contain excess wood which adds unnecessary deadweight. Long straight trees of large diameter are scarce. Wood is engineered or modified to overcome these deficiencies.

The term engineered wood is used in the trade usually to refer to glued laminated heavy timbers. In this text it is used to denote wood which has been modified from its natural state. Some engineered wood products are: plywood, laminated timbers, chipboard, wooden I beams and wood trusses. Each presents the fire service with specific problems.

Plywood

About a hundred years ago, it occurred to someone that one of the defects of wood, lack of shear strength along the grain, could be overcome by slicing the wood into thin layers, placing the layers at right angles to one another and gluing the entire mass together. We know the product as plywood, just about equally strong in all directions. This was the first of many ways in which natural wood was modified to produce engineered wood.

A basic problem of plywood is that it delaminates when exposed to fire. The layers separate. This increases the surface area and thus the rate of heat release (RHR).

Like lumber, plywood can be pressure-treated to render it fire retardant. Also, like lumber, the letters "A," "B," or "C" in connection with a laboratory listing indicate that in a tunnel test, the flame spread, fuel contributed, and smoke developed are less than 25. The letter "A" signifies that only the flame spread was under 25. The notation "30 minutes" indicates that the material passed the 30-minute tunnel test.

At a fire where treated plywood has been installed, do not be surprised at dense smoke. From *Fire Technology*[7] it is learned that "the two most effective fire retardant chemicals, monoammonium phosphate and zinc chloride, greatly increased the smoke density index values for the plywood at retentions above 2.0 pounds per cubic foot."

Plywood can be used as an interior finish, or as a building sheathing without structural value, or it can be used as a structural material in floors, roofs, or walls. Such construction is often described as "stressed skin" or "diaphragm"; this means that by design, the plywood provides some of the structural strength of the building, particularly in providing resistance to shear stresses. The cutting or burning of such plywood can have more serious consequences than the destruction of plywood which was used for simple sheathing, without substantial structural value.

Laminated Timbers

The shortage of big trees from which solid timbers could be sawn led to intensive efforts to develop "laminated" timbers. Plank-like sections of nominal 2-inch (or even thinner) boards are glued together under pressure to produce large arches, beams, girders, and columns. Such timbers are also known as "glulam." (Sometimes bolts are used to supplement the glue.) When highly finished, they are most attractive. Combined with wood plank they can provide the necessary structural strength together with an aesthetically pleasing interior finish.

In the days of wooden ships, shipwrights sought out trees that had grown in the special shape required for certain structural members. Today, almost any shape can be fabricated by gluing.

We usually think of arches as having a characteristic "arch" shape. "Two hinge" arches of laminated wood are available, combining in one member both column and girder, and providing a straight-walled structure with a flat roof and a clear floor area. The Forest Products Laboratory at Madison, WI is constructed of arches that provide a floor area the equivalent of five stories in height with a 60-foot span.[8]

Laminated timbers and other finely finished wood are shipped wrapped in a protective paper cover. This cover is kept on as long as possible during the construction period. This paper is hemp-reinforced and coated with a bituminous moisture repellant. It ignites readily, has a high flame spread, and could contribute to a severe loss in a building under construction. This paper was responsible for the extension of a grass fire to piles of "packaged" lumber.

Tongue and groove roof planks have been fashioned in the past by milling

Laminated timbers are shipped in hemp-reinforced bituminous impregnated paper which has a high flame spread rating and can contribute to high fire loss in buildings under construction.

away and wasting undesired wood. Such planks are now often fabricated by gluing three boards together with the center board protruding on one side and indented on the other. Thus a tongue and groove plank is fabricated without any waste. The author wondered whether these would separate like plywood — allowing the boards to fall from the overhead. The sample burned like a solid piece of wood. Since there are no standards, it might be wise for a fire department to run its own test on materials used locally.

Chipboard

Wood chips are glued together to make flat sheets. These are sometimes used for the floors of trailers. The chipboard is water soluble. Be cautious at trailer fires. Walk on the beams.

Wooden I Beams

Look at the end of a steel I beam. Since the steel is extruded, the design can provide the most efficient shape. The top flange and bottom flange are wide to cope with compressive and tensile loads. The web, which separates the top and bottom, is quite thin, literally just sufficient to keep the top and bottom apart. A sawn wooden beam can be thought of as "containing" an I beam with "surplus" wood along the sides. It is not possible to mill it away. This surplus wood is what makes it possible for fire fighters to stand and operate on a burning structure. As long as only the "fat" is burning, the fire fighter is relatively safe.

Economy in the use of wood is accomplished in another way. Recently, in some parts of the country, wooden I beams have appeared. The top and bottom flanges are 2-inch by 4-inch lumber, solid sawn or laminate. The web is thin plywood. As installed, these beams are drilled for holes to accommodate

There is no excess wood or "fat" in a plywood I beam. Its structural strength immediately succumbs to fire.

utilities. Fire that reaches such a beam finds plywood, a material which burns with a high rate of heat release. The plywood has holes in it so the fire spreads rapidly to both sides. Just as soon as the plywood starts to burn the beam begins to lose strength. There is no reserve. There is no margin for safety.

These beams and wood trusses will force a fundamental change in fire fighting tactics, or lives will be lost. The fire suppression forces must have detailed institutional* knowledge, recorded and retrievable, of the construction of each combustible building, including additions and alterations to determine whether inside fire fighting is even feasible once any part of the structure is involved. It comes hard to those trained in the tradition of "get in, take the punishment, and get the fire" to accept that buildings are being built for which "surround and drown" is the only tactic. One way or another, the tactics will change. Either fire officers will recognize the problem and amend tactics, or deaths, injuries, and judgments against incompetent fire ground commanders will force the issue.

In some cases, such floor beams are cantilevered out to support balconies. Fire in the floor could cause the balcony to collapse under the weight of fire fighters making what they think is an outside attack.

Wood Trusses

Wood trusses have long been used in industrial and large-span buildings such

---

*Captain Truehart, who "knows all about those buildings," may be long gone when the fire occurs. Knowledge is "institutionalized" when there is a systematic method of obtaining and recognizing information so that the department as a whole possesses it. The ability to retrieve and use the information on the fire ground is vital, or the information will be useful only to plaintiffs and their attorneys.

as churches and halls. After World War II, lightweight gusset plate triangular trusses became commonplace for gabled roofs on single and multiple dwellings and commercial structures. Recently, there has been an explosion in the use of wooden, and wood and steel parallel chord trusses in flat roofs and floors of wood and ordinary construction buildings.

All wood trusses share the characteristic of trusses — the failure of one part entitles the truss to fail. At times, a distinction is made between heavy "bridge type" trusses and other trusses. This is dangerous. The collapse of a heavy bowstring truss quite early in a supermarket fire killed six New York fire fighters. All trusses are suspect. The massive timbers are not the key; the metal connections may fail and the heavy timbers just become more dead load to come crashing down.

Heavy Wood Trusses

These may be so-called bowstring trusses (humped up like a segmental arch), triangular, or parallel chord trusses similar to those used for early wooden bridges. Since the truss makes use of the basic rigidity of the triangle to achieve stiffness, the manner of connecting the members at the joints is most important. In earlier practice, the timbers were "jointed" together by mortised joints; more recently, side-to-side connections with through bolts were developed. Modern practice is to use split-ring connectors or shear plates. Metal splice plates are used to join short sections to form the top and bottom chords. Metal tie rods are used for vertical struts, which carry tensile loads. (A much smaller cross section of metal than of wood will provide the necessary tensile strength and the connection is simpler.) In addition, the tie rod can include a turnbuckle to allow for adjustment of the tension. In a typical 19th-century town hall, the roof trusses are of heavy timber but the tie rod or "king bolt" is of wrought iron. In a fire, it is likely that the iron rod would fail first, before the fire weakened the timbers, but the failure of the rod would cause the collapse of the truss and the resultant collapse of the roof.

As will be seen later, when unprotected steel is discussed, if the fire department plans to prevent a collapse, plans must be made to use hose streams to keep vital metal cool. Do not worry about "sudden contraction" or "shattering," which are old wives' tales; your problem is elongation and then failure of the steel; you may not have much time to reduce the temperature of the metal.

In describing trusses, *Radford's Cyclopedia of Building*,[9] which provides many insights into construction practices of the late 19th and early 20th centuries, refers to truss members of 1-inch by 6-inch fencing and uses terms such as "light and inexpensive" and "strong and cheap" to describe such lattice trusses. Many large-span buildings, such as churches and social halls, were built by volunteer labor following such do-it-yourself "cookbook" directions. Many light-trussed roofs have stood for decades, successfully coping with snow, rain, and wind loads, but there is no indication that they will successfully and simultaneously withstand the fire load from underneath, the live load of fire fighters on the roof, and the impact load of a ladderman's axe. The metal parts of the truss are vulnerable to heating and resultant failure. Never very large, they have little mass to which to distribute heat received. Lightweight wooden members may burn through long before the heavy timber main members. The truss is only as strong as its weakest part.

The scissors truss is used for steep roofs, most often on churches. Scissor trusses often have steel tie rods that are under tension, and their failure will collapse the truss. A number of different designs of parallel chord trusses are used.

One-story commercial garages and auto repair shops are typical of buildings requiring large, clear floor areas and a lightweight roof that need carry only the snow load. Today, such buildings are usually spanned with steel trusses, but a generation ago, the wood truss was universally used. A fire of record in a vacant one-story, wood-trussed, brick-bearing wall garage, which necessitated a second alarm, shows the danger of wooden trusses. The building was vacant. The roof was the only fuel. The roof collapsed about 15 minutes after the first alarm, bringing down the walls. Photographs taken moments before the collapse showed no sign that the wall was about to come down. The wooden trusses were counteracting the eccentric load of a heavy sign on the brick wall. When the trusses burned, the eccentricity was too much and the wall collapsed.[10]

When trusses support an open roof, such as a garage or gymnasium, the space required for the height of the truss is of no particular consequence. If a truss is used to carry the upper stories of a multistory building, a space problem presents itself. This is usually solved by incorporating the truss in the partition walls of the floor above. If you are opening up a wall, and you find diagonal structural members, be careful; they may be parts of a truss.

Modern long-span trusses are likely to be space frames, or three-dimensional trusses. A college in the northwest has several buildings with timber space-frame trusses yielding a 120-foot clear span and capable of carrying a 5-ton moving crane. Since the laminated timbers in the trusses meet the definition of heavy timber, the steel connectors might prove to be the first point of failure in a fire. The description the author read of the buildings was silent about ceilings, but if ceilings were installed (fire resistant or not), these forests of timber would be protected from hose streams and a fire in the truss areas could rage unchecked.[11]

Lightweight Trusses

Until recently, the roof structure of an ordinary one-family house or garden apartment house was supported on 2-inch by 12-inch or 3-inch by 12-inch or similar rafters about 16 inches apart. (A rafter is a joist set at an angle to support a sloping roof.) The rafters carried the wind and snow loads, provided the necessary rigidity, and supported the weight of the few mechanics who might have business on the roof.

Because it takes time to burn away from rafters that portion of the wood that gives them rigidity, experience has taught us that in the ordinary dwelling fire, there is still enough reserve strength in rafters to permit adding to a burning roof the weight of several fire fighters and the impact load from repeated axe blows without excessive risk of collapse. Thus, standard fire school instructions discuss ventilation of such a roof by men standing on the roof with at best slight mention of the possibility of failure of the roof.

The light weight and wide span of this construction illustrated the advantages of truss construction and shows the high collapse potential which may offset the advantages.

In order to build faster, use less lumber, and provide wide spans without the necessity of interior bearing walls (as required in a conventionally roofed building more than about 25 feet across), lightweight wood trusses are used. These trusses are assembled of 2-inch by 4-inch and 2-inch by 6-inch lumber. The members are all in one plane; there is no overlapping. In such trusses, bolt and nut connectors are not used, but gusset plates hold the truss together. These gusset plates are sheets of galvanized steel from which teeth are punched out at right angles to the plate. (These trusses are sometimes called "Sanford Trusses.") Sometimes, sections of plywood are used as gussets. In a typical garden apartment roof, the trusses are spaced 24 inches on centers, and the roof sheathing is usually plywood. This roof is perfectly capable of carrying its normal load without any trouble.

When a fire occurs, however, the situation may be quite a bit different from what it would be in the case of an old-style raftered roof. A 2-inch by 12-inch rafter has a lesser surface-to-mass ratio than a 2-inch by 4-inch truss member. (Compare the perimeter of each to the respective cross-sectional areas.) All other factors being equal, a piece of wood of greater surface-to-mass ratio will ignite sooner and burn faster than one of lesser surface-to-mass ratio. The lightweight truss is a fast burner.

The problem is compounded by the fact that if one member of a truss fails, the entire truss fails. The failure of one rafter does not cause the failure of any other element.

Metal gusset plates* can act as heat collectors; the heat is transmitted along

---

*Framing anchors, similar to the gusset plates used on trusses, are sometimes used to attach tail joists (joists not bearing on a wall or girder) to another joist. These present the same possibility of heat causing pyrolytic decomposition and resultant collapse.

the prongs, the prongs heat the adjacent wood fibers, and the fibers are destroyed by pyrolytic decomposition. The only force holding the truss together is the friction and compression of the wood fibers on the teeth of the gusset plates. If the fibers adjacent to the teeth are destroyed, the truss will collapse. Plywood gusset plates (usually $\frac{3}{8}$ of an inch thick) can burn away.

The term "trench cut" has come into use to describe a long narrow cut in a roof, typically of a tenement house, to cut off extension of fire in a cockloft. The trench ,cut was developed by units working on sawn joist roofs, with their inherent "fire-time." The concept must be thought out carefully before being applied to trussed roofs. For only one hazard, note that the saw may cut right through truss members.

All of this is not to condemn the gusset plate truss out of hand. Any device that conserves natural resources and reduces the cost of building certainly has intrinsic merit. From the overall fire protection point of view, the early failure of such a truss may well be beneficial in that it may open up the roof and thus ventilate the fire. The building will not collapse; the collapse will be a local collapse, not a general one.

Until recently, one might be wary of a gabled (peaked) roof but could be confident that a flat roof was built of sawn joists. This is no longer the case. It appears that wood trusses with their wider span have taken over entirely. In the past several years, while photographing buildings under construction across the country, the author has not seen a flat-roofed building under construction of any substantial size, with a roof of other than wood trusses.

These trusses are of two general types. There appears to be little to choose between them from the point of view of fire failure. In one type, the top and bottom chords are 2-inch by 4-inch lumber; the web is steel tubing. Wood is cut away to admit the tubing and a steel pin is inserted to hold the flattened ends of the tubes, which have a hole punched in them. The bottom of a truss is under tension like a rope. Wood is removed leaving little at the panel point. What little wood remains is attacked from outside and inside (by conducted heat in the pin.) One failure, and the truss fails.

Recently, the author has seen trusses in which the top and bottom chords were fabricated of two pieces of wood with the metal web members held between the pieces of wood. This provides an even greater surface-to-mass ratio for the wood members, and probably means even earlier failure.

In the other type, the web is of 2-inch by 4-inch wood like the chords. Gusset plates join the units. In one variation, a piece of steel serves as a diagonal web member and both ends are gusset plates.

Such trusses, or wooden I beams, are often used in rehabilitation of, or additions to, existing buildings. What you find in the older section isn't necessarily the same as the construction of the newer section.

Through prefire planning, the manner of construction of every wooden-roof building in your area should be known to you. The author offers you the following rule: "In the case of a trussed roof, if there is enough heat and fire to require ventilation, the roof is unsafe; men ventilating the roof should be supported independently of the roof." Trussed roofs may well justify an aerial platform apparatus, even if there are no tall buildings in the response district.

It is just as hazardous to be under the trussed roof as on top of it. Several fatal fires seem to have the same scenario. Fire fighters arrive and find smoke but little or no visible fire. Inside operations commence. The "ceiling" collapses without warning. If you have smoke and no fire, be wary. The fire may have possession of the trussed roof void. It is possible to have serious fire in the roof void with little or no smoke visible in the building. Fire can build up with explosive suddenness.

Trussed Floors

Trussed roofs are a serious problem, but even more serious is the trussed floor. Parallel chord wood trusses have found wide acceptance as floor joists. These literally introduce a "cockloft" into every floor of the building. The early collapse potential is enormous. There is no characteristic difference in the appearance of the building. The only way to know the hazard is to take notice when the building is being built. The floors are used in commercial buildings, apartment houses, and residences.

In the author's opinion, once fire reaches the floor cockloft, the building should be abandoned. Be especially wary when there are quantities of heated air or pushing smoke, and little or no visible fire. The fire may be literally raging in the trussed floor. Anticipate very early collapse, and a burst of fire.

In an article in *Fire Engineering*, March 1981 titled, "New Types of Construction," Dick Sylvia, the editor (who is also a fire fighter and a fire science teacher), gives a clear exposition of the dangers of hidden fire in trussed-floor wooden buildings. He points to the necessity for a rapid attack with enough water to darken down the fire.

In addition to the collapse potential, trussed floors and roofs deprive the fire fighting forces of another advantage of joisted construction. Each sawn joist or roof beam serves as a firestop, slowing down the growth of the fire across the roof or floor. The trussed floor is one big void. All the air in the void, and that admitted through openings, is available to the fire that starts in or reaches the void. Air is the governing component in fire growth; for each cubic foot of oxygen delivered to the fire, 537 Btu are generated. The fire will be more intense and move much faster in the trussed void.

Those accustomed to lightweight triangular trussed roofs realize that in this construction there is no usable attic space. If there is an attic, the conclusion is drawn that the roof must be raftered. Not necessarily so. In Alexandria, VA, the author noted a gabled roof. The roof "rafters" were, in fact, parallel chord trusses. Thus, there is a cockloft void throughout that roof.

Certain factors are sometimes offered in mitigation of the hazard. The ceiling is "fire-rated" gypsum board. This fallacy is discussed at greater length in Chapter 5, Garden Apartments. Suffice it to say here that the test structures were fully firestopped, and the test procedure does not contemplate fire originating in the void or extending downward through the plywood floor from the floor above.

Brick nogging installed in row houses to prevent fire extension is deficient when attic and floor voids are left unprotected.

Firestopping

Elsewhere in this chapter the problems of firestopping are discussed. It will only be repeated here that even if the structure is divided into areas of some required size (possibly 500 square feet) by firestopping, this is of no assistance to the fire fighters standing on or under the fire-involved section.

At times, those who comment on the hazard of wood trusses are attacked as trying to slow down progress, run up the cost of housing, etc., by those well-known publicists, Rant and Rave, who appear in any firesafety debate. Without apology, this text is devoted to the interests of the fire suppression forces. Fire fighters standing on a burning combustible structure are entitled to the best advice that is available, not "sympathy-soothing syrup."

The Los Angeles Fire Department conducted tests of lightweight roofs supported by wooden I beams, open web (wood chords and tubular metal web) trusses, metal gusset plate trusses, and roof panels of plywood on 2 by 4 joists and 1 by 6 boards on 2 by 4 joists.[12]

The roof sections carried no superimposed load. The fire was of 4 gallons of thinner and wood scraps, apparently about 2 feet below the roof beams.

The only "roof" which could be walked on after 6 minutes was the "roof" of 2 by 4 joists with 1 by 6 boards. All the others failed in 1 to 5 minutes.

Certainly the standard for residential sprinklers, NFPA 13D, does not contemplate sprinkling any underfloor voids. This is not to argue that it should. Life safety (of the occupants) is the big thrust of residential sprinklers, and an attempt to cover all contingencies would be detrimental to keeping down the cost of

Careful attention to detail is necessary for the successful sprinklering of wooden structures. Note the sprinklers on either side of this beam. This illustrates the problem of reaching all the fire with proper sprinkler distribution.

residential sprinklers. In commercial buildings, adherence to NFPA 13 would require sprinklers in the voids. Fire officers should be aware of the local practice.

For those who live or work in such structures, the author offers this advice: Be fully insured so the property will be replaced, have excellent early warning alarm coverage, react to the fire alarm immediately, and don't have anything of irreplacable nature in the structure.

Sheathing

Sheathing is the covering which is applied to the studs or framing. The exterior surface covers the sheathing. Tongue and groove boards laid diagonally to provide shear strength were used in older houses. (Many old houses were built without sheathing; fire spreading though walls can come out through joints in the siding.) In recent years, low-density fiberboard that is moisture- and vermin-proofed and black in surface color has been used in residential construction because of speed of construction and the relatively high insulation value of the fiberboard. (Celotex, which is a registered trademark, is often used as a generic name for any low-density fiberboard. In fact, the Celotex company manufactures many building products, and other companies manufacture low-density fiberboard.)

This low-density fiberboard material carries the warning "Combustible. May burn or smolder if ignited." The common method of ignition is the plumber's torch.

A similar material is often used for sound conditioning. It was applied directly over wood studs with a gypsum board cover in a school. A propane torch was used to sweat copper pipe. The flame went along the pipe and found the fiberboard. The fire extended vertically to the metal deck roof. The brand new school was destroyed. It was uninsured. It was believed to be "fireproof."

Plywood is also used for sheathing. In some cases, the house is totally sheathed in plywood. In other cases, plywood is used at the corners to provide shear strength.

Gypsum sheathing is found in some localities, particularly where combustible sheathing is not permitted. It is similar to the gypsum wallboard used in interiors.

Foamed plastic is now widely used for sheathing. It may or may not be flame-inhibited. In any case, if exposed to fire, it will degrade and may give off noxious fumes.

Siding

The siding on a building is the outer weather surface installed over the sheathing. Many different sidings are used on frame buildings.

Wood siding is usually novelty siding, laid on horizontally. (Novelty siding is often called clapboard, but true clapboards, cut from a log like thin pie slices, are rare.)

Board and batten siding consists of boards laid on vertically, butt to butt, with strips nailed over the joints. Sometimes plywood is used, and battens are nailed on 1-foot centers, to simulate board and batten. Battens have been used on concrete block walls to simulate board and batten siding.

Shingles and shakes (longer, thicker shingles are called shakes) are used for siding. As siding, they do not present the extreme conflagration hazard of wood-shingled roofs.

Asbestos cement shingles are used both in new construction and as a replacement cover over deteriorated wood siding. They are noncombustible, but the presence of wood trim, and often the old wood siding, makes such a building as vulnerable to a grass or exposing rubbish fire as a wooden building. Asbestos cement shingle can explode when heated, and flying particles can cause eye injuries.

Asphalt siding, usually made up as imitation brick, is used as replacement siding over wood. It burns readily, and produces dense black smoke. Asphalt siding is very similar to asphalt roofing which meets certain roofing standards. At times, this leads people to believe that it has some fire resistance value as siding. It has none. Roofing standards will be discussed later in this chapter.

One old house the author knows of in central Maryland showed an interesting series of changes. Originally a log cabin, it was covered with novelty wood siding. When that deteriorated, asphalt imitation brick was used. When the author last saw it, the imitation brick was being removed, and the house was being veneered with brick.

Metal siding, when used on residences, is usually made up to imitate some other material. Currently, aluminum siding is made up to look like clapboards. Embossed sheet metal is made up to look like imitation stone.

Corrugated metal siding is used on industrial buildings of wood or steel framing. Metal siding can present severe electrical hazards, both from stray electrical currents and from lightning. Thorough bonding and grounding of the siding, and a metal roof, if any, will do much to eliminate the hazard.[13] This is rarely done, and the possibility that metal siding is energized should be a factor in fire fighting. Corrugated metal siding, protected with asphalt, is discussed in Chapter 7, Steel Construction.

Stucco is a thin concrete surface. Stucco can be used over any surface such as brick, block, hollow tile, or wood. In the case of wood, it is laid on metal lath. It is used both in initial construction and in rehabilitating old buildings. At times, the owner of the building is given a sales pitch to the effect that the stucco makes the building safer from fire. This is untrue as far as life safety is concerned, but the use of stucco does improve the exposure situation over that of adjacent wood-surfaced buildings.

In some cases, a coat of gray stucco is applied. A coat of red stucco is laid on over the gray. Red stucco is then removed in lines leaving the gray visible. The effect is that of brick masonry. In Florida, the author observed concrete block buildings stuccoed to resemble clapboard. The houses were then painted and look like wood.

Brick Veneer

Brick structures have what the publicity experts would call a "good press." Brick veneer siding is very popular for wood frame residences and garden apartments. The brick is not structural; it carries no load except itself. While initially more costly than wood siding, the savings on painting over the years may make brick veneer economically attractive.

Brick veneer walls are laid up from the foundation in one wythe (a wythe is one thickness of brick in a wall). Such a wall, if not attached to something, would be very unstable because it is thin. In a veneered house, galvanized steel anchors are nailed to the studs. (Nailing to the sheathing is poor work, but it is found.) The anchors are bent at a right angle and embedded in the mortar between two courses (a course of masonry is a layer). If there is a fire in the wall, the nails may detach from the studs due to pyrolytic decomposition. The resultant freestanding wall is quite unstable. It would not have the stabilizing effect provided by the compressive load of the structure as does a bearing wall. On the other hand, its collapse will not affect the stability of the structure. The collapse of a veneered wall of an upper story may be an impact load to a first-floor extension.

Whether the wall is a veneer or a structural wall is immaterial to the fire fighter struck with falling bricks. Some seem to place too much reliance in newer helmets. No matter how strong the helmet, it is worn on a human body of limited impact resistance. The wearer of the super helmet may find that the frontispiece becomes a belt buckle!

Until a few years ago, it was easy to identify a brick-veneered wall. All the bricks were laid as stretchers (bricks laid lengthwise of the wall). There were no headers (bricks laid across the wall; the ends show). Headers were formerly required in a bearing wall. More recently, the use of metal truss masonry reinforcement, discussed in Chapter 4, Ordinary Construction, has made it possible to omit the headers from a solid masonry wall. If the brick wall is all stretchers, suspect brick veneer, but you cannot be positive.

In the event fire involves the sheathing and studs behind the brick veneer,

Brick veneer will give this structurally complete house the appearance of masonry construction. Oddly, the garage at right is of true brick construction.

there is no point in extraordinary exertions to save the veneer. It must be removed to resheathe the house, and to reattach the veneer.

Brick veneer is also applied to buildings other than wood. The appearance of metal buildings is often "upgraded" by applying a brick veneer, often only to the front and easily visible sides. Many buildings are partially solid masonry and partially brick veneer. When finished, there is no apparent difference. This can be determined only by observation when the building is under construction. It is vital that fire departments develop systems for recording and retrieving such data.

Wood Shingle Roofing

Wood shingles or shakes are split pieces of wood used for roofing or siding. Shakes are larger than shingles.

Some of the greatest fire disasters in history have been due to wood shingle roofs. Table 1-5F in the NFPA's *Fire Protection Handbook*, 14th edition, [14] lists conflagrations from 1900 to 1967. The author suggests you go down the list and mark each conflagration in which wood shingles were a factor. Note that the incidence of "wood shingles" as a conflagration breeder decreased markedly after 1937. By World War II, the effect of legislation outlawing wood shingles and the availability of substitutes, particularly asphalt shingles, had been felt.

Despite the hazard, wood shingles have an attractive appearance, which is often compellingly appealing to architects and owners. Note the following from The Building Code Committee of the Department of Commerce:[15]

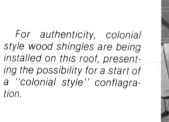

For authenticity, colonial style wood shingles are being installed on this roof, presenting the possibility for a start of a "colonial style" conflagration.

Probably no type of roof covering has caused more comment and discussion than the wooden shingle. The great danger of the wood shingle roof is from chimney sparks and flying brands from burning buildings and bonfires. The danger from chimney sparks is largely confined to wood and soft coal fires and the sparks resulting from the burning of chimney soot.

The wooden shingle has various well-recognized merits. It is light in weight, has excellent insulating value, thus promoting comfort by equalizing temperatures, can be easily applied, furnishes attractive architectural effects, and high grade shingles properly laid produce a roof of excellent durability.

The main objection to wood shingle roofs is the fire hazard. Sparks or flying embers are more likely to roll or blow off from the smooth surface of a newly shingled roof than an old roof with weather-worn shingles having curled and broken edges. For this reason, any treatment of shingles, such as staining or creosoting, which will tend to maintain a smooth surface incidentally improves their fire resistance. Few if any of the compounds used for treating shingles directly increase their fire resistance.

Note that there is no mention of the conflagaration hazard. If consideration of a fire is limited to the problem of the building in which the fire originated, the wood shingle roof might be more desirable than another type, for by burning through, it vents the fire, but the conflagration hazard should be the dominant consideration.

In recent years, wood shingles and shakes have made a strong comeback. Each generation must learn anew. In 1959, the NFPA warned that a major conflagration risk existed where large areas of wood shingle roofs existed.[16] The Los Angeles (Bel Air) conflagration of 1961 proved the prediction to be accurate. There are many areas now where large numbers of houses have wood-shingled roofs. In some areas, they are permitted wherever frame buildings are permitted.

Some insurance rating organizations have removed any penalty for wood-shingled roofs because there hasn't been any bad experience in their jurisdictions. The combination of fast fire department response, one-story buildings, wider spacing of buildings than in bygone years, and the fact that we have, as yet, nothing like the square miles of shingles which existed 50 years ago, have combined to keep the conflagration rate low. But given the coincidence of hot dry weather, brush fires that engage a large part of the available fire fighting forces, high wind, and a hot fire in a wood-shingled home or church, a conflagration will almost surely result.

On the afternoon of July 31, 1979, the City Council of Houston, TX was considering an ordinance placing some relatively minor restrictions on the use of wood shingles. The ordinance was tabled. The day was hot and dry. At about the same time, the fire department responded to an alarm for "fire coming from an apartment" at the Woodway Square Apartments. By evening, 30 apartment buildings were destroyed, hundreds were homeless, and an estimated $44,000,000 in damage was incurred. The fire spread because of the wood shingle roofs. Ironically, the owner was in the process of replacing the shingle roofs with composition. The next day, the council passed the shingle ordinance.

Underwriters Laboratories Inc. rates roofing on flame exposure, spread of flame, and resistance to burning brands (flying pieces of burning wood). Roofings are classified "A," "B," and "C" with "C" being the least resistant. In addition, for wood shingles, a flying brand test is required to determine if the roofing will produce brands. Several companies produce wood shingles that meet Class "B" or Class "C" standards. For many years, there have been many suppliers of asphalt shingles that carry Class "C" ratings. Asphalt shingles, of course, burn and generate dense black smoke, but conflagration hazard, not the combustibility of an individual roof, is the important consideration.

One of the complaints against Class "C" asphalt shingles is their appearance. Recently, several suppliers started offering Class "A" glass fiber reinforced shingles, which have a very attractive appearance and, though they will burn, are very resistant to flying brands.

Wood shingles are permitted in many areas for the currently popular "eyebrows" which are mansard false fronts being erected as cornices on all sorts of business establishments, in an attempt to make them look more attractive. See the discussion of cornices and mansard roofs in Chapter 4, Ordinary Construction.

Know the building code provisions governing wood shingles in your area. If UL-listed shingles are required, bear in mind that only bundles of shingles are labeled, not individual shingles; you can't tell treated from untreated shingles by just looking at them.

For a good history of the whole shingle problem, read "Some New Answers to the Shingle and Shake Problem," by Rexford Wilson, NFPA MP66-1.[17] For a detailed treatment of the tests for wood shingles, read the September 1967 *Fire Journal* article, "A New Class C Treatment for Wooden Shingles and Shakes," by R. H. Bescher.[18]

Row Frame Buildings

In older parts of many cities, frame buildings were often erected in rows; that is, the structures are contiguous. Often they have a common attic or cockloft and may even have party walls (division walls between buildings which provide support to both buildings). In older construction, there may be no fire barrier at all between buildings, or a crude attempt may have been made at a fire barrier by brick nogging (brick and mortar filling between studs). The brick nogging does not close the floor voids or the cockloft.

Brick or stone nogging is also found in old individual houses. It served as a heat sink for winter warmth. This presents an additional hazard in collapse.

In more recent construction, the row house is dignified with the name "town house." In any case, unless there is an adequate masonry fire wall between the separate buildings, all the way through the roof, the entire structure is all one building and should be so described in stating the size for planning or fire report purposes. It matters not one bit to the fire that these houses are separately described on the land records and are separately owned. As long as fire can move from unit to unit totally, or incompletely impeded, the structure is one building.

The change in description may be of real value. Consider a row of frame houses, ten in number, each 20 feet by 40 feet. The row of houses is actually one building 200 feet by 40 feet. A fire involving one house might require a greater alarm, and it is much easier for the fire ground commander to justify sending a greater alarm for a building 200 feet long rather than one described as 20 feet wide.

Closely Spaced Wooden Buildings

In any area where there are closely spaced wooden buildings, the prime tactical consideration must be to confine the fire to the building of origin. This makes it mandatory that units be prepared to place heavy caliber streams in operation immediately upon arrival. Preconnected small lines serve very well in the majority of situations, but the really competent fire officer is one who can recognize a big fire and respond to it *immediately* with big fire tactics, not by working upward through small fire tactics because of (bad) habit. Since many fire departments lack the manpower to handle big handlines, there is a clear need for equipping apparatus with preconnected deluge guns, combining heavy caliber streams with minimal personnel requirements.

Wooden Cooling Towers

From time to time, a fire version of the "man bites dog" story appears in the news; a wooden cooling tower burns. Cooling towers are used to disperse the heat from air-conditioning systems, and when the tower is functioning, cascades of water pour down the inside. Despite this, much of the wood remains dry, and

Where wooden buildings are closely spaced, fire department tactics must be based on confining the fire to the building of origin.

if the tower is not operating, the wood is all dry. Electrical short circuits, fan-bearing friction, and even birds carrying lighted cigarettes to their nests in the towers are but a few of the possible fire causes. Many cooling towers are sprinklered, particularly where interruption of cooling would be costly; others are built of noncombustible materials. When erected as part of a modern high-rise building, the tower is often concealed by masonry walls or constructed in a "well" accessible only from the roof. Wooden water tanks have also been known to burn; one was reported as a "fire in the fence around the water on the roof."

A fire in Toronto, Canada, illustrates the vulnerability of wooden cooling towers. A fire in a trash disposal container was fanned by high winds. Embers were scattered for several blocks. A third of a mile away, a wooden cooling tower atop a 26-story building was ignited and virtually destroyed.

Heavy Loads on Wooden Structures

Thus far, the discussion of wooden structures has been directed primarily to lightweight wooden structures typified by single-family or multi-family residences.

Much ingenuity has gone into the development of heavy wooden construction capable of sustaining massive loads or of spanning great distances. Structures of wooden interior construction, with masonry walls, are classified as "ordinary construction" and are discussed in Chapter 4, Ordinary Construction. A special type of heavy timber and masonry construction, which is designed to control ease of ignition and eliminate void spaces, is called "mill construction."

Here an attempt will be made to confine the discussion of wood principally to its combustibility, leaving the structural problems that flow from this characteristic to Chapter 4, Ordinary Construction.

"The great enemy of the truth is very often not the lie, deliberate, contrived and dishonest, but the myth, persistent, persuasive and unrealistic."[19] This quotation is rather appropriate to this discussion of heavy timber construction.

A typical statement of the "advantages" of heavy wooden construction is: ". . . it should be mentioned, however, that large sections of wood, over 6 inches thick can actually prove to be more resistant to fire than exposed steel. Although steel does not burn, fire temperatures in excess of approximately 600°F cause steel to soften and bend like a pretzel, whereas the surface of the heavy wood merely chars, limiting further burning."[20] The effects of fire on steel will be examined in Chapter 9, Flame Spread. The statement, with respect to wood, is an oft-repeated myth.

In Daytona Beach, FL, a jai alai fronton of laminated timber arches burned furiously and collapsed. The owners looked for some special reason for the collapse. The reason was simple. The arches fell apart at the connections, which were simply unprotected steel straps. These were adequate for the normal building load, but failed very early in the fire. The "slow-burning" timber became just additional deadweight to collapse.

Don't fall for propaganda; look at the connections.

Another typical statement is: "Heavy timber has two all important properties that are too often overlooked when rating its fire resistance: Heavy timber is 'slow burning'; and 'Wood is a poor conductor of heat.'"[21]

The term "slow burning" is advanced uncritically as if it represents an unqualified advantage. In fact, slow burning is an advantage only as long as the fire department can maintain inside positions. If the fire cannot be extinguished, the slow-burning characteristic merely prolongs the air pollution.*

Wood construction has many merits. Properly protected with automatic sprinklers, or subdivided into manageable fire areas by fire walls, it can hold its own in a dispassionate discussion of the pros and cons of various materials. In "The Fire Performance of Timber," a good discussion of the merits of timber as opposed to other building materials, Thompson points out the necessity for sprinkler protection, since no building material can detect or extinguish a fire.[22]

It is a bit discouraging, therefore, to find in an article describing the newly-built laminated timber warehouse of a lumber company that "in order to comply with local fire regulations, the building will be equipped with a sprinkler system."[23] Thus, again, the myth that fire protection requirements are a burden imposed by local authorities, and by inference, not really necessary, is perpetuated.

---

*Each building material has certain advantages and certain disadvantages under fire conditions. An advantage which might tend to limit loss might make the fire fighting operation more hazardous, for instance, and often the question is: "What advantage are you talking about, or, perhaps, whose advantage?" The proponents of each building material work tirelessly to advance their own interests and may volunteer little or no information which might be considered derogatory. So, if at times this text seems to be overbalanced in presenting dangerous or undesirable characteristics, the balance can be redressed, without cost, by writing the appropriate trade association.

For a number of years, fire protection experts, spearheaded by the NFPA, have carried on a campaign, which has been quite successful, to eliminate the use of the word "inflammable," as dangerously misleading, and have urged the substitution of the word "flammable." The time has come to retire the term "slow burning" for the same reason. The term "ignition resistant" would be more accurate, where true. Even the slight change from "slow burning" to "long burning" might be more expressive of the principal characteristic of heavy timber.

The argument for timber structures is made principally on the basis of structural stability during long fire exposure, but in many structures of modern design, the finished surface of the wood is very important to the owner as a principal feature of the interior decor of the building.

A case in point is a fire which occurred in St. Paul's Methodist Church, Kensington, MD. The church was built of laminated timber arches; the ceiling was the exposed underside of the plank roof. The interior was finished in a combustible clear finish. A robing room of lightweight wood stud, plywood and combustible tile construction had been erected in the chancel. A burglar started a fire, and though it spread rapidly, it was speedily controlled, but not before every inch of the interior wood of the church was charred to a depth of ¼ to ½ inch. The following day, the building authorities pronounced the building structurally sound, but in effect, the church was destroyed.

A fire in St. Charles' Church, in Killarney, FL, is another case. The interior surface fire was quickly controlled with three heavy caliber solid streams which were kept moving to sweep the entire surface. (Fog streams would have been "eaten up" by the fire ball and no water would have reached the surface. Only sandblasting was required to restore the surface.) This fire also demonstrated that ventilation of plank on timber arch structures is hardly practical. In this case, it took 45 minutes and three circular saw blades to open a 12-square-foot hole in the roof.

The rate that flame spreads over the finish on the wood may be the determining factor in the total amount of the loss in a heavy timber building. Readily combustible contents, or lightly constructed partitions, or other structures within the building (such as the robing room in the church fire cited previously), may provide the fuel for the ignition of a wood finish with rapid flame spread characteristics.

It is fairly common in model building codes to require that any steel structural member that is used in a building classified as "heavy timber" be provided with 60-minute fire protection. Note the carefully worded question and answer in "Fire Facts": [24] "Q. What is the fire rating on heavy timber construction? A. It has not been rated formally, but many model building codes require that materials used in lieu of heavy timber in heavy timber construction have a one-hour fire resistance rating." If the fire resistance of a material is questioned, and it is "not rated," the reason is usually simple. It can't pass the test.

The substitution of unprotected steel for heavy timber construction, or the use of steel to reinforce aging or overloaded timber structures should be noted. Fire

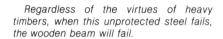
Regardless of the virtues of heavy timbers, when this unprotected steel fails, the wooden beam will fail.

departments should study alteration permits and investigate when building materials are being delivered to a completed building.

The Gibson City Methodist Church, of Gibson City, IL, is a building of masonry exterior walls and interior timber framing. It became necessary to remove a timber column 6 by 8 inches carrying an 18-ton load. A one-story-high steel truss of steel tubes and solid steel tie bars was inserted in the building to permit the removal of the column. A description of this engineering feat is silent as to any fire protection provided to the steel.[25] Unless the steel was protected, it is at least possible that the truss could fail in a fire prior to the time the wooden column might have failed. Since the wooden column was never assigned a fire resistance rating, it was reasonable to replace it with a structure which also had no fire resistance rating. A fire department should avoid attributing fire resistance qualities to structures which do not possess them, because of the myth of "slow-burning" timber.

A chain of supermarkets has adopted laminated wood arches as a distinctive feature of their store design. In at least some observed cases, the arches are supported by three unprotected steel columns. It is not within our expertise to predict the consequences of the failure of such columns, but we can conclude that they would not have been installed unless necessary.

Beams are often penetrated for electrical conduit, and in older buildings for gas pipes. Long bolts and nuts are used to attach timbers to other members. Any of this metal can provide a path for heat to reach the interior. Destruction of wood by pyrolytic decomposition may cause failure. Long solid timbers are hard to obtain. Watch for spliced timbers with overlapping joints and metal connectors. These would be subject to early failure at the joint.

Fire Fighting

If you have responsibility for a large, open-area, heavy timber building, try to keep out the "kindling." If a structure is to be built into the building, such as a room, a ticket office, a hot dog stand, or whatever, urge that it be built of noncombustible material or at least of fire retardant wood. Though many condemn partial sprinkler protection, the author is personally of the "half a loaf" school and would recommend sprinkler protection, even a couple of sprinklers off the domestic system if that is all that is available, to protect small enclosures in heavy-timbered buildings. Don't call it partial sprinkler protection, call it a super extinguisher. If there is no "kindling" to set it on fire, the open-area timber structure should be a good bet to stand a long time.

The structure at Busch Gardens at Tampa, FL, in which trained primates perform, is a wooden dome. It is not sprinklered, but the dressing room is sprinklered.

Fire fighting plans should provide for heavy caliber streams with adequate reach to cover the whole inside surface. Operating points should be determined in advance and "dry run" drills undertaken. A deluge set inside the building (truck-mounted if access permits) with a relatively small solid stream nozzle, to guarantee reach of the stream to all points as fast as possible, might be best. Speed in getting all the interior surfaces wet might be the difference between a refinishing and a rebuilding job.

To sum up, heavy timber structural elements may provide a fair measure of structural stability during a fire if the connectors and any supporting elements are equal to the timber in fire resistance. If not, as is usually the case, anticipate collapse due to failure of the weaker element. This is particularly true of timber trusses.

In a Florida country club, huge wooden parallel-chord trusses are featured in the dining room. The fire department has taken notice that the trusses are supported on an unprotected steel girder. They are aware that the girder may rotate when attempting to elongate and drop the trusses, or simply collapse. Any serious fire here is a planned defensive operation; the building is "designed to collapse."

If there are concealed spaces which cannot be reached with hose streams, if there are other obstructions to successful manual fire fighting efforts, if the contents' fire load has a fast, high-heat-release potential, and if the building is unsprinklered, expect a massive fire. If it is built of wood, it can burn. If it can burn, the fire can be put out by putting water on it. The key point of the prefire plan is to determine how the water is to be applied in sufficient quantity to stop the production of heat.

The objection to sprinkler piping in the open decorative wood structure is easy to understand. The exposed piping "destroys the decor." In some cases, the sprinkler piping is run close to beams, and sidewall sprinklers are used. A competent painter, or better still, a "grainer," charged with making the pipes "disappear," can work wonders.

Imitation Timber

Styles in interior decor seem to run in cycles. In contrast to the severely modern interiors popular a few years back, current styles feature wood in hundreds of ways. Very often the wood is made to appear structural and is not. Often it is not wood. Few if any of the practices are effectively regulated by code. Some deceptions noted are:

- *Unprotected steel beams or columns* boxed in wood to look like wooden structural members.

- *Unprotected steel* encased in plaster surfaced to look like wood.

- *False wood beams* actually hollow wooden boxes of wood boards. In one case, concrete beams were formed in plywood which was left in place and painted black to simulate wood beams. In one case, the wooden box imitation beams were glued up. The connections and hangers should be examined for fire vulnerability. The temperature up at the ceiling can be hundreds of degrees higher than it is at the floor level.

- *Real wood beams* serving no structural purpose.

- *Polyurethane imitation wood beams and fittings* such as brackets. These are readily ignitable and burn furiously. Such imitations are often found in conjunction with plywood veneers. The hazards include the sometimes massive fire load overhead, and the problem of collapse.

Years ago, the desire to deceive was demonstrated in the opposite manner. Epiphany Church in New York City, which burned in a spectacular fire, provides a good example. Built 20 years after the first arrival of famine-stricken immigrants from Ireland, it attempted to demonstrate the success of the parishioners in the new world. In construction, 12-inch by 12-inch wooden columns were used, but there was no "status" in admitting that the interior columns were similar to those used in factories. The columns were rounded out using fitted strips of wood. They were then covered with fabric, and the fabric painted to imitate marble. Painted wooden "marble" columns adorn the front of the Custis-Lee mansion which overlooks the grave of John F. Kennedy at Arlington Cemetery. "Wood graining," a craft which specializes in providing deceptive surfaces, has been revived.

It is now becoming fashionable to make a masonry building appear to be of wood construction. The Teton Lodge, in Teton National Park, is made of concrete. The wood formwork marks were left evident and the entire building stained brown. Brick or block buildings are covered with a wood veneer. This is mounted on furring strips making it easy for the wood to burn on both sides. In one case, a concrete block, one high-story store was covered with shingles stapled to plywood sheets and mounted on vertical wood trusses. The shingles were said to be treated but apparently the plywood and wood trusses were not.

Wooden Suspended Ceilings

The popularity of wood for interior decor has led to the development of suspended wooden ceilings, often in otherwise fire resistant buildings. These can range from a few timbers, suspended from the ceiling of the sporting goods department in the department store, to rather sturdy assemblies providing a large suspended load. In some cases, such false ceilings interfere with the coverage of automatic sprinklers. In other cases, the sprinkler coverage is maintained by openwork construction or by dropping sprinklers below the ceiling. Fire on the upper surface of the ceiling might be untouched by water from sprinklers below.

The fire resistance of the supporting system of an extensive, heavy suspended ceiling is worth investigating. It would not be surprising to find it hung on aluminum wire, or supported in some other equally fire-vulnerable way. It might be interesting to inquire as to the provisions of your local code governing the fire protection of supports for suspended structures. Mere compliance with "flame spread" requirements is not sufficient. The supports are up high where heat concentrates, and the fact that the suspended ceiling is about to fall may not be apparent to fire fighters operating on the floor level.

Strength of Wood

Lumber is a natural product, and two pieces of lumber of similar size can vary greatly in strength. This may be apparent to the skilled eye because of defects, such as a large or loose knot, or other weaknesses, or it may be determined only by test.

In February 1967, the trussed roof of St. Rose of Lima church in Baltimore, MD collapsed, injuring 48 worshippers. The *Washington Post* reported that a consulting structural engineer found that a single 2 by 4 of inferior grade had failed, triggering collapse of the entire roof. The *Post* bought samples of wood in the Washington area and had them tested for strength. In a subsequent article,[26] it was reported that in one case, the wood should have had a modulus of elasticity of 1,760,000 pounds per square inch. Instead, the samples had values of

This beam quit when it tired of being pushed on top (compression) and pulled on the bottom (tension).

410,000 and 1,240,000 pounds per square inch. The samples tested graded half above and half below the required figure. In some cases, averaging values may be satisfactory, but not where failure of any single unit can precipitate a general collapse. Bear in mind the sad fate of the man who drowned in a river, the average depth of which was only 6 inches.

As explained earlier, actual lumber sizes are less than nominal due to planing. In recent years, the lumber industry has reduced the actual sizes another 7 percent (2-inch lumber, formerly 1⅝ inches thick, is now 1½ inches thick). This, of course, reduces the fire fighter's margin of safety further, as there is less stiffening wood to burn away before the fire reaches wood that is providing the designed structural strength, even assuming the new size is normally sufficiently strong.

Wood is a combustible material. It has served man for ages and will continue to do so. It is a uniquely renewable resource. In structures of any consequence, the problem of its combustibility can be dealt with only by complete automatic sprinkler protection, properly designed, with adequate water supplies, and competently maintained. Given this protection, a wooden structure is superior from the life safety point of view than too many unprotected fire resistive structures. A case in point — visitors to Teton and Yellowstone Parks may see two hotels. One is of concrete but it was lined with plywood when last seen. There were no sprinklers. Another is an old wooden hotel but is fully sprinklered. The author would much rather stay in the sprinklered wooden hotel.

References

[1]Huntington, W. C., *Building Construction*, 3rd ed., Wiley, New York, 1965, p. 355.
[2]Radford, W. A., ed., *Radford's Cyclopedia of Construction*, Vol. III, Radford Architectural Co., Chicago, 1923, p. 16.
[3]McKinnon, G.P., ed., *Fire Protection Handbook*, 15th ed., National Fire Protection Association, Quincy, pp. 5-31, 5-32.
[4]Anderson, L. O., "Wood-Frame House Construction," *Agricultural Handbook*, No. 73, Rev. July 70, U.S. Dept. of Agriculture, Washington, DC, p. 33.
[5]"Manual for House Framing," *Wood Construction Data No. 1*, AIA 19B, National Forest Products Association, Washington, DC, 1968, p. 6.
[6]*The Volunteer Fire Department of Old New York 1790-1866*, 1962, Americana Review, Scotia, NY, Back Cover.
[7]Eickner, H. W. and Schaeffer, E. L., "Fire Retardant Effects of Individual Chemicals on Douglas Fir Plywood," *Fire Technology*, Vol. 3, No. 2, May 1967, National Fire Protection Association, Boston, pp. 90-104.
[8]"Arches are High, Wide, Handsome," Engineering News-Record, Feb. 23, 1967, p. 101.
[9]See Ref. 2, pp. 115, 120, 314.
[10]O'Hagen, J., "Wall Collapse, Serious Injury Averted," WNYF, Vol. 29, No. 3, 1968, NYFD, New York, p. 19.
[11]"Timber Truss is Economical," *Engineering News-Record*, Feb. 22, 1968, pp. 46-47.
[12]Mittendorf, John, "Lightweight Construction Tests Open Fire Service Eyes to Special Hazards," *Western Fire Journal*, Vol. 34, No. 1, January 1982, p. 23.
[13]Stone, W., "The National Electrical Code and You," *Fire Journal*, Vol. 64, No. 3, May 1970, p. 33.
[14]McKinnon, G.P., ed., *Fire Protection Handbook*, 14th ed., National Fire Protection Association, Boston, p. 1-34.
[15]See Ref. 1, p. 572.
[16]Bugbee, Percy, "Conflagration Warning," *NFPA Quarterly*, National Fire Protection Association, Vol. 51, No. 4, April 1958, pp. 216-220.

[17]Wilson, R., "Some New Answers to the Shingle and Shake Problem," MP 66-1, National Fire Protection Association, Boston.

[18]Bescher, R. H., "A New Class C Treatment for Shingles and Shakes," *Fire Journal*, Vol. 61, No. 5, September 1967, pp. 52-56.

[19]Attributed to President John F. Kennedy.

[20]Zuk, William, *Concepts of Structures*, Litton Educational Publishing Inc. by permission of Reinhold Publishing Co., 1963, p. 12.

[21]"Fire Facts," American Institute of Timber Construction, Washington, DC.

[22]Thompson, H., "The Fire Performance of Timber," *Forest Products Journal*, Vol. VIII, No. 4, pp. 31a-34a.

[23]"New Timber Techniques Cut Costs," *Engineering News-Record*, Aug. 27, 1967, p. 25.

[24]See Ref. 21.

[25]Gurfunkel, G., "Built in Place Truss Aids Remodeling Job," *Engineering News-Record*, Feb. 23, 1967, p. 25.

[26]"Lumber Misgrading, Builders Beware," *Washington Post*, Nov. 19, 1968, p. 1.

CHAPTER 4

Ordinary Construction

The term "ordinary construction" describes an infinite variety of buildings. The chief common characteristic is that the walls are of masonry.[1] The structural stability of such exterior walls and interior support systems will be discussed in this chapter. Void spaces and masonry walls as fire barriers will be examined, as will automatic sprinklers and fire preplanning. Mill construction is somewhat similar to ordinary construction but does have some important differences. It, too, will be discussed in this chapter.

Masonry Construction Terms

We should be familiar with some of the principal terms and practices used in masonry construction. Among them are:

> **Course**   A horizontal line of masonry.
>
> **Stretcher Course**   Bricks laid lengthwise.
>
> **Header (Bond) Course**   Bricks laid endwise.
>
> **Wythe**   Vertical section of a wall one masonry unit thick.
>
> **Concrete Masonry Unit**   A precast structural block made of cement, water, and aggregates. It may be hollow or solid.
>
> **Terra Cotta**   Tiles made of clay and fine sand, fired in a kiln. Terra cotta is both structural (clay tiles) and decorative, as for ornamental facings. Structural terra cotta has been replaced to a large extent by concrete block.
>
> **Rubble Masonry**   Masonry composed of random stones.

Rubble Masonry Wall A wall composed of an inner and outer wythe of coursed masonry. The inner space is filled with random masonry sometimes mixed with mortar. Such walls are unstable to a lateral thrust.

Ashlar Masonry Stone cut in rectangular units.

Solid Masonry Unit Unit whose net cross-sectional area in every plane parallel to the bearing surface is 75 percent or more of its gross cross-sectional area measured in the same plane.

Hollow Masonry Unit A unit whose net cross-sectional area in every plane parallel to the bearing surface is less than 75 percent of its gross cross-sectional area measured in the same plane.

Solid Masonry Walls Consist of masonry units (either solid or hollow units) laid contiguously with the joints filled with mortar.

Hollow Masonry Walls Consist of two wythes of masonry with an air space in between; the wythes are tied together (bonded) with masonry.

Cavity Walls Hollow walls in which the wythes are tied together with steel ties or masonry trusses.

Veneer Wall A wythe of masonry attached to the masonry bearing wall but not carrying any load but its own weight.

Composite Wall Made of two different masonry materials such as brick and concrete block, designed to react as one under load.

Masonry Columns Masonry walls are sometimes braced by masonry columns incorporated into the walls and called variously "piers," "buttresses," "pilasters," or "columns." These may be built inside or outside the building. Where visible, they indicate where the wall is strongest, often where the concentrated loads are applied, and where not to attempt to breach the wall.

Cross Wall Any wall at right angles to the wall in question; it provides support.

This composite wall shows the diversity of vernacular construction. How many different materials can be identified?

Flying Buttress   A masonry pier at a distance from the wall and connected to it. Such buttresses resist the outward thrust of arches. They are used in Gothic architecture.

Cantilever Wall   A freestanding wall, unsecured at the top, is a cantilever beam with respect to lateral loads, such as wind or a hose stream. Precast concrete walls are very dependent on the roof for stability. If the roof is affected by the fire, the walls are likely to fall. Refer to Chapter 2, Construction Principles, for a discussion of the hazards of cantilever walls as found in masonry buildings under construction.

Hollow and cavity walls are used to limit penetration by rain. Though there are no observed cases to demonstrate, it is at least possible that carbon monoxide from a fire could accumulate in the hollow space and explode disastrously. In the New London, TX school disaster, March 18, 1937, in which 294 persons lost their lives, unodorized natural gas accumulated in the voids in a masonry wall.

Gas from a leaking main outside the building found its way into the hollow of a cinder block wall in a Florida supermarket. An employee flicked his lighter and the gas exploded. In seeping through the earth, natural gas can lose its artificial odorizer.

It is becoming an accepted practice to place sheet or foamed-in-place plastic insulation in hollow walls. The plastics used are of varying degrees of ignitability. Burning plastic produces large quantities of smoke. If the source of smoke cannot be found, it would be wise to check for plastic insulation in the walls.

Brick masonry walls were, for centuries, built only of bricks. In recent years, the composite wall, using concrete block to the maximum to save on brick and labor, has been developed. When such walls were first developed, the conventional system of bonding the wall together by inserting brick headers according to various design practices was used.*

The common practice was to provide a row of headers every seventh course. Uneven settlement often caused the header bricks to crack and so the masonry wire truss was developed. This wire truss is bedded into the mortar in specified courses. As a result, the header course is no longer necessary, and the appearance of a masonry bearing wall may be no different than that of a veneer wall (all stretchers). In some veneer walls, bats (half bricks) were inserted to give the appearance of a bonded wall. Thus, it is impossible to tell a bearing wall from a veneer wall by external appearance alone. We must study the building. (Brick veneer is also used as a facing for wood construction. See Chapter 3, Wood Construction.)

Not all brick-and-block walls are composite. Brick may be veneered onto the block or onto concrete using the same ties that are used in brick veneer on wood construction.

What Is Ordinary Construction?

There are several definitions of ordinary construction. Any description of a specific building, intended for legal purposes, should, of course, describe the building in the terms used in the applicable code.

Together with wood construction described in the preceding chapter, ordinary construction usually fits Condit's[2] term "vernacular construction," i.e., construction which develops out of hand-me-down methods with little or no engineering. The opposite term to vernacular construction would be "engineered construction."

Codes and standards attempt to divide types of buildings into various classes, mutually exclusive, like pigeonholes. Unfortunately, many of the buildings with which the fire fighter must cope were built by people who used the material which seemed best suited to their purpose, or was available, in a manner which seemed to do the job, without any reference to the niceties of distinctions between types of construction as classified by building codes, the NFPA, or insurance underwriters.

Most building codes have a provision for so-called "fire limits." Within the fire limits a structure may not be built unless the outer walls are of masonry.**

---

*Terra cotta tile was used in early composite walls. The pattern produced by various combinations of stretchers and headers gives a name to the wall, as "English Bond," "Flemish Bond," etc.

**New York City has very few pre-Revolutionary buildings. With much fanfare, an ancient farmhouse was relocated to prevent its demolition. A company officer put a violation on the structure, "It is illegal to relocate a frame building within the fire limits!"

Characteristics of Ordinary Construction

Ordinary construction is "Main Street USA." The single unifying characteristic of ordinary construction is that the bearing walls are of some type of masonry, but even here an exception will be made to include buildings in which at least some of the walls (usually only the front, or street-facing walls) are of cast iron. This is logical because the interiors are similar to masonry wall buildings, and there are only a few cast-iron buildings left.

The simplest ordinary construction building is built of masonry bearing walls, with wood joists as simple beams spanning from wall to wall. The joists usually are parallel to the street frontage of the building. The roof may be similar to the floor in construction or it may be provided with a peak, by the use of rafters or trusses.

The masonry bearing walls may be of brick, stone, concrete block, terra cotta tile, adobe, or cast-in-place concrete. The wall may be all of one material; different materials may be used in discrete areas, or different materials may be combined into composite construction (expected to react together under load).

Bearing and nonbearing walls are often identical in appearance and in material of construction. In the typical downtown business building along Main Street, the side walls are the bearing walls, the front and back are nonbearing. Bearing walls are stabilized by the weight of the building elements being carried. Nonbearing walls lack this stabilizing effect.

The simple wood beam floor is satisfactory for buildings up to a practical limit

In typical ordinary construction, such as this four-story building with the front wall removed, wood floor joists up to 25 feet long was the standard. Today, wood I beams or trusses of wood or steel or both may be used. Fire tactics developed for buildings such as this are not adequate for trussed floors.

of about 25 feet in width. For a wider building, or a building of irregular plan, some sort of interior column, girder, and beam system must be provided. There is no limit to the ingenuity of builders. Every possible combination of building materials is used. Columns may be of wood, brick, stone, concrete block, steel, or cast iron. Different materials may be used for columns in the same building. Interior masonry bearing walls may take the place of columns. In light-floor-load buildings, interior balloon frame stud walls may provide the intermediate support. Girders may be of wood or unprotected steel. The connection systems by which the beams are attached to the girders and the girders are attached to the columns are of infinite variety, and it is in the weakness of connections that the principal collapse potential of the building during a fire is often found.

Interior bearing walls may obstruct improvements, such as combining several rooms into a restaurant. The wall is opened. The wall above is picked up on a column-and-beam arrangement. This arrangement will have less inherent fire resistance than the original brick wall. Such alterations were major factors in the Empire Apartment and Vendome collapses described later. Be very wary of the building that has been altered during its lifetime.

As in wood construction, void spaces are an inherent part of ordinary construction. Some fire protection measures (such as "tin" ceilings), intended to prevent the extension of fire from the usable space to the void space, prove to be barriers to the fire department's efforts to reach the fire, once the fire penetrates the void space. Modernizing may make the building one big void space by eliminating windows.

As a general rule, there is no effective fire separation within the building, either from floor to floor or within floors. Even where fire separations exist up through the regular floors of the building, they often are imperfect or nonexistent in attic spaces.

Height of "Ordinary" Buildings

There is an inherent limit to the height of masonry buildings; it is the necessity for increasing the thickness of the wall as the height of the building increases. The usual rule is that the solid masonry walls shall be 12 inches thick for the uppermost 35 feet of height and increase 4 inches in thickness for each additional 35 feet or fraction in height. Hollow masonry walls are more restricted in height; solid walls reinforced by masonry cross walls or piers can be thinner.

The tallest old style masonry bearing wall building in the United States is the Monadnock Building in Chicago. It is 15 stories high, and the masonry piers at the ground level are several feet thick. About the time it was built, steel frame construction was developed, and masonry soon was being used only for fire resistant panel walls, supported on the steel frame, and not load bearing. Not all masonry walls in low-rise, steel-framed buildings are carried on the frame; some are freestanding and merely tied to the frame for stiffening.

In recent years, high-rise brick or concrete block buildings with no wall thicker than 12 inches, and medium-rise brick buildings with no wall thicker than 8 inches have been developed, supplanting the traditional practice of ever-increasing wall thickness. Since these reinforced masonry structures depend in great measure on integration with reinforced concrete, they are discussed in Chapter 8, Concrete Construction.

The Scope of Ordinary Construction

Some buildings constructed in recent years may approach the pure definition of "noncombustible" construction; however, in many cases the difference between them and buildings of ordinary construction, from our paramount point of view of fire suppression, is minor, and, thus, such buildings are properly discussed in this chapter. A noncombustible void space can accumulate carbon monoxide gas as readily as a combustible void. A noncombustible void can contain combustible wiring and thermal insulation. Metal roofs can provide self-sustaining fires because of the use of bituminous vapor seals (see Chapter 7, Steel Construction), and a wall pushed out of line by an expanding heated steel truss can be just as lethal as a wall pushed out of line by collapsing wood joists. In addition, the use of wooden "eyebrows," cornices, canopies, "colonial" belfries, combustible interior wall and ceiling finish, and even wood veneer over masonry leaves us few truly "noncombustible" commercial or institutional buildings. Unprotected steel industrial buildings, which often are truly noncombustible, are discussed in Chapter 7.

Similarly, the boundary line between ordinary construction and early "fireproof" construction (Chapter 11, High-Rise Construction) is not at all clear-cut since building development is evolutionary. For instance, one early skyscraper consisted of seven stories of metal-framed construction, topped by four stories of masonry wall-bearing construction.[3] Many of the hazards of early "fireproof" buildings can be found in ordinary construction; unsupported marble stairs is a good example.

Portions of a building of ordinary construction may have been provided with some degree of fire resistance, either initially or as a result of legal action, as for instance, the installation of properly enclosed, fire resistive stairways in an old school, or the provision of a rated fire resistive barrier around a special hazard such as a boiler room. Rarely does this piecemeal provision of fire resistive features alter the fundamental nature of the building.

Automatic sprinklers may be retrofitted to improve the fire characteristics of the building. It is most important that the entire building be sprinklered, that all spaces are covered. Sprinklers omitted from certain areas for aesthetic or other reasons are an invitation to disaster. If a deficient system is installed, the fire department should be aware of the deficiencies and plan to compensate for them.[*]

---

[*] The author was pleased to see that a hotel in Alaska had been sprinklered, down to a sprinklered shower stall. The unconnected sprinkler riser, however, brought the revelation that, "the contractor never finished the job." The siamese was so located that a connected hose line would make it impossible to open the exit door.

An old office building in Memphis, TN has been extensively modified over the years, including the installation of a light-hazard sprinkler system. The sprinklers were obstructed by air ducts, and the system was not properly maintained. A fire originated in a void space. It took scores of fire fighters 3 hours to literally "dig" the fire out of the walls, cocklofts and other concealed spaces.[4]

A combustible building was sprinklered except for a 3-foot high cockloft. Fire originated in the cockloft and burned for 11 hours. The loss was $600,000.

The sprinkler protection of a dinner theater was impaired by the construction of a mezzanine without sprinklers underneath. The fire occurred underneath the mezzanine and gained headway until ceiling sprinklers operated and sounded the alarm. The loss was much greater than it would have been had proper sprinkler protection been maintained.

The builder or alterations contractor may use fire resistant components, or those similar in appearance, in some part of a building which by law need only be of ordinary construction. Be especially alert for this condition; it may lead to unwarranted assumptions as to fire resistance, which the building may not possess. For instance, the law may not require a finished ceiling in a store. The owner may choose to use a low-flame-spread suspended ceiling for decorative effect in the sales area and omit it in the stockroom. Such a ceiling is similar in appearance to a listed fire resistive roof and ceiling assembly, but there are vital differences.

An office building recently observed under construction is a case in point. The floors are of bar-joist construction with left-in-place corrugated metal forms and concrete topping. The ceilings are of "lay-in" tile. A casual inspection might conclude that the building is of fire resistive construction, using floor and ceiling bar joist assemblies. In fact, the building is of ordinary construction. It was the designer's choice to use the described floor. It need not be fire resistive because the code does not require a fire resistive building. There is an unprotected steel ridge beam running the length of the units. Given a fire in the wooden roof, enough heat would be generated to elongate the beam and push out the brick gable ends of the building.

Alterations through the years, including reconstruction from fires many years ago, may cause one part of a building to be quite different from another. What you see on first inspection may not be what you get. Interior alterations and finish may make it difficult or impossible to determine the nature of the building.

A case in point is the dining hall at the National Fire Academy. When the building was first taken over from St. Joseph's College, it was not too difficult to trace the growth of the building over the years by examining the basement and the attic. The interior improvements have made it much more difficult.

In short, this chapter must be read in conjunction with the chapters on every other type of construction. Void spaces, balloon frames, tile arch floors, marble stairs, forests of timbers, unprotected steel, plain and reinforced concrete — all may be found. Some elements may masquerade as being fire resistive. Some additions may be truly fire resistive, particularly in public buildings. Often the fire resistive and non-fire resistive sections are not effectively fire-separated.

When this structure, an old canal lock inn, was remodeled, a steel girder (arrow) was added for support. Elongation of this girder could bring down the wall.

Problems of Ordinary Construction

The problems presented to the fire department by ordinary construction can be divided into the following headings:

- The structural stability of the masonry wall.

- The stability of the interior column, girder and beam system.

- Void spaces which are inherent in ordinary construction and which can be increased by alteration.

- The efficiency of the masonry wall as a barrier to the extension of fire to the next building.

These are not hard and fast divisions. They are all interrelated and are separated just as a device to organize the material. An exterior collapse will usually cause an interior collapse. Conversely, the elongation of steel beams or an interior collapse may cause the walls to come down. A carbon monoxide explosion in a void space can demolish a building if the carbon monoxide-oxygen mixture is close to the stoichiometric (most efficient) mixture. A minor failure of a wall may permit fire to extend to the next building.

Early in the study of this subject, two books came to hand: *Construction Failures* by Feld, and *Building Failures* by McKaig. Both of these excellent texts cover similar ground. The authors are consulting engineers, often retained after a collapse to fix the cause. If an investigation after the fact could determine that the clues to impending disaster were evident before the disaster, we in the fire service could examine buildings beforehand and determine at least whether there is a likelihood of a particular building failing in a fire. This appears to be a reasonable assumption.

Many, if not most, of the ordinary construction buildings in a typical city were built before any present member of the fire department was born. There is ample opportunity to study the buildings ahead of time and establish suitable pre-planned tactics to minimize risk.

In this text, many clues to potentials for disaster are given. The list could, of course, be endless and is in no sense a check list. It is a guide to help form your own estimation of the hazards of a particular building in which your department must fight fire, and the ways to sidestep those hazards.

Some of the clues are evident from the street; others require detailed examination, which may not be possible. In any event, always be aware that an ordinary construction building, however sturdy and well maintained it may appear, was built without any thought being required as to what would happen to the building in the event of fire. By code definition, such buildings are non-fire resistive; that is to say, they have no designed resistance to collapse in a fire.

Structural Stability of Masonry Walls

The traditional fire service training on the subject of falling walls provides little real guidance. Years ago, an old-time buff related to us the dictum of the legendary 19th century Chief Bonner, "When you see smoke coming out of the walls, puff, puff, puff, it's time to run." In an otherwise wonderful book, *Fire*, which

In many cases, buildings depend on one another for support. The loss of one in a fire may cause collapse of the other. When the building between them was destroyed, buttresses were added to these for the necessary stability.

probably kindled the author's interest in fire protection, Chief Daugherty describes an officer who saw that a wall would fall, but the position was "advantageous," and the crew stayed in a location where a window fell around them! Other books wisely speak of walls falling this or that fraction of the width of the street. Perhaps so, but the bricks fly in all directions. If a single brick gets you, the fate of the rest of the wall is immaterial to you.

Is it probable that a masonry wall will collapse during a fire? Many do; many do not. To the best of the author's knowledge, no study has even been conducted to determine if there is a common pattern in fire collapses. There are, however, certain indicators of probable collapse, some of which can be observed beforehand and noted in a prefire plan. Other indicators may be observed during a fire.

Collapse may be due to a variety of causes; among them are:

- Inherent structural instability, aggravated by the fire.

- The failure of a non-masonry supporting element upon which some portion of the masonry depends.

- Increase in the live load due to fire fighting operations, specifically retained water.

- The collapse of a floor or roof with consequent impact load to the masonry wall.

- The impact load of an explosion.

- The collapse of a masonry unit due to overheating.

- The collapse of another building onto the building in question.

It is good to bear in mind that the right to own a building and to maintain it as one pleases is a fundamental right, well anchored in our legal concepts. Bit by bit, the state has whittled away at the untrammelled right of the owner to do as he pleases, but rarely without legal challenge. One of the rights is to maintain the building in a dilapidated condition, and usually only in the presence of clear evidence of existing public danger will the courts order an owner to repair or demolish his property. *

It is certainly not commonly held, if indeed it has been held at all, that the potential collapse of a building in the event of a fire is adequate grounds to infringe on the owners' rights.

---

*Two old tenements tipped towards one another due to foundation failure. Value did not warrant repair; struts were placed between the buildings; and they are home to 20 families, even though the floors slope. [5]

The wall of this California winery is out of plumb, possibly due to the lateral thrust of wine casks. In any event, the roof load must now be eccentric and the wall is susceptible to collapse under stress of fire.

The fire department is left, then, with the duty of examining buildings from the point of view of possible collapse and adjusting fire fighting methods accordingly.

Some of the signs of potential collapse can be observed prior to a fire and should be noted in the prefire plan.

Rotten Bricks

The word "brick" has a connotation of toughness and permanence which can be misleading. Possibly it starts with the story of the three little pigs. The pig with the brick house survived. Bricks, properly made, can be tough and long lasting. Poorly made bricks can deteriorate readily. Look for signs of poor brick. At times an effort is made to cover the defects with a thin coat of masonry called "parging." The parging may be scored to look like brick. The parging itself may be falling off. This all indicates poor brick.

Mortar Bond

The integrity of any masonry wall depends to a large degree on the quality of the bond between the masonry and the mortar. If the masonry bond is weak or nonexistent, the wall may still stand as long as it is axially loaded. The lack of shear strength may cause it to fail from even a slight load applied laterally, as from a floor collapse.

Old buildings were built with sand-lime mortar. All mortar deteriorates, but sand-lime mortar is particularly subject to deterioration. Weather and chemical pollutants are but two of the enemies of mortar. Mention was made previously of a fire fighter who noticed that the stream had washed sand-lime mortar out of the joints. The officer was alerted and the units withdrawn shortly before a collapse occurred.

There are tax advantages available to those who restore old buildings. They are required to follow the Secretary of the Interior's *Guidelines for Historic Preservation*. This guideline indicates that sand-lime mortar must not be replaced with portland cement mortar; the masonry might be damaged by the harder mortar.

Portland cement mortar came into use in about 1880, but many buildings built later were built with sand-lime mortar totally or blended with portland cement.

Look at an old building being demolished. If the bricks come apart easily and cleanly, the mortar was probably sand-lime.

In the very first edition of the *National Firemen's Journal* (a predecessor to *Fire Engineering*) in 1877, the editor commented on the cheap nature of the masonry walls of the then current construction. Many of the buildings now being "preserved" were poorly built to begin with from those aspects which interest us, and neither age nor the rehabilitation process will improve them.

A house mover whose specialty is moving old buildings describes such buildings as "just a pile of bricks."

Mortar which reacts chemically may "grow," thus forcing the masonry out of alignment. Parapet walls (projecting above the roof line) are particularly subject to weather deterioration, and many such walls, as well as some chimney tops, consist of bricks held in place by gravity, with the nonadhering mortar simply acting as a shim (a scrap of material used to wedge a component into position). Hitting such a wall with a heavy caliber stream can convert the loose bricks into

Note the missing mortar in this wall of a massive old church. It is not unusual for builders to cut corners and fire fighters should never take structural integrity for granted.

dangerous missiles, even though the stability of the building is not affected.

Old masonry walls are often constructed of an inner and outer wythe of finished masonry with rubble, stone, or brick in between. The walls of the Capitol building in Washington, DC were built in this manner. Such walls are called "rubble masonry walls." They can stand for decades coping only with the compressive load delivered vertically but fall into a pile of rocks if disturbed. This happened in the building of the Washington subway when an attempt was made to pick up early rubble masonry buildings on "needle beams."*

Deteriorated mortar permits rain to leak through a wall. The wall then may be "pointed up," that is, the loose mortar is raked out and new mortar troweled into the joint. The fact that the wall required pointing up shows that the wall is deteriorated and leads us to question the bond between units in the interior of the wall. Walls are often pointed up with caulking which may stop the leak, but provides no structural value.

Plane of Weakness

When a piece of glass or gypsum board is cut, a line is scored; hit one side and the material breaks along the line. The line is a plane of weakness. It is possible to put a "plane of weakness" into a building by cutting a chase into a wall. A chase is a passageway, usually vertical, cut into a masonry wall for a pipe or conduit. A horizontal plane of weakness was noted in a two-story-high wall under construction. A slot was left in the wall to accommodate a concrete floor yet to be poured. When the floor was poured, if the concrete was not worked well into the gap, a hidden weakness would be left in the wall. Such a weakness was determined to be the cause of a collapse.

Close examination of the rear wall of an old building in a Canadian city disclosed an ingeniously weakened wall. The roof downspouts were apparently often damaged by trucks moving through the alley. A chase had been cut vertically down each end of the face of the wall, from the first floor ceiling line to grade. The downspouts were set into the chases, free from exposure to damage. When chases are built into a wall initially, they represent a weakness, but at least the joints can be properly laid to forestall disintegration. The chase cut into an existing wall is a serious weakness.

The lines above were written in 1971 for the first edition of this book. At that time, the author had no knowledge of any collapse from this cause. In 1973, the Broadway Central Hotel in New York City collapsed. In the October 8, 1973 *Engineering News Record*, it is reported that the investigation disclosed that 30 years earlier an unauthorized 8-inch chase had been cut into the 12-inch wall for a 5-inch drain. The drain leaked water which washed out the sand-lime mortar.

---

*When a wall is to be underpinned, openings are made through the wall. Beams are passed through the openings to support the wall while the new foundation is built below it. Such beams are called needle beams.

This is just one example of how potential collapse causes can be observed and recorded even in the absence of experience.

Downspouts that need protection from vehicles in alleys should not be recessed into a wall, but rather be protected by masonry or by the use of a cast-iron section. These protective practices were noted in Memphis and Denver. It would be good to determine the local practice.

Another potential plane of weakness has been noted in older masonry buildings across the country and in Canada. In order to set the floor beams level, a joist is set flat into the masonry wall and the floor joists are set on it. This makes the floor level without shims. In a fire, the flat joist will burn away and create an unsuspected void in the masonry wall. This is an example of common building practices handed down from father to son, carried out across the country, and written down no place. Look at your buildings.

A vertical plane of weakness was created when a chase was cut in this wall to accommodate this drain pipe.

Bracing

Settlement of foundations, shifting of floors, overloading, or other deficiencies may cause a brick wall to shift out of alignment. The wall may stay out of alignment for years and give no problem, or it may require correction. This is accomplished by tying the wall to some part of the interior of the building or the opposite wall. Spreaders are used to distribute the pull of the tie over a sufficient area of the brick wall. These may be star-shaped, S-shaped, round or just available pieces of steel plate. Sometimes steel channels (U-shaped steel sections) are used to spread the effect of the tie over a long span. The important thing to remember is that the use of any such bracing is often evidence that the wall is in a deteriorated condition and is therefore suspect. In one church, the spreaders on front were artfully concealed in the masonry. A trip to the rear showed the spreaders and an interior inspection showed tie rods from side to side and front to back.

In some types of construction, stars or other decorative devices were used to attach floors to walls. In such a case, the distribution is usually symmetrical and should not be confused with spreaders spaced at random to correct a faulty wall. However, take note of the discussion further on in this chapter, on the subject of "fire cut" joists vs attachment.

The stars on the front walls of these tenements tell that the walls are tied to floor joists. If the joists fail, the wall may collapse.

The bracing of the wall by attaching it to interior components may create one problem while solving another. Some years ago, the end wall of the Majestic Theater in Pittsburgh was bowed out about a foot.[6] A steel channel was mounted across the wall. Long bolts were used to tie the channel, and thus the wall, to the first of the roof trusses. The trusses not only supported the roof, they carried the substantial load of the ceiling structure. This consisted of joists, which spanned from the bottom chord of one truss to the bottom chord of the next, resting on ledgers (strips nailed to the sides of the trusses). A lath and plaster ceiling was supported by the joists.

Attaching the wall to the truss did not halt the lateral movement; instead, the wall pulled the truss out of line, until the truss was no longer supporting the ceiling joists. At this point the joists fell. Fortunately, the affected area of the theater was sparsely populated.

As will be seen later, there is a basic accepted concept that the interior structure of an ordinary construction building should be free to collapse in a fire without bringing down the walls. Often the walls are tied to a floor joist at right angles to the wall or to a joist running parallel. This practice violates the concept given, but bear in mind that we are dealing with vernacular, not engineered, structures, and inconsistencies abound. Often the problem of the moment is solved without consideration of the consequences. The wall, therefore, may be "braced" by attaching it to an interior component which is itself vulnerable to fire. If the stability of the masonry wall depends upon its connection to fire-vulnerable interior components, the wall itself is thereby made fire-vulnerable.

Where one wall is tied to the opposite wall, the fire-caused failure of the tie rod may cause both walls to collapse. Cables have been used to tie walls together to resist the outward thrust of roof trusses. Such cables are cold-drawn steel and fail at 800°F.

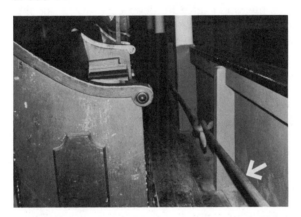

The steel tie rod (arrow), which holds the walls of this church together, passes through the choir loft. No church was ever designed with such a hazard. It was added because of structural weakness.

Cables or ties also may be used in wooden buildings. Steel tension rods may be designed into a building. This is seen in several churches where the rods resist the outward thrust of the roof. In Paterson, NJ, a wooden church was braced with unprotected steel. The steel failed; the roof and wall collapsed. One officer was killed and several others narrowly escaped.

Front walls falling away from the building may be tied on with simple L-shaped straps. One New York cast-iron front is patched in this manner. When it was "rehabilitated," it was painted yellow, but the basic defect remains.

Bracing of the walls should be taken as the equivalent of a sign posted on the building: "These walls are in poor condition — Beware."

Cracks

Cracks in walls indicate possible serious problems; they destroy the homogeneity of the wall. They often indicate foundation failures. If part of the wall is given a lateral thrust, as from collapsing floors, the rest of the wall will not assist in resisting the thrust. Sometimes the cracks are pointed up and patched cosmetically or filled with caulking. This does not correct the fundamental weakness. In one instance, bright sun showed the shadow line of a horizontal crack the length of a long, high wall about 3 feet below the roof. It is very likely that this was caused by summer expansion of steel roof girders, installed without room for movement. During the winter, the beams contract. If the wall has been pushed out of alignment, there is the possibility of a snow load collapse such as the one that occurred in the Knickerbocker Theater in Washington as described a little further on.

Collapse Due to Gradual Changes

Bricks out of alignment or spalling (loss of material from the surface) may indicate overload. Long diagonal cracks indicate uneven settlement. Decay of wood piling due to lowering of the water table by deep excavations, or the demolition of a building whose superimposed load was assisting a foundation in restraining earth movement are only two of the ways in which foundation settlement may occur. Loads may shift from axial to eccentric or torsional. Floor beams may fall from loss of bearing surface (pulling out of the wall).

Floor or roof members can push a wall out of alignment so that the bearing of the floor or roof member on the wall is no longer axial but eccentric. The condition may be borne successfully by the building for several years; then an overload may cause collapse.

The collapse of the Knickerbocker Theater in Washington, DC, on Jan. 28, 1922, which cost 97 lives, is a case in point.[7] There were many factors involved in the collapse, but two deserve special mention. The tile wall had been moved outward by the expansion of steel trusses; when the trusses contracted, they did not draw the wall back. By successive repetition of this process, the wall was moved out several inches, The trusses were no longer bearing axially. A 2-foot snow load, exceptional for Washington, was the trigger.

In addition, the high, thin, hollow terra cotta tile walls of the theater were found to be faulty in several respects. They were too high, some tiles were poor, and there was poor connection between the exterior and interior faces of the wall. Hollow tile is rarely used today as a structural material, but there are many such buildings in existence, and they deserve careful scrutiny. Walls out of alignment, crumbling of tiles under compression (particularly examine for this at the point where the load is applied), and girders, beams or trusses apparently not in their original position are all signs of distress. A handbook shows pictures of Z-shaped tiles designed to be installed in both the inner and outer wythes of cavity hollow tile walls to replace steel ties. If the two wythes pull away from one another, a tensile load is placed on tile, a material with no tensile strength.

Corbels

Portions of the masonry wall may be cantilevered for decorative or structural reasons. This is called a "corbel." In a shopping center built in 1970, a handsome brick chimney provides architectural interest to the building. The author's examination of the building, while under construction, however, showed that the massive chimney was a fake. It was corbelled out from the side wall on both the exterior and interior. The exterior corbelling was a good example of the bricklayer's craft; the interior was another story. Sloppy, incomplete mortar joints and concrete block laid flat gave no confidence that the whole structure would not collapse in the event of any unusual stress, particularly in view of the high center of gravity of the "chimney." A common practice in simple construction is to save on bricks and support the chimney on a wood shelf or bracket, starting the masonry only at the point where the stove pipe enters the masonry. This practice is apparently international. It can be seen in the Russian Archbishop's log cabin "palace" in Sitka, AK.

Russian builders of the Archbishop's home in Sitka, AK saved bricks by supporting the chimney on wood. Unfortunately, such hazardous practices are almost universal, exposing fire fighters to unseen dangers.

Fire Cut Joists

Collapse of floors can cause collapse of walls. For this reason, the ends of wood joists in masonry walls are cut at an angle in hope that when a floor collapsed it would pull the joists away from the wall without their acting as levers against the wall. This is called a "fire cut." On the other hand, every fourth joist or thereabouts is attached to the wall by a metal hanger to tie the floor to the wall; in some cases, these hangers are attached to exterior spreaders, usually star-shaped.

The author knows of no tests which were ever conducted to prove the efficacy of the fire cut requirement, or to resolve the apparent contradiction between fire cutting and attaching of joists to walls. Some architects recognized the fact that inserting joists into walls, no matter how, weakens the wall, so they designed corbelled "shelves" projecting from the wall, on which the joists rested.

In some heavy timber buildings, iron or steel joist hangers or boxes are inserted into or attached to the wall, and the joists are set into the hanger, or box.

Cantilevers and Balconies

Cantilever construction is becoming more and more popular. The cantilever beam projects a substantial distance beyond the point of support. The cantilever is comparable to the seesaw. If the person holding one end of a seesaw lets go, the other end goes down. If the connections in the interior of the building which hold the cantilever fail due to fire, the cantilever beam will fall. Many balconies are cantilevered. In some cases, the beams are continuations of the floor beams. In others, they are spliced to floor beams and sometimes the cantilever beams appear to be held in place only by the flooring. In any case, fire weakening the interior support system, together with the load of fire fighters and equipment on the cantilevered balcony, can cause a disastrous collapse. It appears that one collapse in a southern city, which killed several fire fighters, occurred in this manner.

These lovely New Orleans' balconies are cantilevered. A fire inside can destroy the other end of the "seesaw" and cause the balcony to collapse.

The collapse may not end with the balcony. In one supermarket that was examined, the cantilever beams passed through the walls and delivered their load upward through compression members, to a common header which was laid along the bottom chord of ordinary gusset plate trusses. If there is a fire in the gabled roof, it is probable that the canopy would be swarming with fire fighters. If the trusses fail, as is likely, one end of the seesaw will go up, and the other down. As the beams pivot, they would probably bring the 5 feet of masonry wall down on the fire fighters.

Instead of a cantilever, the designer may use a suspended beam. In such a case, the end of the beam (usually a canopy) is not supported from underneath as is usual, but by a chain or cable attached through the wall to some interior connection.* Such connections are vulnerable to fire. In one case, fire streams iced up a theater marquee and it collapsed, trapping several men. Fortunately, a construction crane was available to lift the marquee and get them out. In another case, an old theater had been converted into a furniture store. Quite late in the fire, the connections holding the old marquee failed and the marquee came down. When it did, it took down the coping along the length of the wall. Six men were killed.

At a fire in the Capitol Hill Theater in Washington, DC, in 1970, fire fighters were alerted to this problem. The marquee was a suspended beam. The outer end was supported by steel tie rods extending downward from the wall above. The marquee was tied through the wall to the roof trusses. In addition, two tie rods ran through the attic, tying the front wall to the back wall. The attic, roof and trusses formed the bulk of the fire fuel. The failure of the interior ties caused the marquee to pull the wall outward almost 3 inches along a mortar joint line. The area was roped off. The owner summoned a wrecking contractor who erected shores under the outer end of the marquee. Thus, the now fully-cantilevered marquee was converted to a simple beam.

Cantilever beams are inherently unstable and depend upon the integrity of the interior connection. All cantilevers should be noted, and the nature of their construction determined as far as possible. The hazard of the cantilever cannot be overestimated.

At a fire, it might be practical to run a piece of apparatus under the free end of the cantilever. Such a support might limit the downward travel of the free end and thus prevent a collapse. A standby arrangement with a local demolition company would be very useful since they have the equipment and manpower to perform the necessary shoring on short notice.

In any event, the fire fighting forces should avoid placing any additional load on a cantilever whose interior support is being attacked by the fire, and should stay well clear of the collapse area.

High Cornices and Roof Structures

Freestanding masonry cornices were often built excessively high. At times the cornice is a full story in height and is, in truth, a false wall to convey the impression of greater height. These walls are vertical cantilevers with respect to lateral loads, such as ladders or fire streams. The exposed masonry in such walls is particularly subject to deterioration from the elements. At times they are supported by metal struts attached to the roof structure. The failure of the roof or the strut may precipitate the failure of the wall.

---

*The suspended canopy of an old Maryland fire station fell off early one morning when the connection failed.

This parapet wall is a vertical cantilever. The sign loads the wall eccentrically. If fire burns out the lag screws which hold the tie rods to the roof rafters, the front wall will probably fall.

A one-story restaurant has a high front coping. There is a sign hanging from the wall. This tends to pull the coping over. This is resisted by two steel angles (tension members), which run diagonally from the top of the wall to the roof. They are attached to the roof with lag screws. A roof fire could release these screws; the rods would let go, and the eccentrically loaded wall would then be free to fall outward. This is in a small bay-side town. Building collapse is a hazard to all fire fighters regardless of the size of the community.

The collapse of a high cornice wall may be an excessive impact load to the portion of the structure it falls upon. If it falls into the interior, the freestanding wall may bring down the floors, which in turn may bring down the bearing walls.

A water tank or air-conditioning unit on a roof is another possible source of an impact load. The collapse of water tanks supported on interior columns many years ago led to provisions in some codes that such loads be supported on two exterior walls.

Walls of Different Materials

It is common practice to build the front wall of a building of different masonry material than the side walls. In older buildings of ordinary construction, the front wall is of face brick, the sides of common; in more recent buildings, the front is brick and the sides concrete block. Brick and tile combinations are also seen.

It is not uncommon in construction practices for the side wall to be carried up a story ahead of the front wall, though it is considered that no wall should be carried more than two stories above another wall of the same building. This recommendation may be much too lenient. The mason leaves a series of gaps in the wall looking like a small child's front teeth. The mason who then builds the front wall is supposed to be able to fit the new wall into the prebuilt section so that there is full bearing and complete adhesion. Watch him work. See how neatly the visible joints are struck. You cannot see the joints within the interlocking wall spaces and the workmanship that went into them. If old buildings are being demolished in your community, look for a wall which was built "blind"; that is, the wall was built against another so that the mason could not see the back of the wall. The blind side often shows broken masonry, missing units, missing mortar, and other defects. Look carefully at masonry walls for signs of separation at the corner.

An examination of one building showed readily apparent separation between the front and side walls up about four floors from grade. The remaining two floors above appeared in good condition. The answer: there was a building on the next lot when the building in question was built. When building the wall adjacent to the old building, the mason was working "blind" and could not make good joints. Above the roof of the old building, he could work on both sides of the wall.

In older buildings, the outer wythe of brick is often a special veneer brick held on by metal ties. The metal ties may be corroded out of existence. The whole veneer may separate and fall. When a heavy caliber stream is used on such a wall, the water can fan out behind the veneer and strip the bricks off the wall.

It is not uncommon to find that the front wall or front portion of the building is of brick and the balance is of frame. Often the frame portion is veneered over with brick in later alterations. The author observed two adjacent buildings being combined into one at Virginia Beach, VA. Split concrete block "brick" was being veneered over a frame building. The veneer wall was continued on the same line across the front of a solid brick masonry building, leaving a cavity between, possibly just the place for a carbon monoxide accumulation.

Building Additions

Many commercial buildings are former residences. Often there was a little garden or areaway (courtyard) in front. In order to make full use of the space, a new front wall was constructed out at the lot line and side walls were extended. What happened to the old front wall? Was it completely removed? Or was just the first floor wall removed, leaving the wall above to be carried on beams provided across the first floor ceiling line? Finished interior walls may conceal the old brick wall. How good is the joining of the front and side walls? It is, of course, possible that the front extension is of superior construction to the older deteriorated original building.

In an eastern city, an 86-year-old brick and wood-joisted building built as a residence had been converted into office space. There was a raging fire on the first alarm. Multiple alarms were sounded and heavy caliber streams were in use. After about 40 minutes, "Suddenly, without warning" (the editors of fire service publications must keep these tragic words set in type), there was an interior collapse. Two men were killed and several injured. The warning is not overt, but it is there. Ask the questions: Did the initial builder or the redesigner of this building have the slightest idea of planning the building to resist the massive combined assault of the fire and fire fighting operations? Is there any reason why the building should not collapse? What is being accomplished to justify, morally or financially, the risk of death and injury?[8]

The modernization of the front of an old business building may be accomplished in several ways. The windows may be filled in with masonry and the whole wall stuccoed. This presents no structural problem but does create another problem discussed later in this chapter under voids. A screen wall may be erected at some distance from the front wall. The screen wall may be of brick, decorative concrete block, terra cotta tile, metal or wood. The manner by which the screen wall is attached should be studied. It may have little resistance to fire and its collapse may be a serious hazard. Such a wall may make window rescues impossible.

The collapse of a building involved in a fire onto an adjacent building may have serious consequences. Nine fire fighters lost their lives in a Chicago factory fire when the fire building totally collapsed onto an adjoining building in which they were trying to rescue two fire fighters trapped there by an earlier partial collapse.[9]

Backdrafts

Deliberate venting or the breaking of a window by the first line may trigger a backdraft, which could bring down any building of ordinary construction. When a backdraft is anticipated, all personnel should be clear not only of the path of the backdraft, but also of the possible collapse of the building. For further discussion, particularly of the fact that violent backdrafts can occur late in a fire, see the discussion of void spaces later in this chapter.

Lateral Impact Loads

Fuel gas explosions have provided a lateral impact load sufficient to collapse walls on many occasions. The entire side wall of the Horton's Brewery boiler house in New York City collapsed, at a cost of several lives, when mixed coal and resin dust used as fuel exploded. A welder was working on the vertical hoist used to charge the hoppers from delivery trucks.

It is a principle of engineering that lateral loads require lateral resistance,[10] and masonry walls are rarely provided with enough lateral strength to resist an undesigned lateral load, particularly if there is a severe impact.

Brick masonry or concrete walls which separate one electrical transformer from another are usually built blast resistant. If the opportunity permits, observe the construction and contrast it with the ordinary wall.

Effect of Fire on Materials

All masonry materials will spall, that is, lose material from the surface in successive layers when exposed to excessive heat. Some stones, particularly granite, possess this characteristic to a marked degree. The loss of material may weaken the wall to the point of collapse. Differences in expansion rates may separate the masonry from the mortar. Often the fire may consume most of the load that the wall was carrying, so that the wall does not collapse immediately, but it still may be very hazardous.

The nature of steel as a building material will be examined in detail in Chapter 7, Steel Construction. Any unprotected steel which is so arranged in relation to the wall that its expansion could exert a lateral load on the wall should be studied to the extent practicable. If the resistance of the wall is stronger than the expansion of the steel, the steel will buckle. If the reverse is true, the wall will move and may collapse, or the steel may just "punch through" the wall. In any event, the basic rule, "cool any heated steel," applies practically universally.

The practical difficulty is that the steel may be concealed and its presence unsuspected. Suspect any altered building. Remember that the building is classed as "non-fire resistive" so that steel members, no matter how important, are not required to be protected.

Openings in Walls

Walls must almost invariably have openings for windows or doors. From time immemorial, there have been only two ways to carry a masonry wall over an opening. The first is by a lintel (a beam). Such an opening is said to be trabeated. The other method is to build an arch. An arched opening is said to be arcuated.

Masonry walls have an inherent capacity to redirect loads. Stress flow diagrams for walls show the stresses "detouring" around the opening.[11] There is a rule of thumb that the actual load on an opening cut in a wall can be described as a triangle above the opening, whose apex is located at a point three-quarters of the length of the span above the midpoint of the span.[12] This rule presumes that the wall is homogeneous and that there is an adequate bond between the mortar and the bricks. An opening breached in a brick wall should be in this triangular pattern, not cut straight across.

Before breaching any wall, examine the condition of the wall, particularly above the point where it is to be opened. If upon opening the wall, you find a wall column or pilaster possibly of brick, solid concrete block, or reinforced concrete, stop immediately and start elsewhere. Such construction not only is stiffening the wall, but also usually is carrying the concentrated load of main girders.

All openings in walls may cause serious weaknesses, whether designed into the wall initially or provided at a later date. They should be carefully examined.

Arches

Arches may be flat or curved, and there are a number of varieties of curved arches.

In the perfect arch, all members are under compression. The careful fitting of stones together to achieve near perfect bearing is a hallmark of the stone arches, which have survived for centuries. In early brick arches, each brick was rubbed to put a taper on it. It was later found to be easier to leave the bricks untouched and allow the mortar to form *voussoirs*, the wedge-shaped pieces necessary to provide the arch. The deterioration of this mortar may cause failure. In one building, it was observed that a stone had fallen out of a flat arch. The masonry was therefore resting on the window frame. Fire burning through the window frame would cause the wall to collapse. In some cases, particularly fireplaces, a steel arch form was used as the falsework (support while the arch was being built) and left in place.

A random sampling of what appears to be brick segmental (curved) arches in 19th century buildings has led the author to the belief that a large proportion of such arches are faked in a most dangerous manner. The outer wythe (or wythes) of masonry is (are) built as a true arch; the inner wythes are carried on a wooden lintel. The reliability of the wall, therefore, depends not on the masonry arch, but on the wooden lintel, as it is impossible to conceive of the inner wythes collapsing without bringing down the entire wall. In an old warehouse in Richmond, VA, a builder used several planks laid flat, one on top of another, not even on edge, for a lintel, a very poor design. Examine even the "sturdiest" arch carefully; in some cases, the deceit was cleverly concealed. Flat arches are also found to be faked in a similar manner.

The outward thrust of the arch can be resisted in only two ways. Either the arch is tied with a tension rod, or a mass of masonry resists the thrust. The former method was used for brick floor arches, and at times for wall openings where the mass of masonry in the wall was insufficient to provide an adequate buttress.[13] The cutting of an opening in an area where the wall is providing the resistance to the arch's thrust may cause collapse.

In some masonry arches, the problem of matching a flat-topped door to a segmental arch was solved by building a masonry core supported by a wooden lintel within the arch. The arch is called a relieving arch. Should this core collapse, the stability of the wall is unaffected, but that of a fire fighter standing in the doorway may be! In other cases, the lintel is inserted into the wall so that the arch springs from the wooden beam. This may cause collapse if the lintel burns away.[14]

Lintels

Every conceivable material has been used as a lintel, a beam to carry the load of a wall over an opening. Stone has limited tensile strength, so stone lintels are of short span. This fact provided us with the beautiful Greek temples with their closely spaced rows of columns. Stone is still used for lintels, and at times steel lintels are used under the stone for strength.

Unashamed wooden lintels are commonly found, once you get into the habit of looking for them. Look very closely at painted or whitewashed buildings; wooden lintels in painted stone buildings are particularly hard to detect.

An intermediate step in the development of cast-iron front buildings was the use of cast-iron columns and wrought or cast-iron lintels at the street floor level carrying masonry above the first floor. This enabled the builder to provide larger show windows, or entrance ways, than would have been possible with masonry. The common lintel used today is a steel "L" or channel section. When such sections were first used, the practice was to arrange the bricks above the lintel as a false flat arch. Later, the pretense of an arch was abandoned and the bricks above the window were set vertically. Today, the common practice is to carry the brickwork horizontally across the lintel. Steel lintels are tied tightly into the masonry wall. When heated they elongate and the masonry can fail.

Precast reinforced concrete beams are often used as lintels in concrete block buildings. In ordinary construction, the concrete lintel need not be fire resistive and most likely is not. Fire can cause failure of the beam, separating the steel from the concrete. This is discussed in detail in the chapter on concrete construction.

Cast-iron Fronts

The small proportion of a masonry wall available for windows led to the acceptance of walls made of prefabricated identical cast-iron sections. Usually a column was cast integrally with the right half of the arch of the window to the left, and the left half of the arch of the window to the right. The sections were bolted together at the crown to form the arch. There are few cast-iron fronts left, and the demolition of one is regarded as an act of barbarism in some circles.

The cast-iron front may be separating from the masonry side walls. Some walls are "tacked" backed onto the masonry with steel straps. The columns, like interior cast-iron columns, may transmit fire vertically.

Building Alterations

Alterations made in a building over the years may have affected its stability or may deceive us as to the true structural condition of the building. Doorways or windows may have been opened in a wall and the load may not have been picked up properly. The Empire Apartment Building in Washington, DC, built in 1893, collapsed in 1947 from a defective bearing wall. The first hole was cut in the wall in the early 1900s. In 1932, the hole was enlarged and a steel beam inserted. This called for a new brick column to support the beam. Instead, the beam was inserted into the old sand-lime mortar wall, and 15 years elapsed before the weakened structure succumbed. Sand-lime mortar, gradually dissolved by a water leak, was an important contributing factor.[15]

It is a common practice to cut openings between two or more adjoining buildings at the first floor in enlarging a store area. The question of how the walls above the openings are carried is important. Was a true arch built? Most unlikely. More likely, the contractor slipped in a wooden or steel beam from front to back, and the whole job was plastered over. The failure of the beam will cause the collapse of the two interior walls (or wall, if a party wall) above the openings. This will bring down the floors and very probably the exterior wall.

This opening cut through the wall without a lintel being installed could precipitate a collapse.

The opening may be of such a span as to require columns. The columns may be of steel or cast iron, and are rarely protected against fire. Thus, the wall lacks any of the fire resistance provided by any masonry wall, and a collapse may take place fairly early in the fire.

The way such work was usually done leaves the wall very susceptible to collapse, particularly if there is any lateral thrust. The wall was picked up on jacks, the masonry was cut away, two steel "L" sections (lintels) were placed under the wall on either side. The column was placed, and the wall let down onto it by releasing the jacks. There is only a gravity connection.

Older buildings have often been altered in unimaginable ways. Nine fire fighters died in the collapse of the Hotel Vendome in Boston. In bygone years, an interior masonry wall had been altered in the manner cited above in order to combine two rooms. When the building was being "modernized," a hole for an air duct was cut through a masonry wall directly below the cast-iron column.

Renovating older buildings is a chancy business and the developer is tempted to cut corners. A summary of the procedures followed in modernizing the Vendome is contained in Appendix A. Could such avoidance of proper supervision take place in your area?

Collapses like the Vendome suggest that fire departments might consider radical changes in procedures. For no real reason, there is the practice of cutting up the structure "to find hidden sparks" and prevent a "rekindle," the most op-

The remains of the century-old Hotel Vendome in Boston in which nine fire fighters lost their lives. The building had been undergoing remodeling at the time of the fire. Recycling old buildings is a current fad. Fire fighters should be wary. Courtesy A.E. Willey, NFPA.

probrious term to the fire service. What is the hurry? The fire in the structure no longer represents a threat to the city as a whole. Why not pull back and call the owner? Tell him that he has a problem. His building is probably structurally unsafe. Advise him to employ a licensed consulting engineer who will advise on the next steps. It might be hard to find one at any price who will even enter the building. That should tell us something.

Develop the habit of looking at buildings from across the street. Before you know it you will be able to spot the existence of a hidden wall.

When wall collapse is anticipated, units should be withdrawn a truly adequate distance. This is farther than most think. Do not forget that an interior collapse may precipitate wall collapse.

In Spokane, WA, a collapse was anticipated at a six-story brick and wood-joisted building. Two fire fighters were in an elevating platform 12 feet from the wall. The bucket was hit with bricks and part of the cornice. One fire fighter died.

Three Montreal fire fighters died when a wall collapsed 3 hours after the first alarm and 2 hours after the fire fighters had been ordered from the building.

More than one case is recorded of fire fighters being killed removing apparatus from the danger zone. Even though apparatus is expensive, the loss of a fire fighter is more costly.

Stability of Interior Support Systems

In this text, interior and exterior collapse are discussed separately in order to organize the material. In fact, they are interrelated. An interior collapse of an overloaded floor can cause the walls to collapse. The collapse of an exterior wall into the structure or onto the structure next door may cause an interior collapse.

The interior structure consists of the floors, any necessary interior support for the floors, and any suspended loads, and the roof. These are all interrelated. The collapse of a heavy interior suspended load may present an excessive impact on the floor below. The contents may become watersoaked and thus overload the structure. A floor may collapse due to burning of the floor members, or the failure of a column or an interior wall may drop the floor. In some cases, the floors are literally "designed to collapse."

Every conceivable method of dealing with the problem of interior support for buildings of ordinary construction has been tried. These systems have but one characteristic in common; no consideration was given to the design of a system of compatible elements that would function together to resist an assault by fire of specific magnitude for a given period of time. In other words, no such system was ever subjected to any test even approaching the tests for fire resistive construction as specified in NFPA 251, *Standard Method of Fire Tests of Building Construction and Materials.*[16] The only consideration given to fire characteristics assumed collapse in case of fire; such detailed questions that arose had to do with whether to design connections to permit collapse of girders without collapse of columns, or collapse of floors without collapse of walls.

In a series of fire resistance tests carried out many years ago, wooden columns were tested. The report of those tests showed that no wooden column was able to survive a 1-hour exposure to a standard fire test.[17]

A newspaper article describing the operations of a contractor who specializes in demolishing buildings with carefully placed explosives quoted him as saying: "We just kick the supports out from under the building and let gravity do the rest."[18] While not as fast as explosives, fire can also kick the supports out, and we can be quite sure that gravity will do the rest.

The fire officer's problem is further complicated by the fact that it is not necessarily simple to determine that a building is of ordinary construction. It was noted earlier that an office building which on first glance appeared to be of fire resistive construction was, in fact, of ordinary construction. The unprotected steel, particularly in the roof, can elongate and push walls out, and can fail and drop the roof. There is no substitute for institutionalizing the fire department's knowledge of buildings in a record and retrieval system.

Floors and Roofs

Floors and roofs are the fire department's working platforms. Procedures are based on experience with sawn wood-joisted floors and roofs. As noted earlier, sawn timbers inherently contain extra material which is not vital to strength. As long as only this wood is burning, the structural stability of the floor is relatively intact. This is also true in the case of roofs. Bear in mind, however, that roofs are often designed for lesser loads than floors.

Recently, as part of the work in connection with Operation Breakthrough (the effort by the Department of Housing and Urban Development to provide lower cost housing by the application of industrial methods or so-called systems building), the National Bureau of Standards subjected some typical, currently accepted floor structures to test. A floor of 2-inch by 8-inch floor joists with ⅜-inch plywood flooring failed due to heat transmission at about 10 minutes, and the test was terminated, but observers reported that collapse was imminent following the standard procedures outlined in NFPA 251.[19]

Some nonstandard tests, that show the rapidity with which such trusses and wooden I beams can fail, are described in Chapter 3, Wood Construction.

It is extremely dangerous to apply the experience learned on sawn beams to other members such as trusses and wooden I beams. These do not have the reserve of sawn beams.

Steel bar joists (parallel chord trusses) are often used. They may last as long as 7 minutes in a fire equal to the ASTM E 119 fire test standard. At times, a ceiling of tiles may be installed. This does not make the floor a fire resistive floor and ceiling assembly.*

Wood trusses and I beams have no reserve. As soon as they are involved in fire they are subject to failure. Do not be deceived by the application of "⅝-inch rated gypsum board" which is sometimes taken to be sufficient to produce "protected" combustible construction. In Chapter 5, Garden Apartments, there is a thorough discussion of the fallacy of considering rated gypsum board as having any substantial merit when it is installed very differently from the way tests are conducted.

Six fire fighters were killed in New York when a wood bowstring truss collapsed at a supermarket fire. The fire originated in a mezzanine area which in fact put the fire close to the trusses at an early stage. There was difficulty ventilating the roof because there were two roofs. A common method of repairing a stubborn leaky roof is to build a whole new roof atop the old one. Time from initial alarm to initial collapse was 37 minutes. The entire roof had collapsed in another 25 minutes.

All trusses are vulnerable to collapse if a single element fails. Trussed roofs involved in fire are not safe to ventilate by current procedures.

In Summit, NJ, fire fighters barely escaped a trussed roof collapse. Again, the fire involved a loft above the floor which put the fire directly into the trusses early. Note the similarity to the Wichita, KS steel truss collapse related in Chapter 7, Steel Construction.

New York fire fighters were fighting a fire in an abandoned garage. The only fuel was the trussed roof. The wall was eccentrically loaded by a large overhanging display sign. When the trusses failed, the wall collapsed, as usual, "without warning." The warning is there but not as it is usually thought of. The warning is in the brain, in the capacity to understand buildings and anticipate how they will react to a fire.

An article about the New York tragedy notes that wood trusses will fail while steel trusses may sag. That is true as far as it goes, but steel trusses can fail rapidly. Also, if they do sag, the fire fighter on the roof may be sliding down a greasy slide (from liquefied tar) into the fire.

---

*See Chapter 7, Steel Construction. If there are any legal requirements at all they would address only the flame spread rating of the tiles. Tradesmen and some building officials sometimes use the term "fire-rated" tiles without discriminating between fire resistive assemblies and flame spread.

The trussed roof may also collapse on fire fighters in the interior. A Tempe, AZ fire fighter was killed and several others narrowly escaped death when a wood-trussed roof of a one-story restaurant collapsed 14 minutes after their arrival. The fire fighters noted fire visible through roof vents and little smoke in the restaurant. When there is heat and smoke and no visible fire, the building is burning. Look at all the restaurants in your area. Wide open flexible space is desired. Trusses are the answer. The wider the space, the deeper the truss, and the bigger the void.

A report of a fire in Ottawa, KS contains this quote, "The preplan showed that the roof was supported with open, unprotected metal trussed supports and that we should anticipate rapid roof collapse... All lines were immediately ordered out of the building." The rear half of the roof collapsed 7 minutes after the arrival of the fire department, the balance, 10 minutes later.

A Los Angeles fire fighter was killed in a roof collapse. In making an addition to an existing building, mortar was picked out from between bricks, and shingles were driven into the gaps. A ledger board was nailed into the shingles. The roof beams were supported on the ledger board.

Different areas of the building may have different construction. The original roof may be joisted; the addition may be trusses. In one old courthouse, the floors are wood-joisted except in one area where an early type hollow tile floor was installed after a fire a century ago. Two fire fighters were injured in New York when a roof collapsed due to steel plates added to the roof to prevent burglary.

When a stairway is relocated, the old opening is usually covered over. The closure is usually much lighter than the floor. In one observed case, 2 by 4s were used, whereas the floor was of 3 by 10s. From above, the floor looks uniform. Such an opening has figured in at least one serious collapse.

Deficiencies of Materials

Wherever materials are combined in the structural support system, the fire characteristics of the poorer performer should govern in assessing the collapse potential. Structures having heavy timber columns and girders and beams are found reinforced with added steel in a variety of ways. In such cases, the failure of the steel may precipitate collapse, as it can be taken for granted that any structural support which was added after the building was built was added because it was absolutely necessary. The addition of unprotected steel to a building with a wooden interior structure is usually readily apparent, but in one Richmond, VA building, known to the author, the new steel was fully concealed by interior finish. In addition, steel bar joists had been installed between each pair of wood joists. Demolition of the building disclosed the additional support.

Brick and stone interior walls may be used to carry structural loads, and they can have all the defects of exterior walls except perhaps weathering. Some very deteriorated mortar joints have been noted (upon demolition) in brickwork which, while technically inside the building, was inaccessible due to the placement of rafters, concealment by inaccessible voids, etc. It was apparent that visible and accessible deterioration had been attended to while the inaccessible deterioration had been neglected. Stone columns and interior walls may lose strength due to spalling during a fire.

Wood beams may have been trussed initially or at a later date. This is usually accomplished by erecting a strut downward from the center of the beam and stretching tension rods from one end of the beam to the other, over the strut. The strut or the rods may fail first. This technique has been used where heavy loads, such as belt drive shafts, were suspended from the overhead.

This wooden girder appears to be supported by a column, but it is also supported by a tension rod anchored to the girder (arrow) that runs up to a support truss in the cockloft. A fire in the cockloft could drop this beam.

A column may be in an inconvenient location or a girder may need strengthening; thus, alterations are made so that the load can be carried upward to a roof truss, to a girder laid across upper floor beams, or to a steel beam supported on exterior bearing walls. The load may be carried on steel rods, which are under tension and very vulnerable to fire.

In one instance, the steel spreader plate on the bottom of a wooden girder and a barely visible rod passing between the joists above provided the clue that the girder was now suspended. An attempt to examine further was defeated by the fact that the second floor was partitioned off into plywood offices, and the suspension system was concealed.*

Overhead Loads

A New York taxpayer** had suffered a multiple-alarm fire. Truckmen were

---

*Note that a girder is a beam supporting other beams.

**The word "taxpayer," applied to buildings, originated in New York to designate a row of cheaply built stores, one story in height, erected to pay the taxes on the property until it was fully developed. (The term "two-story taxpayer" is used in New York to describe what would elsewhere be called a business block.) Taxpayer stores are the scenes of many serious fires. Common cellars and cocklofts (spaces between ceilings and roofs not high enough to be attics) allow fire to spread rapidly.

overhauling the fast food outlet, "seeking out hidden pockets of fire." The roof and air-conditioning units crashed down on them. Three fire fighters died.

There was no air conditioning when many buildings presently in use were built. Air-conditioning units on the roof are added deadweight. There is less of a safety factor in the structure.

A supermarket was converted into a Japanese restaurant of the type where the chef cooks at your table. This necessitates many grills which require many heavy hoods. All this extra weight is hanging from the roof. It is not the same structure it was when the sole purpose of the roof was to keep out the rain.

The author observed a fire in a supermarket. The building was almost literally noncombustible (not just code-defined "noncombustible" which may not be). Health laws require a washable ceiling in the cooler room where meat is prepared, so gypsum board was substituted for the suspended tile used elsewhere. Wood beams were provided for nailing the gypsum board and supporting the overhead refrigeration units. The fire, electrical in origin, never involved any of the merchandise in the store; its fuel was the wooden beams. A truck company was pulling the ceiling. The author asked an officer to have the workers stand back, and cut away the gypsum gently so the damaged support system could be photographed. When the board was touched with the hook, the very heavy refrigeration unit fell to the floor. Had the operation continued as it was going, there probably would have been serious injuries.

The loss was about $500,000 even though no merchandise was burned. When the store was rebuilt, it was sprinklered. The author was informed that the wooden structure within the roof was replaced as it was. It is unsprinklered. If a similar fire occurs, the results will be similar!

A supermarket designer wanted to have shingled canopies over the checkout stands. If they were supported on posts, the arrangement would lack the required flexibility to accommodate future changes. As an alternate, the canopies were suspended. The ends of steel rods are held by two 2 by 4s which extend across the bottom chords of bar joists. A fire overhead, possibly a metal deck roof fire, could either burn the wood away, letting the washer slip through, or cause the restrained bar joist to attempt elongation and twist, thus dropping the 2 by 4s.

The temperature overhead at a fire is often several hundred degrees higher than it is at floor level.

The list of inherent defects of interior supports is endless. Those that are most important to you are those in your area of responsibility. Observe, study, ask questions. Here are some examples:

- Balloon frame wall carrying interior loads may fail very early due to the small cross-sectional area of the studs. (Note that in older masonry buildings the interior walls, if not of masonry, are of balloon frame construction. Masonry walls are usually much thicker than wooden walls.)

- Balconies, mezzanines, half stories, and other suspended loads, including merchandise racks, are very hazardous. Suspended loads are becoming more common as architects realize the space advantages of suspension over columns. In a fire, the loads may be obscured by smoke, and the suspension system may be subjected to temperatures hundreds of degrees higher than at floor level. All such tensile loads must be brought to earth in compression. There may be a number of connections. The weakest (to fire) is governing.

  In a branch library in Maryland, the mezzanine is supported on a main wooden beam. One end of the beam is suspended from the overhead on a slender steel rod. Failure of the connection of the rod to the overhead will drop the mezzanine.

- In one timbered church, boards have been used as false timbers to box in the rods supporting the choir loft to hide the use of an architecturally incongruous material.[20]

- Adjustable steel jack posts or simple steel posts may be used to support overloaded beams. Interior racks are often of very lightweight steel construction. Never get into a position where they can collapse behind you.

- Concrete block columns of excessive l/r (length divided by radius of gyration) ratios (see Columns in Chapter 2, Principles of Construction) have been noted.

Connections

The designers of mill construction realized that a building was no better than its connections and a great deal of attention was given to this matter. By inference, if in no other way, we are led to suspect the fire stability characteristics of connections in ordinary construction. The suspicion is justified. For convenience, beam to girder connections will be discussed first, then girder to column connections.

Beam to Girder Connections

The only way to connect wood beams to wood girders, and to receive the full benefit of the time it takes the wood to burn through, is to set the beams atop the girder. This is undesirable, however, because it adds costly height to the walls. All sorts of systems were designed so that the top of the beam could be level with the top of the girder.

The old-time craftsman was proud of his mortise and tenon joints. The beam was cut down from the top and up from the bottom to form a tenon or tongue. This was inserted through a mortise cut into the girder. A wooden dowel, or trunnel, was driven through a hole bored in the protruding end of the tenon. Lesser craftsmen set notched beams on strips nailed to the side of girders. The weakness of any method that involves any reduction in the size of the wood is readily apparent. The effective strength of the wood under fire attack is determined by the size of the thinnest portion, not the mass of the member as a whole.

Beam to beam connections must be made whenever an opening is provided in a wooden floor. The joists which are cut to provide the opening, and thus do not reach to the girder or bearing wall, are connected to a header beam which in turn is connected to trimmers. All these connections are accomplished in various ways. Mortise and tenon joints are found in older buildings. Notched beams are common. Joist hangers or stirrups and heavy steel or wrought iron straps shaped to receive the joist and nailed to the top of the header are common. Tests made years ago showed that such stirrups are dangerous, charring the joist and softening rapidly.[21] More recently, very lightweight hangers have been developed for lightweight construction. The rapidity with which such hangers will fail in a fire is an important factor. Certainly, the small mass of metal and the dependence upon nails do not inspire confidence.

Some joist-to-joist connections are accomplished by simple nailing, either by end nailing or toe nailing; such connections are very undependable.

Beam to Column Connections

The shear load on a beam is greatest in the portion nearest the column. This is blithely ignored by the amateur builder who wishes to support a load such as a canopy or a back porch or an air conditioner and decides that pipe threaded for water service is just the material to use. The threads remove material just where it is most needed and remove the zinc from galvanized pipe, thus hastening corrosion. The connection usually shears off right through the thread. (In the design of threaded rods, such as used in trusses, this loss of material is taken into account, and if necessary, the rod is upset, that is, made thicker where the threads are cut.) Avoid placing any additional load on such structures during a fire.

Timber buildings can present many methods of connecting girders to columns:

> . . . there are no generally accepted standards of connections for timber structures. . . . no important detail should be left to the discretion of the carpenter. With all due respect to his experience and care, he seldom understands the requirements of any detail but the simplest, and many times in his endeavor to improve on a detail, but hazily indicated by the designer, actually weakens the structure.[22]

Those instructions were written with respect to a building professionally designed; they apply even more forcefully to buildings erected without professional help.

Two conflicting criteria face the designer of any column and beam system. The connections should be firm enough to resist movement yet designed so that they will release to permit partial collapse in the event of fire. Self-releasing floors can be seen in most old timber interior mercantile buildings. The girder sits on a steel shelf which protrudes from the column. A scrap of wood or a "dog iron"* (literally a big staple) connects the girders and imparts some lateral stability under normal conditions. Note in the quotation below the assumption that the building will collapse in whole or part:

> To prevent the girders in falling from pulling the columns with them, in case of fire, standard practice recommends that the attachment to the girders be made self-releasing. The author believes, however, that in the event of a fire serious enough to burn through the girders, the interior posts of the building are almost certain to fall. For this reason, where it is necessary to secure lateral stiffness in a building, he believes it well to design the connections . . . [to be] relatively strong, providing continuity across the columns.[23]

The more the designer of the building followed the advice above, the more there was the possibility of total collapse, once collapse was initiated. On the other hand, self-releasing floors may limit the collapse area, but any system which is designed to precipitate collapse must be weaker and more likely to collapse. In any event, many of the buildings we must cope with as fire fighters are "vernacular" buildings, designed without benefit of professional engineers.

The principle that columns should set directly atop one another so as to carry their loads to the earth as axially as possible was widely violated. Where a column is offset, the girder on which it rests becomes a *transfer beam* and is subject to severe shear stresses unless it was designed as a transfer beam. Improperly loaded structural components are more likely to fail than similar components properly loaded, all else being equal.

Cast-iron columns created many problems. Good practice dictated that the column be topped by a pintle, a solid iron pin of much smaller diameter than the column. The pintle passed through the girder, surrounded by massive wood, and then enlarged to meet the full width of the column above. This practice was much less common (the author has never seen it), however, than the practice of passing the column through the girder at its full width. The ends of the girders were cut away in a semicircle to make room for the column. Thus, a relatively small amount of wood rested on the flat plate cast to extend from the top of the column. The loss of this wood could cause the initial collapse. F. W. Dean said, "This is dangerous construction."[24]

---

*The observant fire officer, on seeing dog irons, will recall that the designer intended the floors to fall in a fire, and will plan to have units outside if he anticipates a collapse.

While shopping in a famous New England department store, the author noticed that the basement ceiling had been cut away at every column to expose the connections, probably for a structural analysis. The tremendous wooden girders, about 24 inches by 24 inches, had been cut around in a half moon to accommodate the column. The loss of very little wood at this point could cause a collapse. There was an upward draft into the void. Though the building was sprinklered, it is possible that a fire could get into the void and destroy the building even before the sprinklers activated.

Over the years, failures of cast-iron columns have been blamed for many serious collapses. It is true that columns can fail, particularly if they are poorly cast so that the material is thin at one point. The often cited "red hot cast iron hit by cold water" is at least questionable. After looking at a number of buildings with cast-iron columns, the author suggests that the chief cause of failure is the typical unsafe connections. Most often the cast-iron column is held in place only by gravity so that the slightest lateral movement can cause it to kick out.

Hollow cast-iron columns stacked above one another form a hidden flue for the passage of fire.

Roof and Attic Structures

In a building of large area or monumental characteristics, the designer faced the necessity of supporting a roof over wide spans and producing a roof line of a design other than flat. He turned usually to the truss. Some trusses are so deep as to provide room for a story, or at least an attic, between the upper and lower chords. Often there is no access to this space for outside streams to protect steel connectors and ties from heat or to stop the loss of substance from wooden members. Collapse of trusses may precipitate collapse of the walls.

The fact that the attic space is usable does not mean that trusses were not used. Recently, parallel chord wood trusses have been observed used as roof rafters. This provides a huge combustible void between the attic rooms and the roof. If there is fire in this void, it is not safe to ventilate it.

Sometimes the trusses are built outside the building above the roof. Deterioration from weather may weaken them.

Working Load vs Fire Load

The fact that a building was built in recent years is no guarantee that its design is any improvement over the practices of the last century. This is particularly true of buildings built in areas where building codes are weak or nonexistent.

If a building is erected and does not collapse under its working load, the builder has achieved a prime objective, but not the only one. All may be well as long as the building is axially loaded. Lateral loads, such as wind, sudden shifts of heavy objects, partial collapse, movement of people, explosion, and last but not least, earthquake, may not be adequately resisted if lateral resistance is not designed into the building.

There is a two-story, circular, owner-designed concrete block commercial building not far from a famous Virginia historical tourist attraction. Because the building is circular, there is a polygonal ring of concrete block columns in the center. The cinder block columns are carried in reduced cross section up to the roof. A ring of wooden girders spans from block column to block column. At the first floor level, steel pipe columns were spaced alternately between concrete block columns. One column was too short, so it was pieced out with scraps of 2-inch by 4-inch lumber stacked one on top of the other with a piece of plank nailed to them. There is no apparent connection between the girders and the column. At the second floor level, the designer's ingenuity really found expression. The girder ends are tied to each other with pieces of reinforcing rod wrapped around the girder and clinched with bent-over nails.

This column with its 2 by 4 extensions carried a major load in a Virginia tourist attraction. A letter to the governor caused it to be replaced before it could collapse.

In another location of the same building, a piece of channel steel atop a steel post provides a socket for a girder and is attached to the girder by four nails driven through holes in the channel. At times, there are hundreds of shoppers in the building; it has stood for several years and has not yet collapsed. *

---

*The author communicated his apprehensions to the State Governor. The owner was persuaded to engage a structural engineer who made a number of changes. The short column, for instance, can no longer be seen.

The point is that the ability of a building to stand up to its daily burden is no indication of how it will perform under unusual stresses, and a fire is an unusual stress. It also consumes the very substance of the building. Long experience in fighting fires in such buildings is not necessarily a guarantee of safety, as each building has its own weaknesses.

Pages could be written describing the astonishing variety of ways men have found to erect buildings that will stand for 100 years under normal loads, but for only a few minutes under attack by fire. There is no limit to the effort the fire department should put into learning as much as possible about the buildings in its area.

Void Spaces

As in frame buildings, void or concealed spaces are an inherent part of buildings of ordinary construction. The voids may be totally combustible, as in a wood-joisted building with wood floors and wood lath and plaster ceiling. Partially combustible voids are exemplified by wood stud construction with gypsum interior finish. A modern commercial building, with steel bar joist roof beams, noncombustible ceiling tile, and a gypsum roof, has a noncombustible void.

The various voids may be interconnected in a manner difficult or impossible to detect. Requirements for firestopping may not have existed or may have been ignored. Even if firestopping has been installed, the execution may have been imperfect, or it may have been removed to permit the installation of ducts or wires.

Void spaces are a major fire fighting problem. The existence of combustible voids has caused many serious losses even in sprinklered buildings. The fire can be extinguished only if it breaks out of the void or if the fire department breaks into the void. If circumstances are favorable, expansion of steam (created by hose streams hitting other burning material) into the void may accomplish extinguishment.

Noncombustible void spaces often contain some combustible material, e.g., insulation, construction trash, wood blocking, etc. Continued smoldering of combustibles can create a smoke problem long after the main fire is suppressed. The fact that a particular void was not as noncombustible as similar voids was responsible for the loss of a major part of a new high school.[25] Low-density combustible fiberboard had been applied to the studs behind gypsum wallboard on interior walls in the school to provide acoustical treatment. Started by a plumber's torch, the fire found a fully combustible void space in a wall and extended quickly to the void space above the ceiling. The roof was an insulated metal deck. (See Chapter 7, Steel Construction, for a discussion of metal roof decks.) The only practical way to control a roof deck fire is to cool the underside of the metal. The noncombustible acoustical tile ceiling proved to be a hindrance to the fire department's efforts to deliver the water to the underside of the roof, the vital place to have absorbed heat.

Interior Sheathing — Vices and Virtues

There is a concept, never clearly stated but acted upon nevertheless, that by providing a protective interior sheathing or finish for a structure, a substantial contribution has been made to the building's safety from a fire in its contents. This is true as long as the sheath keeps a contents fire out of the structure. However, when the fire originates behind the protective sheathing, be it metal, gypsum, mineral tiles, etc., or extends behind it, the protective sheath becomes a detriment to the fire fighting effort. This was clearly recognized by those who developed mill construction, which is discussed later.

At many fires, no flame is apparent upon arrival, but the building is heavily charged with smoke. It is important to determine whether the fire is in the contents or in the inner structure of the building. If the building is of wood-joisted construction, with the usual interconnected void spaces sealed off from the occupied areas by a tight protective sheath, consider the void to be one big rigid "balloon." If the volume remains the same, as it must in a building, and the temperature increases, as it does in a fire, then the pressure must increase. A sign of such a pressure increase is smoke pushing, or even "jetting" from openings. In one observed fire, a plume of smoke, 3 feet long, was noted coming from the lock bolt "receiver" of an aluminum door frame, and smoke was pushing from a covered electrical outlet box on a nearby wall. The smoke in the occupied areas, though heavy, was static. It was immediately suspected that the fire was in the inner structure of the building. The location was determined when a plywood shadow box burned through. The plywood, a weakness in the defense of the structure against a contents fire, proved an advantage when the situation was reversed.

Fire can be burning in void spaces and have a substantial hold upon the building without the occupants being aware of the situation. In the fatal Anne Arundel, MD Oyster Roast fire,[26] the building was a one-story wooden building with peaked roof and combustible ceiling. The fire was observed coming from the ceiling. It was fought with, and apparently extinguished by, several extinguishers; however, the fire continued to gain headway in the void, and when it broke out again, disaster was inevitable.

Ceiling Spaces

Older buildings have much higher ceiling heights than those built today. False ceilings are commonly added to conserve conditioned air and to modernize the building. It is not uncommon for a building to be "modernized" several times in the course of its lifetime with a new and lower ceiling each time. The author watched a full first-alarm assignment search a downtown area for 45 minutes, including raising aerials in the street, with men to look for fire. The fire eventually was found in the many-layered ceiling of a saloon. The smoke had found its way into an old flue; no one in the saloon had any indication that there was any fire at all.

In Wyoming, MI, a one-story, relatively small, ordinary construction building had been built with a combustible tile ceiling suspended 26 inches below the roof on a typical 2- by 4-inch furring strip arrangement.

When the store was remodeled into a restaurant, the remodelers met the code requirement for a place of public assembly by suspending a 1-hour, "fire-rated" (what this really means is often hard to determine) ceiling 38 inches below the combustible ceiling. Fire was burning in the ceiling area. Very suddenly, fire burst from the ceiling, and in moments the restaurant was an inferno. The collapsing ceiling and roof trapped two fire fighters, who apparently ran out of air and then died. The city of Wyoming now requires the removal of the old combustible ceiling.

Even experienced fire fighters can be deceived. It is possible to have a raging fire overhead without it being apparent in the space below.

A third-alarm assignment was operating at a taxpayer fire. The building faced 200 feet on one street and 150 feet on another. The fire was at the east end. There was a sewing machine store at the west end; it had been opened early in the fire and found to be clear. As the author stood fascinated in the doorway, a paint spot on the "tin" (embossed steel) ceiling turned brown, then cherry red, then bright red, then flaming paint fell from the ceiling and ignited the flimsy material draped on display. In about the time it takes to tell, the store was fully involved. There was a raging fire overhead, but there was no evidence in the store that there was anything amiss until fire dropped from the ceiling.

The author was an observer at the scene of the tragic Our Lady of the Angels School fire in Chicago in December 1958. It is his personal opinion that it is at least credible that the fire was in the attic over the heads of the victims, and that it "blew" down on the victims.

The restoration of older buildings is often carried to ridiculous extremes. In a Florida city, an old building with a high ceiling (of sheet metal) was being restored. The new ceiling, also of embossed sheet metal which was carefully removed from a demolished building, was installed several feet below the old. Now there are two voids, and a double barrier between the fire department and the fire.

Joist Spaces

Important void spaces in any multistory, wood-joisted building are the joist spaces. Containing many square feet of exposed fuel, they are protected from hose streams by their construction and the ceiling below. As the fire gains in volume, it spews torches of flaming gas out of every rat hole, pipe hole, and utility service hole in the floor above. The old-timers had a simple technique for stopping this extension; they put about 2 inches of water on the floor above. The fire coming up met the water coming down. Those who never saw a fire when it was out of control, and "modern" fire chiefs, who fell for shibboleths like "put water only on the fire," joined hands to wail over the "unnecessary" water damage. There is, strangely enough, no water damage at a total loss. If a sprinkler system operated on the exposed floor and did exactly what the old-timers' system did, it would be credited, and rightly so, with stopping the extension of the fire. When the chief improvises a sprinkler system because none was installed, he is criticized. Strange are the ways of those who substitute sloganeering for a critical analysis of available alternatives.

Combustible Gases in Concealed Spaces

The ignition of combustible gases, accumulated in voids within the building, may provide the fuel for a devastating explosion, even though the building is ventilated in the accepted manner.

At a stable fire in New York in 1938, the author witnessed an explosion of carbon monoxide that had accumulated in void spaces. The violent explosion took place an hour and a half after the first alarm. It was presignaled by dense clouds of "boiling black smoke," and a longtime observer of fires told us, "it's going to blow." The blast caused the collapse of a side wall and the loss of one officer's life.

There was actually a detonation. Apparently, the gas-air mixture was just in the right proportion. The building had been vented according to standard procedures. It was just unacceptable to the investigating committee that a backdraft explosion could occur 1½ hours after the first alarm in a vented building, and it was held that the wall simply collapsed.

In 1946, New York fire fighters were battling a fire in an abandoned icehouse. Gas accumulated in a 9-foot cockloft and exploded. The back wall came down on an apartment house at the rear. Thirty-nine persons were killed.[27]

More recently, the author observed a fire in a row of stores. The building was vented; the front window of the store was completely out. There was bright fire which was suppressed with hose streams. The wind shifted. Heavy black smoke boiled out. Like the old-timer mentioned above, the author said to a companion, "It's going to blow." It did. Fortunately, the fire fighters were just knocked down.

There was an old saying, "Don't be in such a hurry — get there after it blows." This exemplifies the myth that if there is going to be a backdraft, it will take place early in the fire. A backdraft can take place any time the conditions are right. One sign is heavy, black, boiling smoke. Unfortunately, with the prevalence of plastics today, this is not a sure sign, but if there is no known reason for the smoke, prepare for a blast.

Large Voids

Churches, schools, town halls, and other monumental buildings are often made monumental by the inclusion of vast void spaces. Add a basement full of combustibles, non-firestopped walls, and sometimes hollow columns, and, truly, the building is "built to burn."

At a fire in a Takoma Park, MD college building of stuccoed, hollow tile walls and joisted construction, fire fighters were making a valiant stand in the interior. Darkening of matchboarding on the underside of the wooden column-supported pediment over the front porch, caused by fire above, disclosed the fact that they had been defeated by the designer; he had laid a huge open combustible attic over the whole building.

Writing in *Fire Chief*, Chief Emmanuel Fried of Chicago Heights, IL, summarized the problem of coping with void spaces in buildings.[28] Realistically, he dwelt on the impracticability of interior hand lines; the fire cannot be reached, and the collapse of roof trusses and massive plaster presents serious risks to fire fighters. His final sentence is all too true; he advised setting up snorkels and ladder pipes.

Do not consider void spaces a problem only of old buildings. One church, built in 1970, has the usual high-ceilinged auditorium. The roof is wood decking on wood joists supported on unprotected steel girders on steel columns. The steel is "protected" from a fire in the auditorium by a suspended ceiling; it is not protected from a fire in the wooden roof above. The freestanding walls are composite brick and block, supporting only their own weight, but probably tied to the steel columns for stiffening. A fire in the roof void could cause the distortion of the steel and possibly the collapse of the walls.

A typical fire, which moved through void spaces and destroyed a building with little visible fire until the building was fully involved, is described in the article, "Multiple Alarm Fire Destroys Fraternal Lodge."[29] At one point, the article describes the floor of the second level as "hot and spongy." If fire isn't visible and heavy smoke is present, the building must be burning; its resistance to gravity is melting away.

Ten guests died in a motel fire. Fire raced up a stair tower which was located at the intersection of two non-firestopped trussed gable-roofed attics over two wings. The fire raced both ways through the combustible attics.

In a California hotel, built in the 1920s, fire spread from the guest room of origin through pipe chases to the attic and a large crawl space between the first and second floors. Fire fighters held the fire to one of the two wings of the hotel.

A four-story business block, which housed six businesses, was damaged by fire. The two top floors had been removed. A common undivided cockloft was thereby created over the entire building. Fire in the basement was stopped from coming up the stairs by the fire department. However, fire reached the cockloft via interior stud walls. The loss was about $500,000.

The law library of Temple University Law School was destroyed by a fire which originated in and extended through concealed spaces. The building had been converted from a synagogue. The full article[30] is worth bringing to the attention of those who permit unique values to be in danger in unsuitable buildings. The only adequate protection for reference libraries is a properly designed automatic sprinkler system. The new Madison building of the Library of Congress is fully sprinklered and work is underway to sprinkler all of its buildings including the historic main building with its priceless contents.

Interior Structural Walls as Fire Barriers

In large buildings of ordinary construction, masonry walls may provide the interior load-carrying structure. Such walls may be useful in providing a location at which the fire fighting forces can make a stand. Sometimes closures of more or less adequate fire resistance have been provided at openings. These can be useful. However, since the purpose of the wall is not to stop fire but merely to carry the interior floor loads, it would be unusual to find such a wall carried through the attic space. The combustible attic may provide a path by which fire, initially confined to one section, may pass over the top of the wall. If this situation is recognized, prefire planning might call for special attention to maximum removal of the roof over the fire section to permit heat to escape to the atmosphere. It might also be practical to plan to vent certain windows, while keeping others intact, to relieve the heat through a safe path.

A large school, built in the late 19th century of ordinary construction, burned with what was described as an "unbelievable" amount of fire. The building was shaped like an "H." The brick-bearing walls formed natural fire barriers and needed only fire doors at the openings to convert them into quite respectable fire walls. These were provided; thus, the legs of the "H" were separated from the "cross piece." Unfortunately, the walls extended only to the top floor ceiling, leaving an undivided attic over the whole area. The fire spread over the top of the walls through the attic and took the entire building.

Voids in Mixed Construction

Many buildings are composites of older sections, possibly frame or ordinary construction, and newer sections of protected steel or concrete construction. Unless the architect is aware of the hazard, or an alert fire department can present a convincing analysis of the total fire problem to the building department, the new fire resistive addition may be completely at the mercy of the old building, particularly when the architect wishes to make the result look like one building.

An industrial building in a New England city appears to be of fire resistive concrete construction. The view of the building from an upper story of a hotel nearby shows that it was built around an older building with a huge peaked-roof attic on which air conditioners are mounted. It is almost certain that the attic is of wood construction. This would be quite a surprise at a fire. Get up high and look down on your problems.

A fire in the Barbara Worth Hotel, El Centro, CA, in 1962, is a good illustration. The main section of the hotel was four stories, brick and wood-joisted, with voids undivided both vertically and horizontally. A concealed vertical opening between the top story and attic, caused by the arrangement of a wall of the new

A peaked roof is being built on top of this flat roof which already has a cockloft. Such construction dooms the building as it is practically impossible for the fire department to hit fire in the inner voids.

fire resistive section, admitted fire to the attic, which was common over both sections. The fire completely destroyed the original building and badly damaged the annex. The report of this hotel fire in the *NFPA Quarterly* was accompanied by a picture taken early in the fire.[31] It is hard to believe that at that point it could be said: "Because of construction defects, the building is doomed."

Modern ordinary construction includes more noncombustible components than its predecessor, but some practices will cause unwarranted losses.

Cornices and Canopies

Styles in buildings are like styles in women's clothing, except that the mistakes last longer. A good knowledge of styles in buildings is invaluable to a fire officer, particularly in suppressing a fire in a building for which one has not had the opportunity to prepare a prefire plan.

In reaction to the severe functionalism of modern architecture, which permits no ornamentation, the cornice is returning to the commercial building and to garden apartment buildings and row houses. In most cases, a cornice consists of a wooden structure, shingled or board-sheathed, which extends the length of the building. In some cases, the cornice extends around the entire perimeter of the building without firestops. A large combustible void, inaccessible to fire streams, is created.

There are several different methods of constructing cornices. In one method, the entire cornice is built on the outside of the building after the masonry wall is finished. In such a case, a fire in the cornice may present no problem of extension, but cutting into the cornice will give no access to the void space of the cockloft.

In another method, the cornice is actually a false mansard roof, a structure in which the outer wall of the top story is built of material different than the walls, and inclined inward at an angle. In *The Brown Decades*, Lewis Mumford calls the mansard "a crowning indignity" in the '70s. (He was referring to the 1870s not the 1970s; thus, the more things change, the more they are the same.) In the case of a building where the mansard is not an attic story, but merely a roof structure, we may have a useful advantage. Cutting through the flimsy mansard may give access to the entire cockloft area. If combustible construction is burning, or noncombustible steel is overheating, rapid cooling may be accomplished by sweeping a large area through a single opening.

At the worst, a cornice can interconnect with void spaces and false ceilings; it then becomes a path of extension of fire and distribution of smoke from area to area.

A noncombustible motel in Pennsylvania is decorated with a huge mansard which extends around the building. The fuel load of a motel room is sufficient to pour heavy fire out the window. This fire would ignite the mansard and wreck the motel. It would be interesting to ask the architect, "Did the owner give you a criterion that a single room fire should put the entire motel out of business?" The question might even excite the interest of the owner.

Closely allied to the cornice is the sidewalk canopy. In some cases, this is of cantilever construction, thus providing a serious collapse potential. Even when not cantilevered, the cornice and canopy represent substantial loads, usually supported by combustible members. The collapse of a cornice or canopy is quite sufficient to cost several lives.

One canopy, several hundred feet long, which the author witnessed under construction, showed an interesting detail. Of wood-joisted construction, it was lathed on the underside with metal lath on which cement plaster would be plastered. Furring strips were laid at right angles to the joists, and the lath was nailed to the furring strips. The joist channels were therefore all interconnected and fire could spread unchecked throughout the entire length of the structure. Far from being an advantage, the metal lath and cement plaster would be a hindrance to efforts to open the void. Since the void was interconnected with the interior, steam generated within the canopy by fire streams could spread into the building proper, possibly doing damage far beyond the fire area.

Sometimes canopies are firestopped, though the firestopping is often less than perfect. Gypsum board is rarely taped and nail-set, and often there are gaps around the edges.

Voids in Modernized Buildings

Modernizing old buildings is a common practice. In one modernization in Washington, DC, the builder installed false ceilings, closed up the old original stairwell into a blind void, and managed to interconnect many existing voids. In a truly amazing operation, the Washington, DC Fire Department fought a third-alarm fire in this building and confined the fire to the structure of the building. The only extension to contents was to a shower curtain and other bathroom contents via conduction through the bathtub from fire in the wall behind it.

Commenting on the rehabilitation of six-story brick "apartment" buildings, a New York property insurance analyst points out that the usual construction creates a huge three-dimensional void across the ceilings. In the rehabilitation process, the building is gutted and apartments are created by removing and/or adding walls. Dropped ceiling frames of 2-inch by 3-inch beams are hung, and the new interior is lined with ⅝-inch plasterboard panels. The walls are not firestopped at the ceiling line, and the vertical stud spaces are unprotected above the ceiling. Old dumbwaiter and pipe shafts are retained for utilities.

Such buildings burn furiously. Fire fighters have to pull down every ceiling to find the fire and occasionally are met by a vapor explosion, some of enough force to blow them out of the room. Most codes do not adequately address this problem.

To at least some extent, our building codes reflect fire loss experience. The requirement for brick exterior walls in certain types of buildings grew out of the not remarkable observation that cities built of wooden buildings, close together, burned down. On the other hand, some valuable fire protection features were never required by code. They resulted from provisions for some other purposes, and the fire protection benefit, though real, was unintended.

An excellent example of indirect benefit was that windows invariably were provided at the front and rear of a building and skylights in the roof. These wall openings are absolutely necessary to the control of any advanced fire in a building of ordinary construction, but were never legally mandated.

In an effort to modernize the typical downtown store, the designer often closes up the front windows of the upper floors. Air conditioning and fluorescent lights do the job of the windows and skylights. The skylights and rear windows are closed up to foil burglars, and the often heavily fire-loaded upper floors become, in effect, one big inaccessible concealed space, completely impervious to fire department streams. What might have been a slight fire in earlier days becomes a major loss as heat and smoke fill the void and destroy the contents. An advanced fire is an impossible problem for manual fire fighting since "you can't put it out if you can't hit it."

The building essentially becomes a volcano. The attack is simply heavy caliber streams from above after the roof vents itself.

A typical fire in a windowless lounge is described in an article.[32] Fire fighters found that they could not operate inside the building or safely from the roof. This left only an overhead attack from an elevated platform. The stream forced fire laterally into the adjacent pizza store. If not sprinklered, these buildings are a "no-win" situation. The fire department did not create the unmanageable problem.

If the upper floors are not open to the public, the stairs may be removed to provide more floor space on the first floor. The opening may be closed only with a piece of sheet metal, which would collapse under the weight of an unwary fire fighter.

In some cases, building code authorities have recognized the fire departments' problem and required fire department access panels; these are usually of limited size. Since many codes require sprinklers for inaccessible concealed spaces below mercantile buildings (called cellars), it appears quite logical to press for sprinkler protection when the void is built above the first floor. The problem is the same. The sprinklers may prevent the loss of the building, but a substantial fire loss is almost inevitable. A fire in a department store in Hicksville, NY is a case in point. Though it was a modern, fire resistive, fully-enclosed building, rather than a modernized building of ordinary construction, the lesson is fully applicable. Although the fire covered an area of only 25 square feet, heat buildup in the windowless structure caused 147 sprinklers to operate, and the loss was $1,100,000.[33]

A billboard covering the front of a building can serve also to convert upper floors into what amounts to an inaccessible void.

In some cases, modernization is accomplished at less cost (the windows are only painted over); in others, wood or metal is substituted for glass in blocking off windows. In one Richmond, VA warehouse known to the author, windows have been bricked up behind the glass panes.

These "windows" provide no access to the interior, hindering fire fighting and rescue efforts.

There is a chain of popular seafood restaurants in the midwest. On the exterior of their standard building, several windows are seen. Windows are also to be noted inside. In fact, there is no relationship between the two. The exterior windows are where the architect wanted them. The interior "windows" are where the decorator wanted them. Time wasted in "discovering" this readily observable fact could cost lives in a fire where rescue operations were required. This appears to be a common practice in restaurants especially.

Fire Extension Through the Building

In a building of ordinary construction, rarely is any provision made to prevent the extension of fire through the stairways and halls. Even if the stairways are enclosed in response to legislation passed after some terrible disaster, there are often many bypasses. In addition to interconnected voids, pipe closets, utility shafts, and elevator shafts provide fire paths. In the famous Equitable Building in New York, an apparently minor fire spread from the basement to the upper floors via old dumbwaiter shafts.

Prepare to be surprised, and perhaps you will not be.

Masonry Bearing Walls as Fire Barriers

Attention now is turned to the masonry bearing wall as a barrier to the spread of fire. A fire wall, by definition, is erected for the sole purpose of stopping a fire, and the truly adequate fire wall can accomplish this function unaided. Masonry bearing walls will be examined to determine those practices and conditions that make a bearing wall less than a fire wall. When defects are recognized, tactics may then be planned to take advantage of the assistance offered by the masonry wall, however imperfectly it meets the standard for a true fire wall.

"Main Street USA" is made up of buildings of ordinary construction, usually brick and wood-joisted, built side by side. If each building has a 12- or 16-inch bearing wall, unpierced, these two walls together form a barrier to the passage of any fire that can be generated. The fire may go over or around such a pair of walls but not through it. Unfortunately, the walls may not be unpierced.

Old Communicating Openings

In bygone years, adjacent buildings may have been connected by means of doorways at one or more floors. If the openings are in current use, the fire department can be made aware of them and plan accordingly. It may have legal grounds for requiring such openings to be protected, but the fire department may be totally unaware of openings that were closed at some time in the past when the occupancies were separated. The closure at an opening may be an ordinary door, a fire door with merchandise stacked against it (thus ineffective), lath and plaster, or properly constructed masonry to restore the full integrity of the wall.

Examine older buildings carefully for interconnections. In some cases, the originally planned openings were properly protected by fire doors, but additional openings were made later without proper protection. A Virginia city has many restored buildings. In one case, two older buildings were combined to form a restaurant. The opening was protected with a fire door. Alongside it there is an unprotected window so the manager, whose office is in one section, can see the dining room.

Fire originating in one building may first show up as smoke in the connecting building. This was the case in a fire at 12 East 23rd Street, New York City, on Oct. 14, 1966, in which 12 fire fighters were killed. The fire was first reported from 7 East 22nd Street. The two buildings were connected by a basement door. Do not assume that the fire first reported is the original fire.

At one fire that the author observed in New York's "Hell's Hundred Acres," a second-alarm fire in a five-story brick and wood-joisted commercial loft was under control, and companies were taking up. Suddenly, heavy smoke was found coming from the cellar of a similar building around the corner. The second fire also resulted in a multiple alarm. After the fire, an underground tunnel connecting the two buildings, long boarded up and forgotten, was discovered. The fire had extended through the tunnel.

The author recalls two New York buildings that faced on different streets and were joined back to back. There was a difference in grade between the two streets. After a serious fire, a "watch line" crew detected serious hidden fire. It was later learned that a new floor had been built over the floor of the store at the lower grade to level the entire floor. The fire was concealed in the void space.

Another method of interconnection is to cut through the wall so that the stairway of one building can serve the upper floor(s) of the adjacent building. This permits the elimination of the stairway in one building and the enlargement of the most profitable rentable area, the street floor.

Keep in touch with the old-timer, the broker who deals in downtown real estate, the alteration contractor, the building engineer; all are people who may have information of value about the buildings. Look at buildings from across the street. Note the businesses which occupy more than one building. Most likely they are interconnected. Determine the value of the protection (if any) on the opening.

Units operating through a fire dooor, whether overhead rolling, or sliding, should block the door. A sudden burst of heat may cause the door to close. Closed fire doors, even sliding doors, can be very difficult to open.

Protection From Exposures

When adjacent buildings are of identical dimensions, and the bearing walls are unpierced and parapeted, the ideal fire barrier exists. Unfortunately, the situation is often less than ideal. When one building is taller than its neighbor, the owner of the taller building may install windows in the upper stories overlooking the smaller building. Fire coming through the lower roof may extend to the adjacent building via the side windows. Burning material may fall out of the upper windows onto the adjacent roof. Old, hidden windows may provide a surprise path for fire to travel from one building to another. A key factor in a $1,000,000 fire in Gloucester, MA was a window in a brick building that had been boarded up.[34] The window was next to the wooden exterior wall of a sprinklered department store. The fire entered the ceiling of the store above the sprinkler piping; the sprinklers collapsed and the store was lost.

If two buildings are not of the same depth, the rear windows of one, and side windows of the other (beyond the end of the shorter building) may expose one another. In such a case, the rear exposure may require protection, even before the fire is attacked; otherwise, a frontal attack may drive the fire out the back windows into the exposure.

Narrow alleys between buildings present difficult defense problems against exposures. Often sheet steel, corrugated steel, or steel-clad wooden fire shutters protect windows facing the alley. Their worth is dubious, and in any case, they usually are not closed. Wired glass is of limited value against radiant heat. Fire fighters operating lines in the alley are in serious danger. Study the problem in advance. Lines directed downward from the roof of the exposed building or ladder pipes directed into the alley from the street are two methods of attack.

At times, outside sprinklers or spray systems are installed to protect against exposure fires. The fire department should be fully familiar with their operation. In most cases they must be turned on manually. The valve is located inside the building. The system should be tested at least annually. The system will do the work of one or more engine companies in covering the exposure, do it better, and of course, without risk to personnel.

Party Walls

In many cities, it is quite common to erect party walls, i.e., walls that are common to two buildings.*

In examining the building, determine the thickness of the wall, if possible. Land records or the owners or occupants may provide information. Because the joists of both buildings are supported on the same wall, the builder often found it most convenient to support the joists in the same socket. Sometimes the joists were overlapped to provide greater strength. Modern codes usually require several inches of masonry between joist ends, but this falls in the definition of legal firestopping (see Chapter 3, Wood Construction) and is thus undependable.

A typical party wall fire extension occurred at a fire in Ottumwa, IA on Jan. 29, 1962, involving two stores separated by a party wall in which joists were lapped in common sockets.[35] The fire in the original store was under control. No examination was made of the ceiling of the adjacent store. Suddenly, the ceiling of the adjacent store was found to be involved. A backdraft explosion drove the fire fighters out, and the building was lost. The fire had passed into the ceiling of the adjacent store by way of the joist openings.

Extension through party walls is sometimes hard to detect. Examination of the adjacent premises early in the fire may show nothing. It may take time for the fire to work its way through. Smoke, which might ordinarily give cause for alarm, goes unnoticed in the general smoky condition of the fire. Without some evidence, officers are often reluctant to order the necessary opening of ceilings. When the need becomes evident, it is often too late.

---

*In the area embracing the original city of Washington, DC, regulations issued by President George Washington required that party walls be built, and that the builder of the subsequent building pay the first builder for half the cost of the wall. Usually, party walls are established by mutual contract between the owners. To the best of the author's knowledge, the Washington situation is unique.

The right of a party wall to exist goes on until terminated by agreement of both parties. An interesting situation arises when a skeleton frame building is to be erected on the former site of one of the party wall buildings, while the other is to remain. The party wall is picked up on the frame of the new building, as a curtain wall. If the older building is thereafter demolished, the party wall remains as a protrusion beyond the line of the framed building.

Partition Walls

Some modern rows of stores approach the definition of noncombustible construction, but they can present problems of fire extension similar to those of party walls. Partition walls between stores may be of unpierced concrete block. The top chords of steel bar joists rest on the wall. Take a close look at the next such building you see under construction. Note that the design of the bar joists provides a gap about 3 inches in height between the top of the wall and the roof, whether of steel or wood. Determine if there is any intention of closing the gap, which can be compared to a similar gap across the bottom of a dam. Some building departments require it.

Note the gap over the top of the wall caused by the ends of the bar joists resting up on it. Fire can travel through this gap.

Depending on the local building code, there may be no ceiling, a combustible ceiling, or a noncombustible ceiling. The ceiling may be over all or only part of the store. If the ceiling or roof is combustible (ordinary metal deck roofs are combustible; see Chapter 7, Steel Construction) the extension of fire is almost a certainty. If the void is totally noncombustible, smoke and steam may pass to do extensive damage.

When the bar joists are set on a steel girder and the partition wall, usually gypsum board on wood or steel studs, extends only to the ceiling, the same sort of passageway exists.

The code may require fire separations between occupancies extending to the underside of the roof. These are usually of gypsum board on steel studs. Often the bar joists pass right through the fire barrier. It is impossible to close the openings. Often the gypsum board above the ceiling line is untaped and the nailheads uncovered in the mistaken belief that this is cosmetic. It is required for fire resistance. Any movement of the bar joists will displace the wall. Such a barrier can be useful if defended but it is unreliable if left alone.

The solid wall parapeted through the roof is the only dependable fire barrier; however, just because you can see a wall extending through the roof, don't assume that it is a fire wall. It may just be architectural.

In some cases, fire walls are made of combined elements. Consider a high bay steel-framed warehouse. A main girder is put in place with an upward camber. The parapet wall of brick is built above the girder. Concrete block is built up to the girder. The girder is sprayed "fireproofed." There is no proof that these disparate elements will function together as a fire wall. If steel attached to the girder is heated, it may move the girder.*

Do not take anything for granted. The partition wall can also be bypassed by the now popular combustible "eyebrows" or other false fronts. Regard every building as one interconnected fire area, unless you know differently. If our fire officers are trained to regard the entire building as the fire area, rather than just the portion where fire is visible, there will be fewer disastrous surprises.

There are uncounted cases of the extension of fire from building to building in "Main Street USA." The two following excerpts were selected from the "Bimonthly Fire Record" of *Fire Journal*[36] because they exemplify several faults of contiguous, ordinary construction buildings.

CLOTHING STORE

Inadequate Separation   Apr. 27, 1964, Salisbury, NC

The clothing store in which the fire originated stood in the center of a congested block of 100-year-old unsprinklered two- and three-story brick, wood-joisted mercantile buildings. A police patrolman who was checking the alleyway in the rear discovered the fire in the clothing store at 1:40 A.M. and radioed the alarm.

Apparently the fire originated in the rear of the basement of the clothing store and, before discovery, spread up an open

---

*There are a number of untested combinations of elements which are supposed to react together under fire conditions. The author likens them to a Chinese restaurant menu — one from column A, one from column B, etc.

stairway to the second story. When fire fighters arrived, the fire had spread to an adjoining variety store through glass doors in the party wall between the two store buildings. The floors of the variety store were oiled to keep down dust, and apparently the oil was ignited by heat coming through the doorway before the overhead fire doors at the opening operated.

On the side of the building of fire origin opposite the variety store, the second and third stories of a building occupied by a jewelry store were ignited when the building of fire origin collapsed and pulled down a portion of the party wall. The front wall of the third story of the jewelry building, which was recessed from the front wall of the first and second stories, rested on a steel beam that soon weakened and fell. When the steel beam fell, it made a hole in the party wall between the jewelry store and the next building in line, a department store. Fire spread through this opening to the ceiling space in the second story of the department store.

When fire fighters arrived, heat prevented entry into the burning buildings. Hose streams could not penetrate adequately from the front of the building, and the smoke build-up in the narrow dead-end alley zigzagging at the rear barred any effective operations from that point.

The clothing store, the jewelry store, and the variety store were destroyed. The department store beside the jewelry store was severely damaged. A warehouse behind the jewelry store and a shoe store beside the variety store were also damaged. Total loss was $976,000. Approximately 100 persons were temporarily unemployed because of the fire.

DEPARTMENT STORE

Substandard Party Walls   Feb. 18, 1965, Fort Smith, AR

A man in the street saw flames in the second-story window of a two-story brick, wood-joisted department store in the center of a mercantile block and telephoned the fire department. Fire fighters found the fire well advanced in the second story of the unsprinklered building and could not immediately gain control of it. Within a short time, the flames broke through the brick party walls and into the unsprinklered buildings on each side of the burning store.

The walls were old and had been put up with lime mortar which had crumbled a great deal through the years. The unparapeted wall on the south side of this store of fire origin allowed fire spread across the roof to an adjoining two-story carpet store. The fire also entered the carpet store through cracks in a single course of bricks which had been used to seal off former windows in the wall that now separated the two stores.

The fire spread to the second story of a printing shop on the north side of the building of origin by way of open joist holes in the party wall. Apparently, the joists had been removed during remodeling in previous years, and the holes had not been sealed. The building of origin and the carpet store were destroyed, and the printing shop building was severely damaged. Loss was estimated at $481,500.

In a school fire in Omak, WA, the steel trusses of the gymnasium collapsed. The collapsing trusses opened holes in the masonry wall as they pulled loose. This permitted fire to enter the inaccessible undivided attic over the classroom section.

Mill Construction

Mill construction is a special type of heavy timber and masonry construction which developed in the textile mills of New England. Essentially they made what is called today a fault tree analysis of combustible buildings and then modified the building design to eliminate the faults. In many codes, there is a classification called heavy timber construction. Unfortunately, a building conforming to the code definition of heavy timber construction probably lacks one or more of the features which the mill construction designers had learned were vital. Properly built and maintained, a mill construction building can be a structure in which fires can be brought under control before the building is involved in the fire. If one or more of the vital characteristics is absent, either by original design or because of the way the building was modified, the building can become involved in the fire. Once this happens to any substantial degree, the heavy timber characteristic becomes a liability, not an asset. Chapter 3, Wood Construction, commented on the fact that "slow burning" is of no value once the building is involved.*

Features of Mill Construction

Traditional mill construction has certain features that must be recognized in order to understand its strength and weaknesses. Among them are:

---

*It is hard to understand how a living tree in the forest which has not been rained on in three months is a "serious fire danger" while a tree trunk which hasn't been rained on in over a hundred years is "slow burning."

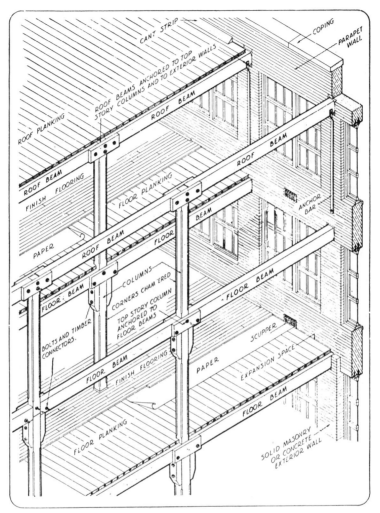

National Forest Products Association.

Some details of mill or heavy timber construction showing some inherent fire resistance. Courtesy National Forest Products Association.

- Exterior bearing and nonbearing walls are solid masonry, usually either brick or stone.

- Columns and beams are usually of massive wood construction supporting floors of thick grooved, splined, or laminated planks.

- Roofs of thick splined or laminated planks are supported by beams or timber arches and trusses.

- Openings between floors are enclosed by adequate fire barriers.

Columns and beams are of heavy timber with cast-iron connectors used to cover joints where the fire might obtain a "hold." The columns are chamfered (edges beveled) for the same reason. Exterior bearing walls are of brick. The ends of girders are fire cut (cut at an angle) so as to release in the event of a collapse without bringing the wall down. Sometimes a cast-iron box is built into the wall to receive the end of the girder. Sometimes the beams are set on a corbelled brick shelf.[37]

There are scuppers (drains) in the wall to drain off water. The combination of waterproof floors and scuppers is provided to reduce water damage on lower floors. If the scuppers do not drain, there may be a serious increase in the floor load due to retained water. The scuppers should be checked to clear debris. A broom handle will hold the scupper open and increase the flow. (In removing the water which flowed from a ruptured sprinkler riser in a huge navy warehouse at Norfolk, VA, fire fighters discovered that the concrete floors had been graded uphill to the scuppers.)

Note also the absence of concealed spaces; the finish is "open" without voids, such as can be created above dropped ceilings or behind wall sheathing. If a paneled office or display room has been constructed within the open floor area of a mill building, it may provide a destructive hiding place for fire. Such spaces can protect the fire from detection by man or automatic detectors located outside the enclosure and from extinguishment by sprinklers or hose streams.

Vital to firesafety is the protection of vertical openings and the division of the building into sections by fire walls. If fire walls are to have any meaning in limiting the loss to the area in which the fire starts, openings in them must be protected by operable, self-closing or automatically closing fire doors suitable to the intensity of fire to be anticipated.

Most important is an automatic sprinkler system with a waterflow alarm and an alarm connection to the fire department. A dependable system of supervisory locks, alarms, and checks are welcome additions to make sure that the sprinklers are continually in service or, if repairs must be made, that special fire prevention and fire suppression precautions are taken.

These are the essential features of basic mill construction. Special hazards should be located in detached buildings. Remember, though, that many of these sturdy old buildings have been around for a long, long time, and that efforts made to remodel them have detracted from their original strength. For example, steel trusses or girders may have been inserted, making it possible to remove some columns. The new trusses, by the very nature of their construction, undoubtedly do not have the inherent resistance of the massive wooden columns they replaced. If the new trusses were of steel, 1-hour fire protection would be required for them in many codes . . . but there has been many a slip twixt the book and the building.

Conversions

Many of these old mill-constructed buildings are no longer used for their original purpose. They are used for tenant factories, storage, discount stores and in some cases, are being subdivided into apartments. In many instances, the basic principles set forth earlier are seriously compromised. Fire loads are often beyond the capacity of the installed sprinklers or sprinklers are turned off. Unsprinklered areas (void spaces) unreached by the sprinklers are created. The result can be and has been disaster.

In many of these conversions, assistance is provided by the local government in order to better the local economic conditions. Great pressure may be put on fire authorities to waive "red tape" and "unrealistic regulations."

The firesafety of a heavy timber building depends on stern maintenance of the features listed earlier. If the public authorities persist in scrapping necessary fire precautions, the deficiencies should be clearly set forth so that there will be no misunderstanding of the potential for a disastrous conflagration.

Fires in Heavy Timber Buildings

An example of a fire where slow-burning characteristics of building construction proved to be meaningless involved a mill-constructed warehouse in New York City in March 1964.[38] The building, used to store household appliances, was about 100 feet by 150 feet and had 20-inch brick walls. Four 12-inch by 12-inch timbers were joined to make up 24-inch by 24-inch columns; beams and girders were 12-inch by 18-inch timbers, and the floors were tongue and grooved planks topped by finished flooring. There was no sprinkler system.

Upon arrival, the fire department found the three upper floors involved, and an immediate second alarm was sounded. The amount of stock in the building, the floor layout (a fire wall with open doors did not block spread of fire but blocked hose streams), and the volume of heat and smoke made it impossible to fight fire from the inside. The fire totally destroyed the building despite the efforts of a fifth-alarm assignment and help from five fireboats. (The fire was placed under control 21 hours after the first alarm; two hours later, the front wall fell in.)

Once the fire department is driven from the building, the "advantage" of "resistance to collapse," which is offered for slow-burning construction, is meaningless.

Shortage of water or fire fighting units was not a factor in the warehouse fire described above. In 1935, in the same area, the author witnessed a fire in a rubber warehouse of mill construction defy the fire department for 4 days, during which 30,000,000 gallons of water were thrown on the fire.

There are other things besides the size of timbers, combustibility of contents, and degree of fire protection offered that affect the rate of burning in mill-constructed buildings. The rate at which a fire burns is greatly influenced by the rate of air supply. Over 500 British thermal units (Btu) are yielded per cubic foot of oxygen delivered to the fire. In the household appliance warehouse fire cited previously, the building was a closed warehouse, the windows were intact, and the air supply was so restricted that there were fears of a massive carbon monoxide explosion or violent pressure release. For some time, the sturdy roof defied the ventilating efforts of two rescue companies with power tools.

In Palmer, MA, a typical 19th-century mill had been divided into tenant occupancies. Employment was provided for over 300 people. The mill was sprinklered to an old standard but there was no waterflow alarm and the water supply was deficient. The huge mill was a total loss despite the efforts of fire fighters from 20 communities using 56 pieces of apparatus. The entire group of buildings was involved in less than an hour. The fire was hot enough to melt brick.[39]

Prefire Planning for Mill-Constructed Buildings

The fire chief of the small city or town in which one or more of these old mills is located has a real problem of several dimensions. A serious fire represents a task far beyond the capability of any force he can muster.

Based on the cases cited and many others which could be presented, reliance on manual fire fighting for the preservation of a mill-constructed building is likely to be futile. This is not to say that determined fire fighters have not made some fine "stops," but the odds are against success. The only fire defense system with a high probability of success must be built around the automatic sprinkler system.

Automatic Sprinklers

The fire deficiences of combustible buildings of wood, ordinary, or mill construction can be largely overcome by automatic sprinklers. To be effective, the sprinkler system must be adequate in design for the hazard, must cover all areas, must have an adequate water supply, and must be kept in service. When it is out of service for any reason, in whole or in part, extraordinary precautions should be taken by the owner and the fire department.

In 1972, a fire destroyed a typical converted mill complex in Wakefield, MA. The complex was sprinklered. The sprinklers were shut off (cause not given) in the building of origin. The fire, pushed by high winds, overwhelmed the sprinklers in other buildings.

To many fire departments, the automatic sprinkler system is a really first-rate, first-aid extinguishing system, but first aid only. Further, they believe that assurance of the reliability of the system is the sole responsibility of the owner of the property. Both of these ideas are erroneous.

On the other hand, there are some fire protection engineers who seem to have the point of view that the provision of sprinkler protection eliminates the need for the fire department in any but a clean-up role. This erroneous concept is supported, albeit unconsciously, by emphasis on the large number of fires suppressed by the operation of one or two sprinklers. Even the fact that the NFPA standard on sprinkler systems[40] allows the fire department connection to be optional helps to create the "either-or" climate. It is a fact that some fire fighters see the sprinkler system as a threat to their economic security. (The economic security of the fire fighter depends truly upon an adequate tax base for the municipality, and the interest of the municipality in an operating business may transcend the interests of the owner, the operator, or the insurer.)

There is an increasing recognition that sprinklers and fire departments are complementary, rather than rivals. A sprinkler system is a fire department tool and it should receive the necessary attention from the fire department to be sure that it will perform as intended. This is most important in combustible buildings, for as has been seen, an advanced fire in a major business is most likely to be beyond the control of all the forces which might be assembled. *

Many buildings were sprinklered to meet what are often somewhat mistakenly called "insurance requirements." A particular insurer who sells insurance at a more favorable rate may make sprinklers a condition of selling the insured a lower cost insurance, but the fact that a building may be easily destroyed by fire is by no means a bar to insuring it, provided the price (premium) is sufficient, and the insurer has insured enough similar risks so that the law of averages will operate.

Over a long term, the cost of automatic sprinkler protection is often reimbursed out of insurance cost reductions, but in the short term it may be cheaper just to buy higher cost insurance. In addition, the cost-benefit relationship rarely benefits a tenant. Furthermore, a sprinkler system may be out of service for an extended period of time before the insurer is aware of the condition. As a result, he is providing the insured with "bargain" insurance.

In some cases, automatic sprinklers may be provided because of legal requirements, particularly where there are life hazards, or as an alternative to exits or area limitations. The author knows of no code, however, that requires that the permitted use of the building be suspended during the time a sprinkler system is out of service. (It is a good bet that when a Christmastime department store disaster happens, it will involve a sprinkler system "temporarily" out of service.)

Over a hundred people died in the fire in the Taiyo Department Store fire in Japan in 1973. The building was unsprinklered. There is no difference between an unsprinklered department store and a sprinklered department store in which the sprinklers have been turned off. It is the water that puts out the fire, not the pipes. There is no difference between the contents of a Japanese department store and an American department store. The lesson is clear.[41]

Macy's Department store in New York suffered a major fire in which one fire fighter lost his life. The area where the fire occurred was the only area lacking water in the sprinklers after the system had been restored to service. The previous day the entire system was out of service to repair a leak.

The developer of a shopping mall characteristically endeavors to minimize the capital investment in the mall and divert as many such expenses as possible to

---

*The practice of depleting the fire forces of a large area to fight a fire that is hopeless from the start is open to serious question.

tenants. One method is for the developer to supply only the sprinkler main. Each tenant is then required to sprinkler the tenant space. This has resulted in sizeable areas, awaiting tenants, being unsprinklered. Such spaces are ideal for storage of supplies and building materials and temporary rental to established tenants for stock storage. When the space is rented, the store constructor often fills it with combustible trim materials and possibly flammable adhesives and plastics. The last item installed is often the sprinkler system because the heads go below the finished ceiling. Fire departments should insist that all spaces be sprinklered. If this requires shutting off the sprinklers to install ceilings, this should be done at night. It is simply not sufficient to accept the fact that the sprinklers will be in service before the store is opened. The contents of the "unopened" store may be a tremendous hazard to the customers of the mall. In a shopping mall, the life safety of all depends on the sprinklers to be in service 100 percent to operate on, and sound the alarm for, an incipient fire.

If the building has a metal deck roof, a well-developed fire in an unsprinklered store will involve the roof. The metal deck roof fire will roll through the ceiling void quite unaffected by the sprinklers operating below. Only streams directed onto the steel roof deck from the underside to cool the steel and stop the production of gas will stop the fire. The ladder pipe is futile. See Chapter 7, Steel Construction.

It is common practice today to "trade off" many fire protection requirements if only sprinklers are installed. A book published by a sprinkler manufacturer lists some of the advantages that can be gained — longer exit travel distances, fewer exits, more flammable surface finishes, and many others. In the typical wide open shopping mall, the safety of all depends on the sprinkler system being in service.

For this reason, it is incumbent on the fire department to insist that the sprinkler system remain in service when the building is occupied. Those who can always find a reason for inertia argue, "We don't have the power." If the exits were provided in accordance with the provisions for sprinklered buildings, the unsprinklered building simply doesn't have enough exits. There isn't a fire marshal worthy of the name who can't take action against a building with insufficient exits.

Make it plain to management what the consequences of turning off the sprinklers are. If sprinkler contractors are licensed, put the burden of notifying the fire department of any impairment on them. Few buildings will be shut down; the work will be done outside of regular hours.*

Many basic fire prevention statutes provide broad powers to the fire marshal or fire chief to find that an "unsafe condition" exists. Why not try this route to keep sprinkler systems in service?

---

*When the author was with the Atomic Energy Commission, one of our university contractors was conducting operations in a very poor mill building which was fully sprinklered. The author required that sprinkler work be done on overtime or that the staff be sent home. When AEC auditors (as an ex-auditor, the author says that auditors often know the price of everything and the value of nothing) demanded to know why "plumbing was done on overtime" it was explained that the alternative would be far more costly. If you make it costly to do it their way, they'll do it your way.

If the property is to be owned or managed by a governmental or quasi-governmental body, get your story in early. Do not let some accountant "balance" the cost of improving, restoring, or providing the sprinkler protection against the cost of fire insurance only — make sure the cost of replacement at today's prices, the loss of tax ratables, the loss of sales and other business taxes, and the loss of payroll are included. Work up your own figures; put the fiscal expert on the defensive.

There is a tendency on the part of governments to "go easy" on themselves in matters of regulation. In addition, there are many restrictions on the power of one government agency to force another to act or refrain from acting. The ultimate excuse is, "Sorry, no funds."

If the property is sprinklered, the reliability of the sprinkler system should be a matter of major importance to the fire department. Keep in touch with the fire insurance engineer who services the property. Work closely with the maintenance superintendent; help him plan changes. Provide effective special watches when the sprinkler system is out of service and see that hazardous processes are shut down. Use the spare pumper, or if you must, borrow equipment, but provide pumping equipment to keep the pressure up on the system if necessary. Do not simply accept the message, "sprinklers out of service," and do nothing. The goal must be: protection 24 hours a day, 7 days a week with every part of the building under the eye of a sprinkler supplied with adequate water.

Automatic Alarm Substituted for Sprinklers

Automatic alarm systems are only another means of summoning the fire department. They do nothing to suppress the fire. Even with prompt alarm, the conditions in the typical rehabilitated old building, which the author has seen, give little hope for successful manual fire fighting. We cannot overlook the fact that arson is a real threat today in all occupancies. The amount of gasoline that one can carry is sufficient to create an unmanageable fire almost instantly.

Preplanning the Fire

This is not a general treatise on fire fighting, so this discussion is not directed at all facets of prefire planning. Insofar as it is practicable, attention is directed here to construction features that can influence fire fighting practices.

An officer preplanning fire ground tactics, at buildings of ordinary construction, should consider the following together with other factors which are not within the scope of this book:

- The probability of exterior collapse.

- The potential for interior collapse.

- The existence of void spaces in buildings and their possible interconnection.

- The potential for extension to or from adjacent buildings.

The four factors listed above are, of course, interrelated. The load on the floors, for instance, may be greatly increased because the current contents possess a high water absorption capability. The point need not be belabored, but it is necessary to start considering a problem someplace, and the divisions given above will serve.

Risk Analysis

There are no relevant statistics, but any observer of the fire fighting scene would probably agree that far too many fatalities to fire fighters occur in combustible construction buildings. When the reports of these fatalities are studied, these questions keep coming to the fore: "Why were those men where they were? What benefit was to be obtained in return for the risks undertaken?" Too often the answers to these questions have been unsatisfactory. If all concerned were able to be completely truthful, we might learn that the real reason was what Latins call machismo or manhood. *

Fire fighters have been injured and died, and the taxpayers saddled with unnecessary expense, because fire fighters were operating in buildings that could not possibly be saved in any economic sense of the word. It may provide a good feeling to say, "We made a fine stop," when we fought to extinguish a fire in already half destroyed material, totally ignoring the fact that "half burned is totally burned." But fire departments are organized to provide a certain service, not to make fire fighters "feel good."

If the fire service is ever to be regarded by management professionals as a truly professional service, some analysis of the relationship between risk incurred and benefits obtained, must be undertaken.

It is certainly no answer to say proudly, "I never send my fire fighters anyplace I wouldn't go myself." This may simply increase the number of lives foolishly risked, both in fire fighting and overhaul operations. There are many unavoidable risks in fire fighting, and they are all the more reason for the fire officer, who has the lives of his fire fighters in his hands, to be well informed about construction features of buildings that have been available for study for many years.

Size-up

Even where prefire planning was not accomplished, much can be learned at

---

*This sentence is repeated from the first edition. Since then the term "macho" has become widely used.

the time of the fire, and deduced from a good knowledge of similar buildings. Walls obviously deteriorated, braced, or tied, certainly indicate both deterioration and the possibility that the walls have, of necessity, been tied to the floors so that a floor collapse will probably bring down the walls.

The volume of fire gives a clue as to how long the floors will last. Ordinary wood-joisted floors are not formally rated by any standard fire resistance test, but it is dangerous to trust them for more than 10 minutes. (The 10 minutes may have expired before the arrival of the fire department.) Heavy volumes of boiling smoke persisting even after the visible fire has been well controlled indicate fire in inaccessible voids.

A serious error is not to prepare to use volumes of water. The old style building of ordinary construction is a combustible structure surrounded by masonry walls. Master streams used in the initial attack may turn down the heat production curve of a fast-growing contents fire and hold the fire out of the structure of the building. Proper training of company officers will prevent a ridiculous initial attack with a booster line on a fire already far beyond the heat-absorbing-capacity of the small line.

Danger to Exposures

In the case of row buildings, expect extension to the adjacent exposures. Get them opened up early and get charged hose lines into position inside the exposed buildings. If a shortage of manpower does not permit such lines to be kept manned, leave the charged lines in place with a patrol. In case after case, the fire department "suddenly found that the fire had extended," and by the time operations could be rearranged, the exposed building was fully involved.

Anticipate the possible extension of fire as a result of collapse. The collapse of a grain elevator extended the fire across a wide waterfront slip (a slip is a basin for boats which "projects" into the land, much as a pier projects into the water), to a fire resistive warehouse at the Mid Hudson Terminal Fire, Jersey City, NJ, in May 1941.[42] The author watched this tremendous radiant heat source force open wire glass factory windows on the warehouse.

Safety for Fire Fighters

Reduce the number of fire fighters in hazardous positions to the barest minimum. All lines which might require sudden abandonment should be in master stream devices. Be sure that all fire fighters have a path of escape and that they are aware of it.

If the fire is not being substantially reduced in volume, the operation is not a "standoff" — the building is being destroyed. Positions which might have been relatively safe at an earlier stage may now be unsafe. In other words, in this type of construction, if you are not winning, you are losing.

Every fire department should have a clear, readily understood signal which means "Everybody out NOW" — without hose lines or other equipment. Drills should be conducted on this tactic, and the use of the signal should not be restricted to the fire ground commander. "Remember Pearl Harbor." *

Under pledge of anonymity, the author often hears stories which reflect on the competence of senior officers, particularly those who have not kept up with tactics but still insist on responding to serious fires and taking command. Stories of several disasters had a common thread. The senior officer, arriving after the fire had been in progress for some time, insisted on tactics appropriate for the first arriving units at a fire which has not had a chance to get a grip on the building.

When a chief's duties as administrator keep him so busy, leaving no time to study, it is time for him to hang up the helmet, promote himself to a position such as "fire administrator," and let competent subordinates fight the fires.

An army general who goes to the scene of a battalion fire fight doesn't take command. He observes. We in the fire service might profit from this example.

When the fire has been controlled, do not be in too much of a hurry to commence overhauling. Make a thorough survey of the building to determine the extent of structural damage. Seek professional assistance from the building department. Use fresh alert units, not men whose judgment may be weakened by fatigue. Wait until daylight.

Writing in *Fire Command*, the author suggested that fire departments might properly make an analysis of the benefits to be derived from inside attacks on well-advanced fires in ordinary construction buildings, versus the cost in probable injuries and deaths to fire fighters.[43]. The result might be formal notice to the owner that the installation of automatic sprinklers is recommended. If, as is often the case, the owner pleads economic inability to comply, the fire department might properly take the position that any advanced fire would be fought from the beginning from the outside. The author believes that someday this will be commonplace, if not initiated by the fire department, then initiated by city managers.

A less dramatic procedure would be for the fire department to set up a "board of building review." Experienced officers, who had studied building construction, assisted by consultants recruited from the building department or private sources, would survey major buildings of ordinary construction and typical buildings of small duplicated types. Considering the available fire load, the type of construction, the nature of connections to adjacent buildings, water supplies, protection provided, and other relevant circumstances, an upper limit in minutes would be put on the time that fire fighters could be permitted to stay in the building. The time rating of the building would be part of the prefire plan. The fire ground commander, of course, would have the power to order the fire fighters out before the expiration of the rated time, but strong restrictions should be placed on authority for keeping them in the building after the expiration of the rated time.

---

*At Pearl Harbor, it is reported that junior personnel operating the radar saw the incoming Japanese planes, but no alarm was given.

The fire ground commander would be substantially relieved of an unfair responsibility, i.e., determining the boundary between stability and collapse, under fire conditions. A building under demolition by fire can be properly compared to a building under construction. In the latter case, the latent uncalculated space frame strength is not available, while in the case of a fire, the space frame is being disconnected. "The boundary between stability and instability, between sufficiency and failure, is a thin line. Ignorance of the boundary is no excuse when a failure occurs."[44]

Such a procedure might limit the number of lives lost needlessly such as what happened in an eastern city some years ago. Fire fighters had been battling a multiple-alarm fire in a large building of ordinary construction for some time. A battalion chief ordered three engine companies, manning lines in an inside stairway, to lash the lines and back out. The 70-year-old assistant chief ordered three ladder companies under command of an acting deputy chief to take over the lashed lines. Lacking any specific guideline, the subordinate had no acceptable basis for questioning the order. The building collapsed shortly thereafter costing the lives of two men and the crippling of several more. The civilian fire commissioner on the scene, who did not have fire fighting command responsibility, but who had observed many fires, was incredulous when told that men were still in the building at the time of the collapse. Only heroic rescue efforts, directed by a skilled fire officer who responded after the collapse, kept the death toll to the figure cited. Must "The past is prologue" be our motto?

The following quote from "Notes on Fire Ground Command" by Chief Alan Brunacini in *Fire Command*, January 1981, is appropriate:

> When the FGC decides to change from an offensive to a defensive mode, he announces the change (as emergency traffic), and all personnel must withdraw from the structure and maintain a safe distance. In such cases, the FGC must maintain an effective organization with adequate, well-placed sectors. This system is designed to control both the position and the function of operating companies (when the FGC says "out," he better have an effective organization already in place). In these retreat situations, sector and company officers assemble and account for all personnel on the outside of the building (roll call time). Interior lines are withdrawn or abandoned, if necessary, and repositioned in a defensive mode. Lines should not be operated in door and window openings, but should be backed away so that personnel are in safe positions.

References

[1]Huntington, W. C., "Masonry Wall Construction," *Building Construction*, 3rd ed., Wiley, New York, 1963, pp. 207-297 (A useful description of masonry wall construction).

[2]Condit, C. W., *American Building*, (The Chicago History of American Civilization, Boorstein, D. J., ed.), University of Chicago, Chicago, 1968, pp. 64 and 131. (This is a most useful resource for

the serious student of the history of construction techniques, as distinct from architecture.)

[3]See Ref. 2, p. 118.

[4]Adelman, R.K., "Sprinklered Building Burns Because of Alterations," *Fire Engineering*, August 1975, Vol. 128, No. 8, p. 72. (An excellent description of the fire problem of the much-altered building.)

[5]Feld, J., *Construction Failure*, Wiley, New York, 1968, p. 56. (Together with *Building Failures* by T. McKaig, referenced below, a treasury of information about failures of all types of structures, both under construction and in use.)

[6]McKaig, T., *Building Failures*, McGraw Hill, New York, 1962, p. 203.

[7]See Ref.6 p. 139.

[8]Canavan, J., "On the Job," *Firehouse*, April 1981.

[9]"Large Loss of Life Fires — Industrial," *NFPA Quarterly*, Vol. 55, No. 3, January 1962, p. 334.

[10]See Ref. 5, p. 14.

[11]Zuk, W., *Concepts of Structures*, Reinhold, New York, p. 28, Fig. 32.

[12]Spalding, F., *Masonry Structures*, 2nd ed., Wiley, New York, 1926, pp. 88 and 379.

[13]Kidder, F., and Parker, H., editors, *Kidder-Parker Architects and Builders Handbook*, 18th ed., Wiley, New York, 1954, p. 321, (Fig. 6 and the paragraph, "Segmental Arches with Tie Rods"). See also p. 326 for ten ways in which a masonry arch may fail.

[14]*Radford's Cyclopedia of Construction*, Vol. 3, Radford Architectural Company, Chicago, 1923, p. 217.

[15]See Ref. 6, p. 202.

[16]NFPA 251, *Standard Methods of Fire Tests of Building Construction and Materials*, National Fire Protection Association, Boston, MA, 1960, 22 pp.

[17]*Fire Tests of Building Columns*, published jointly by Underwriters Laboratories Inc., Factory Mutual Insurance Companies and Bureau of Standards, U.S. Department of Commerce, 1920.

[18]Grubisch, T., "And the Walls Came Tumbling Down," *Washington Post*, Washington, DC, Dec. 24, 1970, p. D5, col. 2.

[19]Private communication.

[20]Hool, G. S. and Johnson, N.C., editors, *Handbook of Building Construction*, 1st ed., Vol. 1, McGraw-Hill, New York, 1920, pp. 588 and 179. (This book provides a great deal of information on the design and construction of ordinary construction buildings, though the term is not used.)

[21]See Ref. 13, p. 890, "Weakness of Wrought Iron Stirrups When Exposed to Fire," also see Ref. 20, p. 394.

[22]See Ref. 20, Dewell, H. B., "Timber Detailing," p. 308.

[23]See Ref. 20, Dewell, H. B., "Floor and Roof Framing, Timber," p. 378.

[24]See Ref. 20, pp. 393-4.

[25]Peterson, C. E., "Fire Delays the Opening of Two New Schools," *Fire Journal*, Vol. 62, No. 2, March 1968, pp. 19-22.

[26]Bryan, J. L., "Psychology of Panic," *Proceedings of Fire Department Instructors Conference*, Western Actuarial Bureau, Chicago, 1956.

[27]Juillerat, E., "The Menace of Abandoned Buildings," *Fire Journal*, Vol. 59, No. 1, January 1965, pp. 5-10.

[28]Fried, E., "Watch Out for Fires in Old Churches," *Fire Chief*, April 1970, p.41.

[29]Bradish, J. H., "Multiple Alarm Fire Destroys Fraternal Lodge," *Fire Chief*, Vol. 25, No. 2, February 1981.

[30]Wiley, E., "The Charles Klein Law Library Fire," *Fire Journal*, November 1972, Vol. 66, No. 6, p. 16.

[31]"Reports of Important Fires," *NFPA Quarterly*, Vol. 56, No. 3, January 1963, p. 278.

[32]Stow, D., "Windowless Buildings, A Challenge to a Small Department," *Fire Chief*, Vol. 19, No. 10, October 1975, p. 44.

[33]Juillerat, E., "Fire Hazards of Windowless Buildings," *NFPA Quarterly*, Vol. 57, No. 1, July 1964, p. 22.

[34]"Large Loss Fires of 1961," *NFPA Quarterly*, Vol. 55, No. 4, April 1962, p. 393.

[35]"Reports of Important Fires," *NFPA Quarterly*, Vol. 56, No. 4, April 1963, p. 404.

[36]"Bimonthly Fire Records," *Fire Journal*, Vol. 59, Nos. 1 and 5, January and September 1965, pp. 43 (Jan.) and 47 (Sept.). ("Bimonthly Fire Records" in *Fire Journal* are brief but often extraordinarily useful accounts of fires. Particularly noteworthy is the attention given in the records to the factors which caused or contributed to the loss, rather than just to the cause of the fire. They provide a valuable source of vicarious experience.)

[37]For good discussions of what constitutes and what does not constitute mill construction, see *The*

Factory Mutuals 1835-1935, Manufacturers Mutual Fire Insurance Company, Providence, 1935, pp. 224-226, or *Able Men of Boston*, Boston Manufacturers Mutual Insurance Company, Boston, and *Handbook of Building Construction*, McGraw-Hill, New York, 1st ed., 1920, Section 3-58.

[38]Love, F. J., "Brooklyn Warehouse Fire," *WNYF*, New York City Fire Department, Vol. 21, No. 2, 1964, p. 4.

[39]Peterson, C. E., "Fast Fire in Mill Buildings," *Fire Journal*, Vol. 63, No. 3, May 1969, p. 39.

[40]NFPA 13, *Standard for the Installation of Automatic Sprinklers*, National Fire Protection Association, Boston, MA, 192 pp.

[41]Taiyo Department Store Fire, *Fire Journal*, May 1974, Vol. 68, No. 3, p. 42.

[42]Hayne, W. "Jersey City Waterfront Fire," *NFPA Quarterly*, Vol. 35, No. 1, July 1941, p. 83.

[43]Brannigan, F., "Building Weaknesses — Do You Know Them," *Fire Command!*. July 1970, pp. 22-27, (reprinted as F 37-3).

[44]See Ref. 5, p. 3.

CHAPTER 5

Garden Apartments

In the two previous chapters, the deficiencies of buildings of wood or ordinary construction were examined. We could now start examining buildings by types, by looking at churches, schools, row stores (called "taxpayers" in New York), warehouses, and so on. In this chapter, the author has chosen to describe the problems encountered by the fire department in combustible low-rise multiple dwellings, emphasizing what are known popularly as "garden apartments."

The term "garden apartment" includes row houses, town houses, quadriplexes, and all forms of low-rise, combustible, multiple dwellings, whether the occupants are owners or renters.* In some areas, they are referred to as "condominiums" (a method of ownership, not construction).

Though the author is without any firm data to prove his point, he believes that there is no type of building for which an understanding of the building's deficiencies is more important. A fire department which has an adequate knowledge of the defects of the garden apartments in its area can truly serve the citizens. The ratio of value saved to value exposed can be many times higher than in the case of a one-family house. Also worth consideration is the fact that the victims of a multiple-dwelling fire are, in the main, the victims of the criminality, carelessness, incompetence, or greed of others. To compound the problem and thus make a "good stop" even more of a benefit to the citizens, a large proportion, often more than half, of the victims of multiple-dwelling rental units have no insurance on their personal property. Overwhelming losses are suffered often by those least able to bear such a disaster.

Firesafety in Multiple Dwellings

"A man's home is his castle" is a long-established principle of our common law. This principle is carried forward into statute law, such as building codes, in that fire protection aspects of building codes bear very lightly on single-family

---

*For ten years, the author investigated garden apartment fires for the National Bureau of Standards to determine building features which were responsible for the extension of the fire beyond the area of origin. Most of the material in this chapter is drawn from personal observations.[1]

dwellings. Regulations calling for masonry exterior walls and spacing between buildings limit one's "right" to burn down one's neighbor, though recent relaxation of rules against wooden shingles have restored this "right" in some areas. Generally, inspection laws do not give the fire inspector access to private dwellings. The principal impact of the statutes is in the regulation of the initial installation of electricity and heating appliances, but there is generally no warrant for reinspection to determine that installations have not been substantially altered after initial approval. Even when requirements are upgraded, usually no attempt is made to apply them retrospectively to existing private dwellings.

It is impossible to house each family in the United States in its own individual "castle." Multiple-dwelling units are necessary.

In the multiple-dwelling unit, there is an interaction between a fire in any unit, and the safety of the occupants in all the other units. In addition, multiple-dwelling units contain common areas accessible to all occupants, and to varying degrees, to any outsider. A fire in the common area threatens all the occupants.

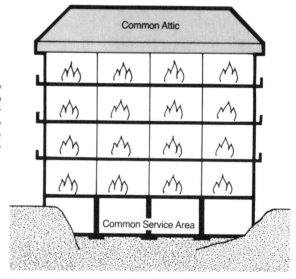

This sketch illustrates how fire in any unit of a multiple dwelling threatens all other units. Fire fighters must be aware of the many ways fire can extend throughout the building.

Historically, multiple dwellings in many cities were created by the subdivision of large one-family homes into smaller units as the former owners moved away and neighborhoods deteriorated. In cities like New York, the supply of deteriorated housing was insufficient to meet the needs of hordes of immigrants, so tenements were built to house several family units. These were structurally indistinguishable from the multistory one-family house, the typical New York "brownstone," of the well-to-do.

Terrible losses of life occurred in these tenements (now "Old Law" tenements in New York parlance). The first official recognition of any special firesafety problem was the requirement that outside fire escapes (vertical ladders, with a platform at each floor level) be provided.

About 1903, as a result of sociological pressures, sweeping tenement house legislation was passed in New York giving rise to the "New Law" tenement. Though not so stated, the principle that underlay the requirements was that no person should lose his life due to a fire which started outside his apartment. The entire thrust was towards easy escape of the occupants, and for many years there was no loss of life in such a building due to extension of fire.

While the fire protection emphasis was almost exclusively directed towards safe escape, requirements for life safety also tended to limit extension of fire. Fire resistive stairways, self-closing fire resistive apartment doors, and similar requirements for dumbwaiters, together with the universal complete plastering of all interior surfaces, served to contain almost all fires to the apartment of origin, for many years. The provisions for easy escape of occupants lightened the burden of rescue operations so that fire fighters were available to swiftly check for, and suppress, any extension of fire. Exterior fire escapes provided direct secondary access to all apartments. *

Unfortunately, building codes are parochial and as communities shift from individual to multiple dwellings, the same error of regarding a multiple dwelling as simply a larger version of a single-family house is repeated. Typically, in 1966, the District of Columbia asked the National Bureau of Standards to investigate possible methods of improving, *in situ*, the fire resistance of existing wooden apartment doors. Tests indicated that the improved fire performance could be obtained by removing the door and frame and substituting metal, a determination made two-thirds of a century earlier in New York.[2]

The May 1973 *Fire Journal* contains an article by M.H. (Jim) Estepp, now Fire Chief of Prince George's County, MD, entitled "Apartment Townhouse Complex Fire Safety."[3] Prince George's County has thousands of garden apartments and has been in the forefront of requiring design improvements. Chief Estepp lays out a comprehensive plan of attacking this problem on all fronts. The present depression in the housing business will lead to a pent-up demand which will probably be met with sub-substandard housing using the excuse, "Let's get rid of the nit-picking regulators and get on with our noble function of housing the homeless." Fire and building officials would be well-advised to be fully familiar with this forthright treatise on the pitfalls of boom construction.[4]

Characteristics of Garden Apartments

The multiple dwellings discussed here are of a post-World War II type known generally as garden apartments. The exterior structural walls are of solid masonry or brick veneer over platform wood frame. In some cases, portions of the exterior surface, typically the spandrel spaces directly between vertical lines of windows, are of wood. Others are completely of wood. The usual nominal

---

*Many of these apartments are now being rehabilitated. Some of the fire problems being generated in the reconstruction are detailed in Chapter 4, Ordinary Construction, in the discussion of void spaces.

height is three stories, but by taking advantage of terrain, the building can be four stories on one elevation.

In the usual case, the apartments are all on one floor; however, town houses and row houses are multi-floor units. Some apartment houses have duplex or triplex units, i.e., rooms in one unit on two or three floors.

Balconies are customary, either extended or re-entrant, and variously of combustible or noncombustible construction. Common gable roof attics usually extend over the entire structure, broken in area only to the maximum permitted by the local code when the building was built, and by barriers frequently as inadequate as the code or inspections will permit.

Typically, garden apartments are from one to three stories high with exterior walls of solid masonry or brick veneer on platform wood frame, such as this structure. Understanding the deficiencies of such buildings is vital to fire departments.

Regardless of exterior construction, the interior construction is almost totally of wood, using identical construction techniques to those used in one-family homes. The addition and stacking of units multiplies many times fire extension potentials which are inherent in current construction techniques.

Vertical and horizontal voids, three or four stories high, are, thus, fully interconnected. In installing plumbing fixtures, it is often the practice to cut and, thus, weaken structural members. The result is that any fire which starts in or penetrates this void could extend rapidly to as many as 16 family units, and the attic above. The weight of fixtures, cutting of structure for accommodation of plumbing, and the air supply due to interconnected voids make early collapse not unexpected. The contrast between the potential ultimate effect of an initially identical fire in a one-family ranch house and a garden apartment is startling, yet no code provision really takes this into account. Firestopping, even if mandated, is usually deficient.

Bathrooms provide an excellent illustration. Years ago, the bathroom was fully plastered and the bathtub stood independently on feet. Glamorizing the bathroom brought about the development of the built-in bathtub. Because of its contours, such a tub interconnects the void spaces of one wall, the void spaces of the wall at right angles, and the void spaces under the floor. These void

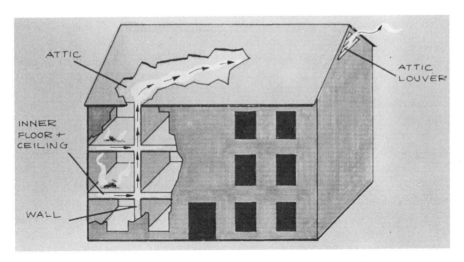

Undress the building to see how fire can extend vertically and horizontally through a garden apartment. When one room is "off," the whole building is "off."

spaces are interconnected with vertical voids which must be of substantial dimensions to accommodate waste and vent piping. In the layout of the apartments, the plumbing facilities are consolidated for economy. This can result in four bathrooms being located in four quadrants around a point.

If one were to ask the designer or the building regulator how a fire in one apartment would be confined to that apartment, and not extend to the apartments of innocent bystanders, the probable answer would be: "by the use of fire-rated gypsum board."

Gypsum board is an excellent material. Gypsum itself is the only common building material which is endothermic. It takes heat from the fire (as much as 580 Btu per pound) when tested in a bomb calorimeter. For gypsum board, the gypsum combined with binders is sandwiched between two sheets of paper. Gypsum board will yield some Btu when tested because of the combustibles added. It is available in sheets 4 feet wide, 7 to 10 feet long and from ⅜ to ⅝ inches in thickness (even thicker boards are used for fire barriers and as shaft liners in fire resistive buildings).

Gypsum board is a part of a number of fire resistive assemblies listed by Underwriters Laboratories Inc. and Factory Mutual Research Corp.

Fire Resistive Assemblies

Of particular interest here are the listings in UL's *Fire Resistance Directory* of fire resisting/combustible wall and floor-ceiling assemblies. At first glance, the terms "fire resistance" and "combustible" would appear to be contradictory. Fire resistance (or endurance), however, is a measure of how long a particular structure will resist collapse or the passage of fire (or both, as the case may be). A structure may be combustible and yet meet the standard. The well-known "tin"-clad wooden fire door is an example.

The assemblies which are tested and listed in the laboratory are generally simple structures. Walls are generally of gypsum on vertical studs without penetrations. Floor and ceiling assemblies are typically of gypsum board and wood flooring nailed to 2- by 10-inch sawn wooden beams tightly firestopped, without penetrations.

The structures which are built are far more complex than the simple structures tested. A chief cause of failure is the fact that there is often an unlimited supply of air to the space on the unexposed side of the gypsum board. In each listing of a combustible floor-ceiling assembly, a "finish rating" figure is given. This is often just a small fraction of the assembly rating time. This figure represents the time that it takes to reach significant temperatures behind the gypsum board. The tight firestopping of the test assemblies limits the damage as combustion requires free-moving air. In the attic or the interior void, when the fire temperature is reached in the void, there is adequate air to support combustion. The assembly is being attacked by the fire from both sides, not just from one side as in the test laboratory. The result, not surprisingly, is "unexpected early failure."

A sample listing is shown in the accompanying illustration.

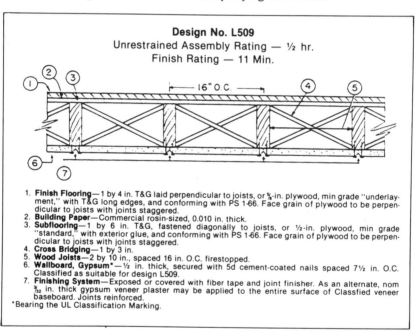

Design No. L509
Unrestrained Assembly Rating — ½ hr.
Finish Rating — 11 Min.

1. **Finish Flooring**—1 by 4 in. T&G laid perpendicular to joists, or ⅝-in. plywood, min grade "underlayment," with T&G long edges, and conforming with PS 1-66. Face grain of plywood to be perpendicular to joists with joints staggered.
2. **Building Paper**—Commercial rosin-sized, 0.010 in. thick.
3. **Subflooring**—1 by 6 in. T&G, fastened diagonally to joists, or ½-in. plywood, min grade "standard," with exterior glue, and conforming with PS 1-66. Face grain of plywood to be perpendicular to joists with joints staggered.
4. **Cross Bridging**—1 by 3 in.
5. **Wood Joists**—2 by 10 in., spaced 16 in. O.C. firestopped.
6. **Wallboard, Gypsum***—½ in. thick, secured with 5d cement-coated nails spaced 7½ in. O.C. Classified as suitable for design L509.
7. **Finishing System**—Exposed or covered with fiber tape and joint finisher. As an alternate, nom ³⁄₃₂ in. thick gypsum veneer plaster may be applied to the entire surface of Classfied veneer baseboard. Joints reinforced.
*Bearing the UL Classification Marking.

A typical listing of a fire resistant floor and ceiling assembly from UL's Fire Resistance Directory. The finish rating of 11 minutes is the time in which an average temperature rise of 250°F is reached within the joist or a temperature of 325°F is reached at the bottom of the joist. Underwriters Laboratories Inc.

Note that the spacing of the nails, the setting of the nails, and the taping and cementing of joints are all part of the listing. Bear in mind that the tested assembly was built by the gypsum board manufacturer and that there are no unnecessary details.

The problem starts, therefore, with the fact that the code permits buildings to be built which are far more complex than the simple assemblies tested in the laboratory.

The gypsum board tested in the laboratory is arranged to present an impervious sheath or membrane to the fire. In the structures built, there is an infinite variety of penetrations of the sheath, any one of which can, and has, admitted fire to the combustible voids. These will be examined in detail shortly.

Some defects are the result of incompetent or venal inspection. Non-rated gypsum board, which is cheaper, is on the job site for use in locations where rated board is not required. It is found to be used where rated board is required. Many believe "it's all the same." It is not. It costs about one cent a sheet to run gypsum board through UL and get it listed. This sum would not justify two inventory and product lines.

Many believe taping and nail-setting to be cosmetic. In fact, it is a part of the fire resistance system. Note whether gypsum board in attic firestops, for instance, is taped and nail-set.

The author has seen instances where the inner gypsum board layer of two-layer systems was just odd scraps which were not even complete. *

Contrast the single layer of plywood with carpeting or tile surface that is often found with the actual floor constructions listed and tested.

The floors which are tested have solid sawn joists. When wooden I beams are used, the fire in the void, instead of scorching sawn beams, is consuming structural strength. Almost invariably the I beams are penetrated by holes for wiring. An electrical wire hole through a piece of plywood will pass fire much faster than a similar hole through a nominal 2-inch sawn joist.

Trusses are most common today. They provide a huge supply of air to the fire which develops as heat penetrates the gypsum ceiling and ignites the truss members. The forest of small dimension lumber is then available as fuel.

Fire Loads

There is a basic fallacy in the test method. The ASTM E 119 furnace test was developed to test fire resistive structures which were also noncombustible. (See Chapter 6, Principles of Fire Resistance, for a description of this test.) It was reasonable to test a floor-ceiling assembly from below only. The author has been

---

*A few years ago, the author was in a Florida hotel giving a lecture on this topic over the phone to a recorder at the Florida State Fire School for use in a program he could not attend. The maid was listening attentively. When the author finished she said, "It's true. My husband is a dry wall foreman and he comes home and cries at what he has to do. He was told to cut $50,000 out of a $200,000 job." The next day she was very upset. Obviously she had talked to her husband. "Everything he does is strictly according to code."

told that when it was proposed to test combustible assemblies, some studies were made of fire temperatures which might be expected at the combustible floor. They were nominal. This is not the case today. The typical fire load in a garden apartment will develop temperatures at the floor which will ignite the floor early in the fire and permit penetration of the combustible assembly from the top. The author has seen this happen.

In 1980, the National Bureau of Standards conducted tests to determine the characteristics of fires in basement recreation rooms.[5] The fire loads used could also represent a living room or bedroom in an apartment.

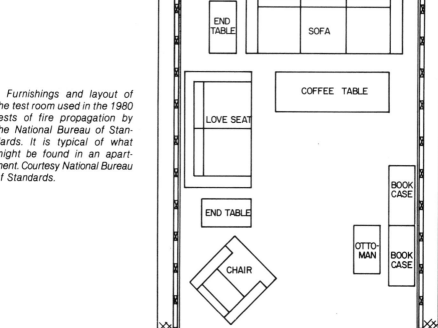

Furnishings and layout of the test room used in the 1980 tests of fire propagation by the National Bureau of Standards. It is typical of what might be found in an apartment. Courtesy National Bureau of Standards.

It is apparent to the author from the NBS tests that the typical apartment room fire can develop an assault on the structure far beyond that provided in the standard ASTM E 119 furnace test. In addition, test data showed that heat radiated to the floor from a contents fire probably would be more than sufficient to ignite the floor, even leaving out the puddles of flaming liquid which are characteristic of many plastic fires. Thus, the concept of "fire resistive combustible," which underlies all garden apartment construction, is, in the author's opinion, very questionable, because the ASTM E 119 test does not envision a fire assault to the *top* of the floor structure. Even the "strictest code" is a broken shield.

In less than 2-½ minutes after ignition, fire in the test room was pushing heavy flames out the doorway. There is no such thing as a "safe" burn time. Courtesy National Bureau of Standards.

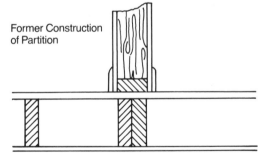

Former Construction
of Partition

Present Construction
of Partition

Past practice had been to build partitions over doubled joists. New lightweight materials permit partitions to go anyplace on the floor. Fire burning down into the void can extend from one unit to the next.

Even under the best of circumstances, therefore, the structures are not very resistant to the potential fire assault from a fire in the contents. In addition, there are many specific paths for entry of fire into the interior voids of the building. Not all are found in all buildings and there are probably others not given here, but a representative sample is provided.

Pinholes in the Protective Membrane

Openings around utilities such as gas pipes and electrical service provide fire paths. In one case, a piece of 1-inch by 2-inch lumber was used to complete the basement ceiling. Holes were drilled to accommodate the electrical cables. The fire burned through the wood. Gypsum was ended 6 inches short of the wall to accommodate gas pipes.

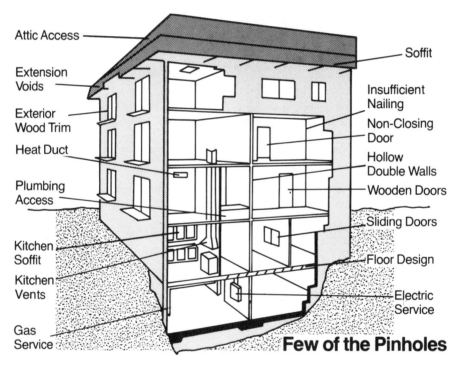

Attic Access

Extension Voids

Exterior Wood Trim

Heat Duct

Plumbing Access

Kitchen Soffit

Kitchen Vents

Gas Service

Soffit

Insufficient Nailing

Non-Closing Door

Hollow Double Walls

Wooden Doors

Sliding Doors

Floor Design

Electric Service

Few of the Pinholes

Though gypsum board has good fire resistance qualities, these often are of little help. This sketch shows how openings ("pinholes") provide fire paths throughout a structure.

The bathtub is built in. If of metal, it is usually permitted to be installed on bare studs. Fire in the void extended into an apartment by conducted heat igniting clothing on the bathtub. In the case of plastic bathtubs, some jurisdictions require the gypsum sheath to be completed. Often the joints and nails are bare because the tub is installed before the gypsum board taper is on the job.

The plumbing access panel is usually plywood or hardboard providing a ready entry to the void.

The heating/cooling system consisted of hot or cold water pipes with convectors and fans. In a living room, the access to the unit was a plywood panel. Fire burned through the panel and extended along the piping out through the plywood panel to the upper apartment which, ironically, suffered much more damage.

The attic service access trap door is usually in the stair hall. At times it is located in a top-floor apartment. It is usually plywood, or even fiberboard. A fire in a bedroom closet may actually be in the attic before the fire department arrives. Know your buildings.

Plywood paneling may be used on one or more walls. If the plywood is installed over taped gypsum board, then it is the equivalent of a combustible piece of furniture. If the plywood is directly installed on the studs, then it is another point of penetration of the void. This can be determined by taking the cover off

Plywood paneling should be installed over gypsum board (arrow) to protect against fire penetration into, or from, voids. Fire fighters should be aware of the actual construction in multiple dwellings.

an electrical outlet and examining the construction. Because air is available to both faces, this plywood will burn with a high RHR (rate of heat release) and produce an intense fire.

When a doorway is not fitted for a door, the gypsum is finished with metal corner bead and cemented. When a door frame is installed, the gypsum sheath is not completed and the wood structure is left bare.

Many apartments have interior kitchens or bathrooms. There must be a mechanical vent. The duct runs up through the void without any consideration of its being a fire path. It is often heavily coated with grease and passes through wood without clearance. The intake is right over the stove. The whole design often looks as if a conscious effort was made to convert a stove fire into a building fire. Be alert for the extension possibilities.

A fire heavily involved an interior kitchen of a garden apartment. The old-time volunteer chief of the department was very opposed to opening walls. "We're

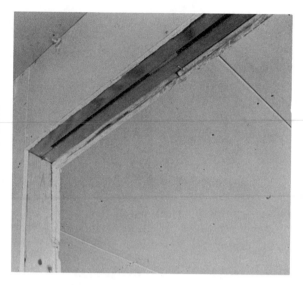

This doorway framing is not protected by gypsum board providing another point of penetration and extension of fire.

not housewreckers," was his excuse. The volunteer lieutenant of a mutual aid truck company knew his business. He disregarded the standing orders, went above the fire, removed the refrigerator, opened the wall, and found the fire roaring up the wall void. At another fire in the same complex, exactly the same extension occurred.*

The following observation from Chief Alan Brunacini of the Phoenix, AZ Fire Department is pertinent:

> Where fire involves concealed spaces (attics, ceiling areas, con-struction voids, etc.), it becomes very important for ladder com-panies to open up and engine companies to operate fire streams directly into these areas. Early identification of and response to fires in concealed spaces can save the structure. FGCs who hesitate to open up because they do not want to beat up the building can be found an hour later trying to hold the fire to the same part of town. Concealed space fires require a large commitment of truckies to per-form access operations. Losses in such fires are typically caused by inadequate or ineffective support functions, not by insufficient water. Fires in concealed spaces can be extremely difficult. The FGC must apply water directly on the fire. He must get forces ahead of the fire, cut it off, and then try to overpower it. He must forecast effectively and operate aggressively.[6]

---

*The author is pleased to note that the lieutenant, one of his first students, is now the career chief of a large fire department.

Vertical Voids

When the fire has entered the void space by one or another of the pinholes provided, it finds readily available vertical paths up through the building.

At this point, it should be noted that not all fires start in occupied spaces and enter the voids. A fire started in a void from a short in armored cable. The fire dropped out of the void into the basement, where it ignited a couch. The first unit on the scene made the basic size-up error, "What you see is what you have," and turned back the rest of the assignment. The fire burned up through the void unnoticed, until it broke out in the attic and a citizen brought it to attention. The "couch fire — we can handle it" had become a multiple alarm.

"Cheap construction" is often equated with fire problems. In at least one instance, the cheaper construction presents no problem. Wall-hung kitchen cabinets, when low priced, are simply hung on the finished wall. In more expensive construction, a soffit is constructed to close the space between the cabinet and the ceiling. The usual result is that there is only the thin wood or hardboard

Soffits (arrow) over kitchen cabinets connect wall voids to ceiling voids. If back-to-back, they permit fire extension from one unit to the next. Fire fighters should be alert to such possibilities.

top of the cabinet between a kitchen fire and the floor-wall void. If the kitchens are back to back, the fire is into two apartments very early, and has access to the vertical plumbing void. In some areas, the builder is required to firestop the void. This is resisted, however, because it requires bringing the dry wall contractor on to the job twice. Most of what the author has seen would be ineffective. Odd scraps are used, without taping and nail-setting.

One set of apartments had attractive sliding louvered shutters closing the pass-through window from kitchen to dining alcove. Fire starting at a toaster entered the structure through the sliding door opening. A fire inspector living in the development heard the alarm, looked out, and took pictures. The pictures clearly show that the fire was in three apartments before the units were out of the firehouse.

The most significant vertical void is the plumbing void. Consider an apartment building with four apartments off each landing. The front and rear walls are bearing walls. Floor joists or trusses run from these walls to an interior bearing wall. This is usually of wood studs, often arranged as a cavity wall. The cavity and double wall structure cut the transmission of sound between units, and provide room for the plumbing pipes. A fire in this void is most serious. The burning studs are the structural heart of the building. In this area are the heaviest (and most expensive) components of the building, the kitchen and plumbing fixtures. At the top of the wall the top plate is cut to accommodate the plumbing vent which passes through the attic to the outside. If the fire gets a hold in the attic, the building is doomed. In a flat roof, the fire will spread laterally. Because the roof boards are shimmed up to put a drainage pitch on the roof, the joists do not act as firestops.

A fire in the central bearing wall area demands a vigorous coordinated attack. In addition to the attack on the main fire, units must get above the fire, open the walls, and extinguish fire. The roof must be opened over the wall and a line must be gotten into the attic. All these actions take time. There isn't any time for the chief to learn about this problem at the scene. Preplans and drills should prepare all units, including mutual aid, for their assigned roles.

Fire can travel upward through even minor vertical voids such as the furring strips which mount gypsum board to a masonry wall.

Exterior Extension

There are a number of potentials for exterior extension.

Heavy fire out of the window and back into the building through the window above is a serious problem. Formerly, it was believed that about 5 feet of masonry between the top of a window and the sill of the window above would be adequate to stop extension. This is no longer true due to the increase in fire load and rate of heat release brought about by modern furnishings. Not only are there foamed plastic mattresses and upholstery, but also furniture pieces made of rigid plastic, including some which imitate wood in appearance.

If the structure is of wood, expect rapid extension over the exterior surface. In the case of brick veneer buildings, be aware of wood panels used between windows in a vertical line.

The exposure should be covered without driving the fire back into the apartment into the faces of the interior attack forces.*

---

*New York's famed Chief Hugh Halligan (who gave his name to the Halligan Tool) said, "There are only three rules. The first is, never work two lines against one another; you'll kill everybody." When pressed for the second, he replied, "I'll never tell you because you guys don't pay any attention to the first rule."

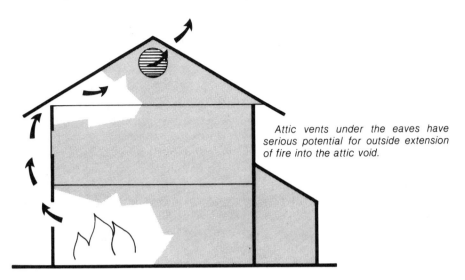

Attic vents under the eaves have serious potential for outside extension of fire into the attic void.

The most serious outside extension potential is due to the requirements for attic vents. The attic is insulated above the top floor ceiling. There are requirements for venting the attic to avoid trapping moisture in the insulation.

These vents usually involve lower openings along the eave line and upper openings at the top of the gable, or along the ridge line. A constant current of air is moving through these openings. The lower openings may be in the fascia above the top floor windows or on the lower side of the roof overhang (soffit). Fire rolling out the top floor window is literally drawn into the attic lumberyard following an already established air flow. There are cases where fire coming out of a window below the top floor entered the attic bypassing the top floor.

These vents could be redesigned to retard this extension greatly. One simple change is required in Prince George's County, MD. The vent opening must be located out of line with the windows. This change was proposed at the recent BOCA code meeting in Virginia. It was voted down. One can only speculate on the reason for voting down such a simple change but it again proves that, if proof is needed, codes are political not technical documents, and even "good" codes have serious technical deficiences. The real solution to this "design for disaster" is to locate the vents on the roof, using metal units designed for that purpose.

If heavy fire is coming out of the window, use a stream to cut off vertical extension.

Balconies may hinder or accelerate outside extension, depending on design. A concrete balcony fitted tight to a masonry wall will retard vertical extension. A concrete balcony against a combustible wall (such as the wooden panels mentioned earlier on brick veneer buildings) will form a "hood" and direct the fire upward. Wooden balconies will, of course, spread fire upward. In one case, the balconies were provided with gypsum board ceilings. The upward extension was fueled by simple dividers fashioned of 1- by 6-inch vertical boards and plywood panels on the front of the balconies. The balconies themselves were uninvolved.

The damage to the innocent party in the top floor apartment was total compared to minor damage in the first-floor apartment where the fire was maliciously started.

In fire training, it is learned that fires extended by "conduction, convection, and radiation." In fact, these are the mechanisms of heat transfer. Fires extend also by the movement of heated or flaming material. Flaming material falling down from a balcony on combustibles below has caused extension, and a serious fire developed below fire fighters who were not at all alert to the problem.

Fire also falls down into kitchen cabinet soffits from the floor above or the attic, through the hollow walls of balloon frame construction and into the stairway from the attic.

Trash bins should be located far enough from the building to be no exposure. Sometimes the bin is located close to the building and "concealed" in an attractive wooden structure. Such an enclosure became involved in a trash fire and resulted in the destruction of the building.[7]

A car fire in a combustible covered carport developed into a multiple-alarm fire which destroyed several units.

The Attic

Lateral fire spread is possible even on flat roofs because of the necessity of providing a pitch which creates a void. Most of the garden apartments have gabled (peaked) roofs. These are created by lightweight, factory-made gusset plate trusses. The attic space is not used for storage, but facilities such as electricity, gas, and heat ducts pass through it to serve units below. Vent stacks from plumbing, and kitchen and bathroom exhaust ducts pass through vertically. The sheathing of the attic is usually plywood. The shingles may be fire retardant or combustible wood shingles.

Keeping the fire out of the attic should be a chief objective of the fire forces. All must understand that once the fire is beyond a mattress or the equivalent, there is no "room off" or "apartment off"; the building is off. The fire has a good chance of taking everything until it is stopped by an adequate fire wall, or open space.

When the attic is involved, only a well-equipped, properly trained, adequately staffed fire department with good water supply and perfect circumstances can stop the fire short of a fire barrier. The first tactical requirement, therefore, is to defend the barrier on the downwind side of the fire. The next requirement is to suppress any fire that drops into the top floor. Ladder pipes take time to place and set up. The setup should be made not with regard to where the fire is then, but where it will be when the stream is operating.

Again, to quote Chief Brunacini:

> The FGC must develop critical decisions relating to cutoff points and must approach fire spread determination with pessimism. It takes a certain amount of time to get water, and while the attack is

being set up the fire continues to burn. The FGC must consider where the fire will be when the attack actually goes into operation; if he misjudges, the fire could burn past his attack/cutoff position. The FGC can't afford to play catch-up with a fire that is burning through a building/area. He must project the setup time, write off the property that will burn during that time, and stay ahead of the fire. The FGC must set up adequately, then overpower the fire.[8]

Stairways can be dangerous for the occupants and fire fighters. One might think of a masonry enclosed stairway with self-closing rated doors as a place of refuge but most codes simply permit the extension of the combustible attic over the stairway. The "protected combustible" construction is being attacked from the upper side (contrary to the code concept that fires come only from below). In at least one case the author reported on, the attic was falling on the escaping tenants early in the fire.

Fire Barriers

The only barriers which can be depended on to stop an attic fire are adequate space between buildings, the absence of wood shingles, or an unpierced masonry fire wall parapeted through the roof and either extended out from the building or meeting a masonry wall of adequate width. See the NFPA's *Fire Protection Handbook*, 15th edition, p. 5-97, for a description of true fire walls.

Any other type of barrier, "rated" or not, is subject to deficiencies. These deficiencies should be known by the fire department so that adequate plans can be made to cover them.

Very often, in multi-unit buildings, the laundry rooms may be located in only one of the units. In order to provide covered access for the tenants, fire doors are cut into the fire walls at the basement level. These fire doors are characteristically blocked open because tenants find them hard to open. Often they are removed.

Utilities should be run parallel to the building and brought separately into each unit. Often they are run through fire walls without proper closures. If such passages exist, they must be guarded. A fire "under control" suddenly blossomed out in the next unit. It had followed the gas line.

When the New York City Tenement House Act of 1903 was written, life safety was the paramount consideration. Masonry separation of units, therefore, was required only to the top floor ceiling line. Several serious fires spread through the cockloft space (these structures characteristically have nominal flat roofs). The law was then amended to require that the walls be brought to the underside of the roof. Fire spread over this "barrier" readily and the law was amended again to require parapeted walls.

The use of parapets is resisted by designers and builders. Not only does it cost money but also it provides a location where only a competent roofer can prevent leaks. Flashing a parapet properly takes infinitely more skill than throwing down shingles.

Masonry fire walls are, therefore, often carried only to the underside of the roof. There are no blocks made to fit the angle of the roof. The fit is often a gap of several inches. Even when fitted tight, the plywood delaminates and the fire passes over.

Quadriplexes, four units divided by masonry walls, were observed in Florida. The top floor trusses were hung on brackets which passed over the top of the wall with a socket on each side for the truss. The plywood roof was nailed over the top of the wall. The thickness of the truss brackets provided a fire passage between the top of the wall and the roofing. The fire would find fuel in the roof trusses and would have a good hold on the next unit before being observed, unless the fire department is alert to the problem, gets into the next unit, pulls the ceiling and cuts the extension off.

All sorts of deficiencies are found.

Masonry fire walls are carried to the top-floor ceiling with gypsum board fire barriers above. The gypsum board is nailed to an available truss which only rarely is directly above the wall so the barrier is seldom continuous.

The gypsum board is rarely taped and nail-set, and is often just odd scraps which do not even cover the wood.

The big fallacy of the code is in assuming that fitting the fire barrier to the roof will stop the fire. The roof is of plywood. Plywood delaminates, and even in the best cases, the fire passes over the top of the barrier. In an attempt to overcome this deficiency, some codes require fire retardant plywood 4 feet out from the barrier on each side. The author has no experience to cite, but he would act to protect this the same as untreated plywood. The treated plywood is supported on an untreated wood truss. When this truss fails, what happens to the plywood? It may well just fall away. Recently, fire walls of heavy gypsum board, two 1-inch layers in steel frames, have made their appearance. Apparently, they have the same problem as other barriers whether masonry or gypsum board on trusses — that of making a truly effective seal to the roof.

At times, structural members or utilities penetrate the barrier. Maintenance employees cut doorways through the barrier. In one case, the openings were cut by one of the occupants of a row of houses. He was a peeping tom.

Often the barrier does not block the overhang, allowing the fire to sweep around the end. "The code does not require it," was the main defense in one lawsuit. Such a defense is not adequate in the face of charges of professional incompetence.

Garden apartment buildings are usually symmetrical. This often places the barrier so as to split the stairway opening. In one design, the unneeded space at the top of the stairway was used as a room for one of the apartments. Fire rolling out the window of this room entered the attic through the vents on both sides of the fire barrier.

Fire departments should be aware of the deficiencies of these barriers and prepare to defend them. In one case, a fire progressed through several multistory units. It was stopped when the stairway ceiling was pulled and a stream directed upward onto the exposed side of the barrier. The stream and the barrier together stopped the fire.

Plan your attack. Get lines to the exposed top floor, pull ceilings, get water protection up into the attic along the barrier. At the same time, do not neglect the fire rolling over the top. The shingles may be fire-rated (resistant to flying brands), but they burn intensely and this spread must be stopped by an exterior line.

Some fire walls end at the inside of the exterior wall. Combustible siding or sheathing may be laid right over the wall. This makes it difficult to determine where the barrier is and where to make the stand. The solution is to know where the barrier is located. A picture attached to the preplan, with the barriers drawn in, might be very helpful.

In some cases, the fire walls are also party walls. Structural members common to both buildings are supported on the wall. The sockets are usually common to both buildings. Watch for this in town houses where the plans have been reversed, placing the doorways next to one another. The stairway headers may be in a common socket, as any alternative is costly.

In such cases, there is the possibility of subtle extension of the fire into concealed spaces of the next unit. The smoke which might ordinarily give early warning might be attributed to the original fire, until the extension is well advanced.

This subtle extension is guarded against by opening up the ceilings of the adjacent units at the points where connections exist.

In some areas, fire walls are not permitted to serve as party walls. The wall is unpierced. Such walls lack the stabilizing effect of an applied load. If a unit is completely burned out, be wary of the fire walls; they may have been leaning on the burned-out unit.

Fire Tactics

This chapter does not pretend to offer a full exposition of details of tactics. There are some guidelines which grow out of the nature of the structures which should be useful.

Know your area. The best practice is for the fire fighters themselves to draw maps showing streets, cul-de-sacs, hydrants, addresses (these are often extremely confused), obstructions to operations such as fences, ditches, and steep grades,* and drivable and non-drivable lawns. In Orlando, FL, for instance, many developments feature lakes between the buildings.

As is typical, the building code makes no distinction between accessible and inaccessible buildings, even though by permitting combustible construction, the code assumes that the major total loss control element will be manual fire fighting. Where manual fire fighting is seriously impaired by design or layout, fire departments should press for automatic protection.

Know your water supply. A serious fire will require heavy caliber streams im-

*In one winter fire, ladders were laid on the ground so fire fighters could get up and down the hill.

mediately. What volume is available at an adequate residual pressure? Can supply be increased on request? Is there standing water which can be drafted?

Make energetic efforts to control parking. Legally designated fire lanes may be required on private property. Insist on police cooperation in enforcement of regulations.

Consider preconnected deluge guns for possible blitz attack on a heavy moving fire.

Plan to get above and on all sides of the fire. Protect deficient fire barriers.

Plan an adequate first-alarm response, including automatic mutual aid if necessary. Assign responsibilities.

If roof overhangs or balconies are burning, use heavy caliber streams to beat back fire and knock off loose wood which is ready to fall. This may buy a few minutes to effect entry with less risk.

Major collapse potentials are brick veneer walls, trussed or wood I beam floors and roofs, the use of unprotected steel girders, and the interior combustible bearing wall which also provides the plumbing voids.

The cantilevered balcony is truly a seesaw. Fire inside may be destroying the other end of the seesaw.

In rescue operations, do not be in too much of a hurry to use ladders. There are many possibilities of injury. The interior stairway is by far the best way to bring people out of the building. If necessary, hold up opening the door of the fire unit. Guard it with a line; attack from the exterior via ladders, until all the occupants have been removed.

If gas is used as a fuel, make a study of the gas distribution system. In Anne Arundel County, MD, a faulty gas regulator increased pressure throughout the development starting many fires simultaneously. Gas meters are located in one place for the convenience of the meter reader. The gas lines run from the meters

Study the gas distribution system. Piping may penetrate fire barriers. It may be hung on crossbars supported by pipe strapping which may fail in a fire.

to the apartments. Twenty or more gas lines may be grouped together and supported on a cross bar. Often this cross bar is supported on pipe strapping nailed to the joist above. In a Silver Springs, MD fire, the strapping failed and the entire group of gas pipes fell to the ground in a sort of catenary curve. Fortunately, the crews were not past the point of collapse. The collapse caused gas leaks which added to the fire. In some units, gas mains pass through the attic, feeding downward into apartments. The method of gas distribution and the location of shutoffs should be noted on the plan.

Salvage work is one operation which tends to prove G. K. Chesterton's "Whatever is worth doing at all, is worth doing badly." Any partial salvage effort is worthwhile. First-class salvage work is a tremendous benefit to the citizens involved and reflects well on the fire department.* The author would be happy to receive a fire department annual report which featured a picture of a good salvage operation instead of one of the year's big mistakes.

If you find burned paper money, advise the owner to bring it undisturbed by the best means to the U.S. Treasury Department in Washington, DC. There, a group of trained persons will carefully separate the money. All identified currency will be paid for. They cannot do anything with mixed ashes.

Chief Edward Spahn of the Orange County, FL Fire Department has itemized some of the tasks beside fire fighting that must be accomplished at a serious garden apartment fire.[9] They include organizing the investigative task force, arranging for an interrogation room, caring for displaced residents and concerned relatives in cooperation with the Red Cross, and establishing a security area for property of occupants discovered in overhauling operations. The latter task could become complicated if the occupants are not at home.

Educating the Tenants

Public fire education should deal not only with fire prevention but also with fire alarming. Recognize that, particularly in low income projects, the manager has the last word. The manager can evict or not evict when the rent is not paid. For this reason, the tenants hesitate calling in the fire department, but run looking for the manager. The manager, in turn, usually goes to the scene to determine that there is a fire. By that time, the fire is often well advanced. Workmen in the development will report a fire to the manager, or attempt to use fire extinguishers before they phone an alarm.

Don't beat around the bush. Try to talk to the manager and staff together. Point out the short time it takes for a fire to get away. Get the manager to tell the staff that phoning the alarm is to be done first. Get them over the fear of "publicity." Point out that small fires do not make the papers or the evening news, but big ones surely do. Work with the tenants' association and as many of the tenants that you can reach. Get them out of the fear of calling authorities.

If you can get to the fire when it still is confined to the room of origin, you have a good chance of success.

---

*Be considerate of the property of the tenants. In rental properties, often less than half of the people have any fire insurance. Sort out partially damaged or undamaged property and set it aside. Think how you would feel if somebody decided for you that your desk was badly burned and so the contents should be dumped in a pile. Be particularly careful of childrens' toys. Make every effort to return dolls and other precious possessions. The child has suffered and will suffer a severe trauma for some time. Help to ease the pain.

There is probably no class of building in which good planning and training can pay off more than in garden apartments. The knowledge the fire department gains from the day the development is planned should be institutionalized. By this the author means that it should become the organized property of the fire department and not be solely in the head of one person or another. The system should make the information available to the incident commander at the fire scene.

Automatic Sprinklers

This chapter is not intended as a treatise on the design of garden apartments but one observation is too important to pass over:

In some areas, it is common practice to install automatic sprinklers in basement storerooms, laundry rooms, and other spaces where there are sizable fire loads and where many fires break out or are started. Without any statistics to back up his opinion, the author is convinced that this is probably the most efficient use of automatic sprinklers, if one were to compare the number of sprinklers operating with the number installed.

Some amazing results have been experienced from even nonstandard systems fed by a small water line. It is better to have a system with proper control valves and an alarm which is at least connected to the building fire alarm. The fire department will be called if only to "get the bells shut off." These utility rooms are the ideal location for the arsonist, usually a juvenile. It may be difficult to prevent the fire from being ignited, but the basement sprinkler system will provide peace of mind both to the tenants and to the fire department at little expense.

Summary

To summarize:

- Garden apartments are very poor fire risks.

- They are really enlarged single-family dwellings in which a non-systematic attempt is made to deal with the hazard which one occupancy presents to all the others.

- Codes are based on a concept of fire behavior which is unrealistic and on misinterpretation of tested components.

- Fire loads have increased in recent years.

- Alarms are often delayed so that fires are well advanced upon the arrival of the fire department.

- The fire department must attack the problem on several fronts.

- The code path is difficult. Builders and material suppliers, rather than a possible fire loss, often have more influence on elected officials.

- Prefire plans and tactics deserve a great deal of attention.

- The fire department should work very hard to assure that a fire will be alarmed as soon as discovered.

- Urge tenants to procure adequate insurance. Many think it comes automatically.

A meeting of builders or real estate promoters is not complete without a speaker who raises the shibboleth, "Buildings would cost far less if we didn't have those terrible *specification* codes. New performance codes could let us use imagination and technology." When this argument is raised, fire departments might offer this performance code: "A combustible multiple dwelling shall be so designed and built that the design basis fire will be confined to the area of origin for 30 minutes."

In 1974, a nationally recognized building code expert made specific recommendations based on this author's study of garden apartment fires for the National Bureau of Standards. The author recommends the publication, "A Study of Fire Spread in Multi-Family Residences,"[10] as being useful to fire service officials in dealing with building departments and with builders to improve fire-safety in garden apartments.

References

[1]Brannigan, F.L., "A Field Study of Non-Fire Resistive Multiple Dwelling Fires," Center for Fire Research, National Bureau of Standards, Washington, DC, 1974.

[2]Shoub, H. and Gross, D., "Doors as Barriers to Fire and Smoke," National Bureau of Standards, Washington, DC, Building Science Series 3, March 1966.

[3]Estepp, M. H. "Apartment Townhouse Complex Fire Safety," *Fire Journal*, Vol. 67, No. 3, May 1973, p. 49.

[4]See Ref. 3.

[5]Fang, J. B. and Breese, J.N., "Fire Development in Residential Basement Rooms," NBSIR 80-2120, Center for Fire Research, National Bureau of Standards, Washington, DC. Every fire department should have a copy of this publication.

[6]Brunacini, A. V., "Notes on Fire Ground Command: Fire Control," *Fire Command*, January 1981, p. 9.

[7]Nailen, R. L. "Fire Feeds on Design Weakness," *Fire Engineering*, Vol. 129, No. 11, November 1976.

[8]See Ref. 6.

[9]Spahn, E., "Newly Created Department Provides Full Services at Fire," *Fire Engineering*, Vol. 135, No. 4, April 1982.

[10]Vogel, Bertram, "A Study of Fire Spread in Multi-Family Residences — The Causes and The Remedies," NBSIR 76-1194, Center for Fire Research, National Bureau of Standards, Washington, DC. In this publication a nationally recognized building code expert makes specific code recommendations based on the author's study of garden apartment fires. It should be very useful in dealing with the building department and with builders to get improvements because it is an "authoritative source."

CHAPTER 6

Principles of Fire Resistance

Toward the end of the 19th century, it became apparent to thoughtful observers that building structures of "fireproof" (noncombustible) materials did not guarantee that a fire would not result in a tremendous loss. As steel frame buildings developed, the necessity for protecting the steel from the heat of fires became apparent. J. K. Freitag[1] gives fascinating information on the early designs of fire protection for steel. It appears that engineers in charge at construction sites had quite a bit of freedom to design "on the job." Fires in buildings built to such innovative but untested techniques brought the realization that adequate standards and test procedures for fire resistive construction were necessary.

Early Fire Tests

Floors received the earliest attention as columns were apparently considered to present no problems. Chapter 4, Ordinary Construction, pointed out that there were two schools of thought on the behavior of wood floors in a fire. One school maintained that if floors were designed to resist collapse, they presented the risk of a general collapse of the building. The other school believed that if a floor was designed to collapse early in a fire, the walls would be preserved. The dilemma would be resolved by designing "fireproof" floors.

In 1890, the first fire test of a "fireproof" floor assembly in the United States was conducted in Denver for the Denver Equitable Building Co.[2] Hollow tile floors were tested. It was determined that porous hollow tiles set in so-called end construction (tile cells at right angles to beams) were superior to dense tiles set in "side construction" (tile cells parallel to beams). The floors were subjected to load, shock, fire and water, and continuous fire tests (24 hours at 1,300°F).

Six years later, the New York City Building Department conducted a series of tests on "fireproof" floors using brick kilns as test furnaces. The inside dimensions of the kilns were 11 feet by 14 feet. There was a 10-foot height from the grate to the top. The floor to be tested formed the top of the kiln. The central panel of the floor was loaded to 150 pounds per square foot (psf). A wood fire was built on the grate and maintained at 2,000°F, as nearly as possible, during

199

the last 4 hours. After the fire, a hose stream was applied at a pressure of 60 pounds per square inch (psi), and the floor was reloaded to 600 pounds per square foot (psf) for 48 hours, with the final load resting on the arch, not on a beam. Other tests were conducted at Columbia University, and the use of gas or oil fires for better control, and requirements that fire not penetrate the floor were developed.

Between 1896-1916, not much work was done on columns and such tests that were conducted showed cast iron to be superior to unprotected steel.

Standards for Fire Resistance

New York's Parker Building fire in 1908 and Equitable Building fire in 1912, and the Baltimore Conflagration in 1904, involving many "fireproof" buildings, convinced all concerned that there was a dire necessity for developing standards and test procedures for "fireproof" construction and that column stability was a factor of increasing importance. The effort to develop standards for fire resistance ratings brought together the National Bureau of Standards, Underwriters' Laboratories, Inc., the National Fire Protection Association, and both capital stock and mutual insurance interests. Tests were conducted during 1917 and 1918 and a standard was developed which has existed substantially the same since.

The path by which fire resistance standards are actually made effective in buildings should be traced:

- A municipality decides that, as part of its duty of protecting the general welfare of the citizens, it should have a building code.

- The municipality examines the several "nationally recognized codes" and considers other alternatives.

- By the appropriate process, the municipality decides to:
  Adopt one of the model codes verbatim.
  Adopt one of the model codes, with its own modifications.
  Write its own code.

- The code becomes law when it is adopted by the appropriate procedure.

- A builder wishes to build a building with steel columns. Among the code requirements is one that requires that columns have 3-hour fire resistance.

- Manufacturers of fire resistance systems have anticipated this need and have had their system tested at a testing laboratory, in this case, Underwriters Laboratories (UL).

- UL has standard columns at its laboratory. The manufacturer's staff working in the laboratory puts the fire resistive material on the column.

- When the column is ready (wet materials such as cement must cure), the test is conducted in accordance with the test standard ASTM E 119.

- The column has been wired with thermocouples. When a limiting temperature is reached in the column, the test is ended.

- The system receives an appropriate time rating, for instance, 3 hours.

- A listing of the system with all pertinent details is placed in UL's *Fire Resistance Directory* along with other systems which have passed the test.

- The builder indicates to the building department which system (listed by the laboratory) is to be used in the building.

- The building department is not obliged to accept any particular system. If the department objects to a system, it need not accept it for use in its community (subject to the customary appeal system).

- Tests for floors and walls follow a similar path except that the entire assembly is built in the laboratory by the manufacturer.

It is important to understand what fire resistance intends to provide and what it does not intend to provide:

- Fire resistance of columns is concerned with resisting collapse.

- Fire resistance of floors is concerned with passage of fire and collapse.

- Fire resistance of walls is concerned with passage of fire and collapse.

- Fire resistance of fire doors is concerned with passage of fire.

- Fire resistance is not specifically concerned with life safety, and many lives have been lost in fire resistive buildings.

- Fire resistance is not specifically concerned with smoke control. For example, metal-clad wood fire doors generate toxic carbon monoxide as their wood cores burn. Some earlier fire resistance systems, incidentally, did provide smoke containment, and some fire procedures were based on this fact.

- Modern systems, very often, do not provide smoke containment. *

- Fire resistance is not concerned with the dollar loss due to fire. There have been huge losses in fire resistive buildings.

Materials are classified as to their fire resistance capabilities according to procedures set forth in a nationally-recognized NFPA standard, NFPA 251, *Standard Methods of Fire Tests of Building Construction and Materials*. The text of the standard is the same as that published by Underwriters Laboratories Inc., with the designation UL 263. The American Society for Testing and Materials issues a similar standard, designated E 119, which is recommended by the ASTM Committee on Fire Tests. The same standard is designated A2.1 by the American National Standards Institute.

Fire Resistance Testing

The essential provisions of the standard are requirements for a standard reproducible test fire, a consistent method for conducting tests and classifying results, and specific instruction on the selection and preparation of test specimens. The object is uniformity in testing.

The reproducible standard fire used in the testing procedure follows a time-temperature curve based on the times and temperatures shown in the following chart.

Samples of typical structural elements are exposed to the standard exposing fire. The minimum sizes for test specimens are: columns, 9 feet; beams and girders, 12 feet; partitions and walls, 100 square feet; and floors and roofs, 180 square feet. The load-bearing structural elements are loaded with the prescribed superimposed weight for the assembly being tested. The unit passes the test as long as it successfully resists its superimposed load and, in the case of partitions, walls, floors, and roofs, resists the passage of fire; full details of test requirements are found in NFPA 251.

The complete standard should be studied by those with a particular interest; here the principal requirements are summarized.

The temperatures listed as follows are temperatures of the exposing fire.

---

*A dangerous error has been detected in the thinking of some persons responsible for construction which is required to maintain its integrity and confine noxious materials, such as radioactive or pathogenic, during a fire. Fire containment is equated with hazard containment. Nothing in the standard or the test procedures justifies this assumption.

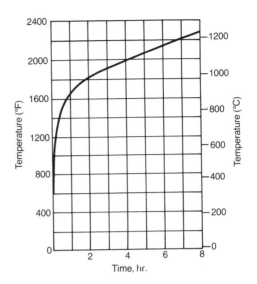

5 minutes after start of test 1,000 °F
30 minutes after start of test 1,550 °F
60 minutes after start of test 1,700 °F
120 minutes after start of test 1,850 °F
180 minutes after start of test 1,925 °F
240 minutes after start of test 2,000 °F
480 minutes after start of test 2,300 °F
75 °F per hour thereafter.

The time-temperature curve.

Columns have several thermocouples peened into the column. When the average temperature reaches 1,000°F or any one thermocouple reaches 1,200°F, the test is ended.

Floor structure must not develop conditions under applied load which would ignite cotton waste on the unexposed surface or permit an average temperature rise on the unexposed surface of 250°F.

Under some circumstances, a hose stream is applied to the assembly after the test.

Most tests are conducted at Underwriters Laboratories Inc. Fire protection systems or assemblies which pass the test are rated in units of hours, as for instance "1-hour," "2-hour," etc., depending upon the time the test unit survived the test fire. Rated units are "listed" in the UL's *Building Materials List* published annually with semiannual supplements. Based on such listings (not approvals) local authorities may permit use of specific materials and assemblies.

It should be noted that there is no point in achieving a rating for 2 hours and 20 minutes, since the assembly will receive only a 2-hour rating. In short, the listed assembly is the *minimum* for passing the test. Therefore, all components are important. Any departure from the listing means that most probably the assembly installed is not the equivalent of the listed assembly.

The unit being tested may be seriously damaged or effectively destroyed in passing the test; thus, fire resistance ratings are often misinterpreted in that it is thought that "2-hour fire resistance" means that the building will be relatively undamaged for 2 hours. This is just not true; the fire damage can be quite severe for short exposure. However, if the building reacts to the fire in the same manner as the specimens did for that particular type of construction in the test furnace, and if the fire load does not exceed the test fire load, and if the duration of the fire does not exceed the time span of survival specified for the components involved, then the structure should survive the fire.

A Word of Caution!

Fire resistance ratings for building materials must be carefully distinguished from fire hazard ratings. Fire resistance ratings, as has been seen, are concerned with the length of time a particular building assembly will continue to perform its function as a barrier to the passage of heat in the face of an assault by a standard fire test. Fire hazard ratings are concerned with the rate at which fire spreads over the surface of a material, the smoke the material develops and the fuel it contributes to the fire (see Chapter 9, Flame Spread). The two terms are not interchangeable, but in some cases, a single material may be evaluated for either its fire resistance rating or its fire hazard rating, depending on the way it is used. Take acoustical ceiling tiles as an example. When the tile is part of a fire resistant (fire-rated) floor and ceiling assembly, its ability to act as a barrier to passage of heat is of prime importance. However, when acoustical tile is used solely for its sound-deadening and decorative qualities, as when tiles are directly adhered to the underside of a concrete floor, then its fire hazard rating is of prime importance. As part of the interior finish of a building, its combustibility is a potential threat to the life safety of occupants.

Estimating Fire Severity

It is erroneous to relate a building's rated fire resistance to a projected collapse time. The following concept is not correct: "The building has a 2-hour rating. If the fire is not controlled in 1 hour and 45 minutes, prepare to withdraw the men and operate from outside." Depending upon the fire load, the rate of fire development, and many other factors, such as possible failure of a key structural element or fire barrier, the building may be in distress, or the fire may communicate beyond a fire barrier, in much less than 2 hours. The converse is equally true.

Since there are so many variables, the limit of confidence that one can have is more nearly the concept that a 4-hour fire resistive building is more fire resistive than a 2-hour fire resistive building and a 1-hour fire resistive building is less fire resistive than a 2-hour fire resistive building. It is not simple to relate the building ratings to an anticipation of what will occur during a fire, either preplanned or actually in progess.

There is a rule of thumb which can be useful in estimating the severity of a building fire. It involves an estimate of how much of a fire exposure certain amounts of fire loading (the amount of fuel per square foot of floor space in a building) will produce. For example, it has been estimated that a fire loading of 5 pounds of ordinary fuel per square foot of floor space in an office building has an assumed potential of releasing 40,000 Btu which would be the equivalent of a

30-minute exposure under the time temperature curve used in fire tests. Higher amounts of combustibles per square foot of floor space produce higher corresponding equivalent fire exposures.

The following table relates combustible contents to heat potential and to equivalent fire severity. Note that the "pounds" are of ordinary combustibles estimated at 8,000 Btu per pound. For plastics, rubber and flammable liquids use 16,000 Btu per pound.

Estimated Fire Severity for Offices and Light Commercial Occupancies

Data applying to fire-resistive buildings with combustible furniture and shelving

Combustible Content	Heat Potential	Equivalent Fire Severity
Total, including finish floor and trim	Assumed*	Approximately equivalent to that of test under standard curve for the
Lbs per sq ft	Btu per sq ft	following periods:
5	40,000	30 min
10	80,000	1 hr
15	120,000	1½ hrs
20	160,000	2 hrs
30	240,000	3 hrs
40	320,000	4½ hrs
50	380,000	7 hrs
60	432,000	8 hrs
70	500,000	9 hrs

*Heat of combustion of contents taken at 8,000 Btu per lb up to 40 lbs per sq ft; 7,600 Btu per lb for 50 lbs, and 7,200 Btu for 60 lbs and more to allow for relatively greater-proportion of paper. The weights in the table are those of ordinary combustible materials, such as wood, paper, or textiles.

It is possible, therefore, to estimate the fire loading of a given occupancy by roughly calculating the equivalent fire exposure for that amount of loading, and then comparing it with the fire resistance rating of the building. The result will be at least a rough determination of the possible effect on the building of a severe fire in the contents. A study of file rooms and supply rooms in a 1-hour rated office building might provide a real shock.

Combustibility and Fire Resistance

Fire resistance does not necessarily mean noncombustibility. There are listings of assemblies which include combustible elements. These have the word "combustible" in the UL listing immediately after the rating. In the case of a "combustible" rated floor and ceiling assembly, fire does not penetrate through a qualifying assembly but it may penetrate into it. Combustible rated ceiling and floor assemblies should be given careful attention during a fire. In some codes, fire resistance ratings that include the word "combustible" are not permitted.

In the case of such combustible floor assemblies, the lower surface of the assembly, the ceiling, is exposed to the test fire. Apparently, an assumption is made that that is the only way the fire will attack the assembly. Unfortunately, fires do not necessarily conform to test procedures. The tests described in the Bureau of Standards NBSIR 80-2120, "Fire Development in Residential Basement Rooms" (discussed in Chapter 5, Garden Apartments), showed that in a typical furnishings fire, the radiant heat flux to the combustible floor, which is part of the assembly, is apparently quite sufficient to ignite it. In addition, the combustible floor may be attacked by a liquid pool fire, as from melted plastics.

These tests should also make plain that the widely accepted belief that some real degree of fire resistance is achieved simply by nailing gypsum board to the underside of a combustible structure is erroneous. The limit of confidence is that the structure is somewhat better off than if the gypsum board were not applied, if the fire chooses to attack from that side.

Note the discussion in Chapter 7, Steel Construction, of the column failure potential where the floor-ceiling assembly "protects" the column.

Fire Intensity and Duration

Since in reality there are no "fireproof" buildings, the measure of the resistance of a structure to fire must involve determinations of both the intensity of a fire and its duration. The difficulty lies in anticipating how big a fire can be expected and how long will its life be. In The Severity of Fires in Steel Frame Buildings, Seigal points out that a high-intensity short-duration fire may be more severe than a low-intensity fire that lasts a longer time.[*3] [Small increases in temperature cause large increases in intensity, because radiant energy is proportional to the fourth power of the absolute (abs) temperature — temperature measured from absolute zero which is at -460°F. Thus, a 2,000°F absolute fire lasting 20 minutes could have eight times the radiant energy potential of a 1,000°F absolute fire lasting for 40 minutes.]

Theory vs Reality

The difficulty of translating theoretical calculations and fire tests to actual hostile fires is demonstrated when one considers that the smoke emitted by the fire at times seriously interferes with the transfer of heat by radiant energy within the fire building. Test fires use smokeless natural gas, so radiant heat transfer is important in tests.

It is argued by some that the fire load anticipated, based upon the type of occupancy, should govern the required fire resistance rating of structures. Under the procedure presently followed in building codes, fire resistance requirements

---

[*]This contradicts the long held belief that if the "degree hours" in different cases are equal, the severity of the fire is equal.

are very imprecisely related to possible fire load. "Warehouses must be of 4-hour construction" or "mercantile buildings must be of 2-hour construction," are typical of the sort of provisions found in codes, regardless of the fact that the warehouse may contain anchors and the store may sell foamed plastic bedding.

While it may be many years before the projected fire load concept is adopted into law (the possible changes in occupancy over the years would militate against it), we should certainly compare the fire load to the fire resistance rating in planning for a fire in a building. The choice of words in the foregoing sentence was deliberate. Before we can preplan fire department operations intelligently, we must intelligently estimate the size and severity of possible fires.

References

[1]Freitag, J. K., *Fire Prevention and Fire Protection as Applied to Building Construction*, 2nd ed., Wiley, New York, 1921, pp. 369-70. (See also Ref. 4, Chap. 6.)

[2]Shoub, H., "Early History of Fire Endurance Testing in The United States," National Bureau of Standards, *STP 301 Fire Test Methods*, American Society for Testing and Materials, Philadelphia, PA, 1961.

[3]Seigal, L. G., "The Severity of Fires in Steel Frame Buildings," *American Institute of Steel Construction Journal*, October 1967, p. 139.

CHAPTER 7

Steel Construction

Steel is the most important metal used in building construction. Without it we would be confined to massive all-masonry buildings with arched floors, or masonry wall-bearing buildings with wooden floors. Steel is tremendously strong. Its modulus of elasticity is about 29,000,000 pounds per square inch (psi), its compressive strength is equal to its tensile strength, and its shear strength is about three-quarters of its tensile strength. It is available in supply generally equal to demand, and it is relatively inexpensive.

Steel has three characteristics of concern in considering its behavior in fire. The balance of this chapter will be concerned with these characteristics, how they affect the fire performance of buildings, and the deficiencies, if any, of methods used to overcome these characteristics. An excellent treatise on the benefits and advantages of steel construction is *Fire Protection through Modern Building Codes*, 5th ed., published by the American Iron and Steel Institute. Here, without apology, we deal with the problems steel used in construction gives to fire fighters.

First, the coefficient of expansion of steel is such that substantial elongation can take place in a steel member at ordinary fire temperatures (about 1,000°F). Elongation may cause the disruption of other structural components, particularly masonry abutting the ends of the steel.

If the steel cannot elongate because of restraint, it will buckle or overturn. This can be particulary significant when other components rest on a steel member. Often a few steel members, usually beams, are inserted into an otherwise combustible building to provide a greater span than a wooden beam could provide or to support an opening in a brick wall. Such unprotected steel beams may be the origination of a collapse.

Second, when heated to higher temperatures (above 1,300°F), which are common at serious fires, the yield point of steel drops drastically; steel members may fail, bringing about a collapse of the structure. Steel is truly a thermoplastic material.

Note that the temperature cited refers to structural steel. Neither steel tendons used for tensioned concrete and for excavation tiebacks nor elevator cables can be depended on above 800°F.

Substantial elongation can take place in steel structural members at ordinary fire temperatures. A fire in what is to be a wood roof on this masonry-walled office building could elongate the steel ridge girder and push out the masonry gables.

Third, steel transmits heat readily. It is a good conductor, a characteristic that is beneficial under some circumstances. A large massive steel member is a heat sink of considerable capacity and thus is capable of redistributing heat received locally. This capacity is not unlimited, and many destructive fires have demonstrated that failure temperatures can be reached at one point in a steel structure when the heat input rate exceeds the rate at which the heat can be redistributed. The conductivity characteristic is also important in that heat can be transmitted by conduction through steel to combustible material that otherwise is unexposed to the fire. This characteristic is important both in the development and extinguishment of metal deck roof fires (discussed in detail later in this chapter).

The fact that steel is noncombustible leads to unwarranted confidence in its "fireproofness" and suitability for all applications where fire is a problem. "I never would have believed it," is a common reaction by otherwise well-informed people when they see the ruins of an unprotected steel building. These brief comments are expanded later in this chapter.

Definitions

Some necessary definitions are given here. A much fuller treatment is provided in Chapter 7 of the book *Building Construction*.[1]

> *I Beams* These are beams shaped like the letter "I."
> The top and bottom of the "I" are called flanges and the
> stem is called the web.

Wide Flange Shapes   These are I beams which have flanges wider than standard I beams. Some are "H"-shaped and being square (with respect to the four extremities), are thus more suitable for columns. The letters WF are part of the designation of wide-flange shapes.

Channels   These are steel structural components which have a "U"-shaped cross section.

Angles   These have two legs at right angles to one another. They are "L"-shaped in cross section.

Tees   A standard I beam cut lengthwise through the web forms two beams with "T"-shaped cross sections.

Bulb Tees   These are tees with a thickened portion at the tip of the stem.

Zees   As the name implies, these are members with a "Z"-shaped cross section. They are not often used in structures. *

Tubes   Steel structural members that are rolled in cylindrical, square and rectangular shapes are called "tubes." They are most often used as columns.

Plates   Flat pieces of steel are called plates.

Bars   Plates less than 6 inches in width are called "bars." Bars are also made square or round.

Box Columns   These are large hollow columns built up out of steel plates.

Box Girders   These are large girders, hollow like box columns, and often used for highway bridges.

Weight   The weight of steel sections is usually given per running foot. Thus, a "730-pound column" weighs 730 pounds per foot. (This is a very heavy column.)

---

* Short lengths of Z sections are used by New York City Rescue companies. Fitted over the edge of a subway platform, they provide a base for a jack to shift a subway car, on its springs, away from the platform.

Rolled or Built-up Members

Steel structural members can be either rolled or built up. Rolled members are one solid piece of metal; built-up members are made up of a number of different sections riveted, bolted, or welded together. As steel technology developed, mills were built which could roll larger and larger sections. Therefore, built-up structural members, of a size for which rolled beams are used today, will be found in older buildings. No matter how big, sections can be rolled; however, designers will need larger members. A girder built up out of steel plate with angles riveted on each side of the top and bottom to form flanges is often called a "plate girder."

There are some terms which are really structural in nature, but are usually associated with steel.

Purlins are beams, usually channels, that are set at right angles to the trusses or roof rafters to provide support for lightweight (usually galvanized corrugated steel, alone or in combination with corrugated reinforced plastic sheets) roofing.

Spandrel Girders are girders that tie the wall columns together in a framed building. They also carry the weight of the panel wall, which can be considerable when masonry is used. In addition, they sometimes are specially connected to the columns and stiffened to help resist wind shear.

Steel as a Construction Material

The first structure which usually comes to mind when steel is mentioned as a construction material is the skyscraper. The development of steel framing as an engineering technique, and of steel fabrication capacity to match, made it possible to erect buildings as tall as desired without unacceptable economic penalties in terms of space taken up by the components of the structure itself. The example of a steel-framed structure which most clearly shows its nature is the universally recognized symbol of Paris, Gustave Eiffel's remarkable tower.[*]

Not so well known is the fact that New York City's equally recognized symbol, the Statue of Liberty, is its first steel-framed structure. Eiffel designed the interior steel frame which supports the copper skin. The strength of steel, the consistency of its structural characteristics, and its ability to be connected so that loads can be adequately transferred, are all important to the use of steel as a building material.

It is useful at this point to note some of the principal uses of steel as a structural material. Some of these uses, and the fire problems which flow from them, are discussed in more detail in other chapters.

Steel is used for connectors of wooden structural members. In Chapter 4, Ordinary Construction, note that there are many buildings standing in which wood-

---

[*] Not all Parisians appreciate the tower. One critic eats his lunch in the tower restaurant daily, saying: "It's the only place in the city I can't see it."

working connections, such as mortise and tenon joints were used, but these are a tiny minority. Connectors, such as nails, screws, gusset plates, lag bolts, threaded bolts, joist hangers, dog irons, rods, wire and cable, struts, and ties, are all made of steel, and in a given fire, the fire characteristics of the steel may be of much more importance than the characteristics of the wooden members.

In ordinary construction can be found steel lintels, channels, and spreaders and tie rods bracing insecure walls, and struts supporting high masonry parapets. Interior floor, beam, and column systems may be completely or partially of steel, either by initial design or by reason of alterations to strengthen a building weakened by age or overloading.

Steel reinforcing bars and stress tendons are vital to concrete construction in providing the tensile strength that concrete lacks. Steel is also used in concrete flooring systems; corrugated steel provides "left in place" forms, which are often designed to react together with the concrete under load, thus forming a composite.

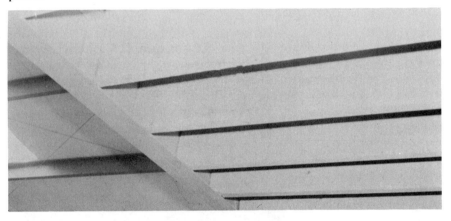

These wooden beams are resting on a steel girder. They will fall if the girder twists from overheating.

Steel framing is used for many commercial and industrial buildings, usually of one-story construction. The framing may be columns and beams with triangular trusses to achieve a peaked roof. The columns may support deep parallel chord trusses spanning wide areas with smaller trusses (often called bar joists) spanning between the main trusses to support a flat roof. Space frames (three-dimensional trusses) provide huge clear spans in some modern buildings, and there are a variety of framing systems available to achieve them. The important thing to remember, though, is that the steel is almost universally unprotected, and at best, such buildings can only be classified as noncombustible.

Steel is rarely protected from the effects of fire until the building is of such height or size as to be required by law to be fire resistive. "Protected noncombustible" construction is found occasionally. In such buildings, major structural elements are provided with some degree of fire resistance, but not enough to qualify the building as fire resistive.

In "protected noncombustible construction" some structural members, such as this column, are given some measure of fire resistance.

Rigid frames are a cousin to the arch and are used to achieve wide clear spans. In the rigid frame, the column is narrow at the base and tapers to its widest point at the top where it meets the roof girder, which is also tapered from narrow at the ridge to wide where it joins the column. This wide "haunch" resists the outward thrust of the roof. Rigid frame bents often are tied together under the floor. If there is a basement, and the ties are exposed to a basement fire, failure of the ties may cause the failure of the building. Rigid frames can provide clear areas of about 100 feet, which can be increased indefinitely by using Y columns.* If such a column is distorted by fire, the area of damage may be doubled.

Large rigid frames, such as these spanning a skating rink, are tied together to resist wind load. Distortion of one frame by fire can cause torsional loads that could collapse other frames some distance from the fire area.

---

*Rigid frames can also be built of reinforced concrete or laminated wood.

When huge spans are achieved by rigid frames, trusses, or space frames, collapse can be sudden, general, and tragic. In rigid frame buildings, the adjacent bents are tied together to resist wind load. The tying together of the steel units means that if one part of the building is distorted by fire, torsional or eccentric loads beyond the designed capacity may be placed on the balance of the building, starting the progressive collapse of the building, often far beyond the area involved in fire. As a matter of fact, the better the building is tied together to resist wind load, the more likely it probably is to suffer progressive collapse due to fire distortion.

Many steel-framed buildings are prefabricated to standard designs. The Butler Company is a prominent manufacturer of steel "prefabs" and such buildings are often referred to as "Butler buildings" regardless of the manufacturer. Originally, the lightweight steel-framed building was exclusively for industrial use, but in recent years, schools, churches, and many other types of buildings have been built to similar designs.

Designers are working wonders in developing wide-span trusses made up of very light sections. The push to design large open areas with a minimum of mass is fraught with the potential for terrible disaster. The collapses of the C.W. Post Auditorium in 1978, a 171-foot span steel and aluminum space dome; the Hartford Coliseum in 1978, a 300-foot span steel space frame; and the Kemper Auditorium in Kansas City, MO in 1979, a 324-foot steel truss roof suspended from a space frame, providentally caused no fatalities. Considering the thousands of people who had occupied these buildings on occasion, it is easy to see that there was a disaster potential which, by contrast, would make the terrible problems faced by rescue personnel in the Kansas City Hyatt Regency Hotel skywalk collapse miniscule.*

Walls Of Steel-Framed Buildings

A steel-framed building can be provided with a variety of walls. Some common wall materials and their fire characteristics of interest are:

Cement Asbestos Board. (Transite, a trade name, is often used as a generic name.) It is noncombustible and is often used for friable construction. Friable construction is provided where an explosion is a possibility. It will break away readily, relieve the pressure, and not itself provide missiles. Some sandwich boards of cement-asbestos "bread" and low-density fiberboard "meat" are used. Fire can burrow into the fiberboard and smolder.

---

*On July 18, 1981, a two-level suspended walkway collapsed at the Kansas City Hyatt Regency Hotel, killing 113 persons and injuring 186. The National Bureau of Standards report said that the walkway's connections were able to support only 27 percent of the total dead and live load they should have been designed to hold.[2]

This steel-framed building has asbestos cement siding. In an explosion, the siding will shatter and break away, hence the name "friable construction."

Glass-fiber Reinforced Plastics. While the glass fiber is noncombustible, the resinous binder most often is readily flammable. This is not generally known, and the material is often thought of as noncombustible.

Aluminum Metal. Aluminum is noncombustible, but it has a low melting point and little mass per unit of area, so it disintegrates rapidly in a fire. This is by no means always a disadvantage, since in some instances it may provide needed venting and access for hose streams to the interior of the building.

Precast Concrete Panels. Panels of this type are erected in large sections, and collapse can be particularly hazardous to fire fighters. Study the relationship of the steel frame to the wall. If the steel expands under fire conditions, will it deflect the wall section from the vertical? If deflected, will the wall section fall freely or will it be restrained from falling? Precast concrete panels formed the side walls of McCormick Place, the big Chicago exhibition hall destroyed by fire on Jan. 16, 1967. Distortion of the roof steel in the fire pushed the wall sections out of alignment and several of them collapsed.

Wood Siding. Wood siding is not common, but partial wood siding may be used for aesthetic effect. Wood siding and any other combustible siding must be considered when estimating the effect of heat on the steel.

Masonry Walls. Probably the most common walls for unprotected steel-framed buildings are of concrete block or a composite of concrete block and brick. Natural or artificial stone walls are also seen. Usually, the walls are only curtain walls (the exterior surface may show unbroken masonry; the interior wall may show panels of masonry between steel wall columns whose interior surface is approximately flush with the masonry). Occasionally, in a building of generally framed construction, one wall, or part of one wall, is found to be a bearing wall. In either case, it is important that an analysis be made of the effect of the expansion of the steel frame on the wall. For instance, if a girder rests on a bearing plate, it may be free to slide, and thus disturb only masonry above. On the other hand, a girder tied to a beam that is incorporated longitudinally into a wall may, by pushing on the beam, bring down all or a substantial part of the wall.

The author has seen several sheet steel-walled buildings "improved" by covering the steel with a brick veneer. Any movement of the steel would likely bring down the brick since its stability depends entirely on the steel.

The signs of masonry deterioration, which were mentioned in Chapter 4, Ordinary Construction, are, of course, equally important in the steel-framed building, but the absence of deterioration does not decrease the possibility of collapse. The most probable cause of collapse is a steady lateral push by expanding steel.

Galvanized Metal Walls.   These are a common industrial standby where heat conservation is not important. Sometimes the metal is weather-protected by an asphalt coating. This is asphalt asbestos protected metal (AAPM), often called RPM (Robertson Protected Metal, the name of one proprietary type). AAPM is of varying degrees of ignitability, but in any event, the coating is composed of combustible material.   •

Metal Panels.   Prefabricated metal panels are often made up in a sandwich construction to provide one unit combining thermal insulation and interior finish. In earlier construction, low-density fiberboard will be found; later, mineral and glass fibers were used. Today, plastics will most often be found. Even when the outer layers are noncombustible, the insulation, or vapor seal, or adhesive in the panels may be combustible. It may be necessary to open such panels far from the main fire area if the interior of the panel becomes ignited. The author had this problem in a fire in a military warehouse many years ago. The panels were crimped around a sheet of low-density fiberboard. Each panel had to be chopped open and the fire extinguished. The result was that the building looked as if it had been sacked by Atilla the Hun, whereas, in fact, we had made a fine stop. Unless the panels present a serious extension problem, the author would recommend that those overhauling work slowly and patiently, cut with the circular saw, rather than with the axe.

In one type of construction, polyurethane insulating panels are protected by gypsum board and stainless steel sheathing. It is quite possible that, many years hence, alterations will be undertaken by persons not aware of the insulation. A cutting torch will be used, and a smoky, destructive fire may result.

In the 1920s, the Rockefeller Institute was engaged in research into tropical diseases. Prefab steel buildings, incorporating corrugated metals which sandwiched combustible fiberboard insulation, were developed for tropical use, and some were erected at Princeton, NJ. Years later, a welder was told to cut a hole for a door. No one was aware of the insulation, and a serious fire resulted.

Many years later at another scientific research establishment, a similar fire occurred. A steel tunnel-like structure was lined with polyurethane insulation. A crew installing a door cut a hole with a torch and again, a serious fire resulted.

Aluminum sandwich panels with foamed polyurethane are also available. Some are listed by Underwriters Laboratories Inc. for low flame spread ratings but "smoke developed" ratings may be quite high.

Metal wall panels can be used on any framed building and in fact, are used on many concrete buildings.

The design of panel walls, the method of installation, and the degradation of insulation or expansion of metal under fire conditions are some of the reasons the closure of the wall panel to the floor slab can fail and permit the extension of fire from floor to floor. This can be particularly significant in high-rise buildings beyond the range of effective ground attack.

Skyscraper Steel Framing

For many years, steel framing stood unchallenged for high-rise buildings. In recent years, concrete is finding more use for framing high-rise buildings, and the two compete strenuously. The principles of construction are the same, and much of what is discussed here applies equally to concrete-framed high-rise buildings.

When steel-framed buildings were first built, builders were hesitant about featuring that it was the frame which supported the building. By long tradition, most people associated strength and permanence with solid masonry; consequently, tons of brick, stone, and terra cotta, all structurally unnecessary, were loaded onto framed buildings. Even the height of buildings was suspect. Rather than arrange the exterior so that vertical lines led the eye upward, as is common today, buildings were "belted" with masonry corbels or trim; different materials were used on successive floors, and all possible architectural devices were used to reduce the apparent height of the building.

The masonry served another purpose. The spandrel space (the distance between the top of the one window, and the bottom of the one above) was of masonry and sufficient in depth to prevent the extension of fire from floor to floor on the outside.

Today, the pendulum has swung the other way, and the exterior extension potential should be noted, as demonstrated by the fire in the ten-story Playboy building in Los Angeles County, CA. The fire spread from the third floor to the seventh floor via outside windows, and was finally controlled by outside master streams, a method which is unavailable for an upper-story fire.

Bracing Against Wind

The steel-framed building must be braced against the shear force of wind. This is accomplished in one or more ways. Diagonal braces connect certain columns and are concealed in partition walls. An arrangement of braces between columns which resemble the letter K is called "K bracing." Heavy riveting of girders to columns from top to bottom of the web is called portal bracing. Some floors are built as diaphragms and in addition to carrying their floor load, actually contribute to the structural stability. In the World Trade Center in New York City, the wind load is resisted by giant vierendeel trusses (see Chapter 2, Principles of Construction) formed by exterior box columns and spandrels. Diagonal braces are connected between exterior columns and featured as part of the visible exterior design in the John Hancock Building in Chicago. Interior walls, necessary for other purposes such as stairway or elevator shaft enclosures, are sometimes made of reinforced concrete to serve as shear walls. If it is necessary to breach an elevator or stair shaft, and reinforced concrete is found, it might be easier to shift around to another wall, which may be only concrete block or gypsum sheets.

Except for the problem of breaching, as noted above, the method of wind bracing the building is of no interest after the building is completed. While the building is under construction, the wind forces must be resisted, as the building is not fully connected. Often, cables are strung diagonally across the steelwork, and at times, these braces are not properly installed. If there are no high winds, nothing happens and the perpetrator of the dangerous "bracing" has another "good experience."*

Carefully examine the temporary bracing before going up to extinguish a fire on a steel building under construction, particularly if the weather is windy. Be aware that it is obviously impossible to make all the connections as soon as steel is placed. Bolts are placed in some rivet holes; thereafter, rivets are driven in the remaining holes, then the bolts are replaced with rivets. The practice is known as "field bolting," and it is good to give a wide berth to field-bolted structures during a high wind.

"Plastic" Design in Steel Construction

For years, there was an accepted engineering principle that structural elements were to be designed so that the applied loads would be within the elastic range of the material. In a further effort to seek economical steel design, the concept of "plastic design" has been developed. The structure is rigid-framed or braced rather than pinned. This is analogous to monolithic construction in concrete. As a result of the rigid framing, a load which might be locally excessive is redistributed over the structure. The result is that the weight of steel in the building can be decreased. Beams are lighter and columns are smaller than they

---

*Experience is the best teacher — of bad habits. In one case of which the author knows, a bracing cable ended in an eyesplice that was to be connected to the eye in the end of a turnbuckle attached to a column. The cable eye was passed through the turnbuckle and "secured" with a scrap piece of wood through the cable eye. The "artisan" who accomplished this engineering triumph then fixed three others the same way.

would be otherwise. This is of more than academic interest. The lighter the steel, the less inherent the fire resistance; thus, the provision and maintenance of adequate "fireproofing" for the steel is even more important in a plastic design building than in one of conventional design.

Fire Significant Characteristics of Steel

At this point, the comments made in the opening of this chapter will be expanded on.

Steel Conducts Heat

The fact that steel transmits heat should be well known. A suit of armor is noncombustible but no one would attempt to fight a fire in it.* Despite this, many building codes required (and still may) so-called "tin ceilings," actually embossed steel, in certain occupancies. These can transmit fire in either direction.

Sheet steel doors were used on vaults, many of which still "guard" valuable governmental and private records. The chief disastrous effect of this characteristic is in the loss of records. Across the country, businessmen and homeowners rest secure with the knowledge that their vital records are "protected" in uninsulated steel files. "Tin boxes" are sold in stores as "fire resistant chests."

It is the author's guess that it is acceptable for a free American to hazard his own records if he wishes; but it is a different case when vital records of others are involved. The next time you go to the dentist or the doctor, ask how your records are safeguarded from loss. Most often you will find them on open shelves or in non-fire-rated files. There are, of course, rated safes and safe files available, designed to protect records against a standard fire for a rated time. A word of caution — if magnetic tapes are to be protected, the file must be so designated. Tapes will fail totally at temperatures much lower than are satisfactory and permitted for paper.

The conductivity of steel can be a factor in spreading fires. For example, heat was conducted through steel expansion joints in a concrete floor at a fire in the New York City Post Office. Mail bags, resting on the concrete floor, were ignited by the heat.

Steel sheets directly attached to combustible surfaces to provide "fireproofing" have in many cases acted to transmit and retain heat. In one case, a steel plate was attached to the wood-joisted ceiling above the end of a rotary kiln. When it was ordered removed, plant personnel were surprised to find that all the wood behind it had carbonized to charcoal.

---

*In World War II, Navy Fire Training Officers were driven almost to drink by Washington public relations photos showing "heroic" navy fire fighters in their "asbestos suits." The "logic" was, "Asbestos is noncombustible, therefore safe."

This uninsulated steel file contained the vital records of a business. The owner was devastated at the condition of the records after a fire — he thought they were safe because the file was made of steel.

While it is generally understood that metal smokepipes must have clearance from wooden members, this concept is often ignored in the case of other metal ducts which may have fire in them, despite requirements to the contrary. Grease ducts are a good example. A grease duct fire often extends to adjacent combustible construction by conduction.

Ships are not buildings, but the practice is growing of converting ships into buildings, as was done with the *Queen Mary* at Long Beach, CA; also, a number of ferry boats have been converted into restaurants. Ships have steel walls. Many people associated with ships confuse noncombustibility with nonconductivity. Typically, welding operations will be performed on one side of a bulkhead (sheet steel wall) without any concern for the transmission of heat through the steel to ignite combustible material on the other side. This can cause a real problem when the other side is covered with wood paneling which is ignited from within the concealed space behind the paneling.

Steel Elongates

The NFPA's *Fire Protection Handbook* states that the average coefficient of expansion for structural steel between room temperature and 200°F is 0.0000065 for each degree. For temperatures of 200°F to 1,100°F, the coefficient is given by the formula: $e = 0.0000061 + 0.0000000022t$, in which e is the coefficient of expansion for each degree fahrenheit and t is the temperature in degrees fahrenheit.[3] In more understandable terms, steel will expand 0.06 percent to 0.07 percent in length for each 100°F rise in temperature. The expansion rate increases as the temperature rises. Heated to 1,000°F, a steel member will expand 9½ inches in 100 feet of length. As will be seen later, at temperatures above 1,000°F, the steel starts to soften and fail, depending upon load.

Expanding steel exerts a lateral force against the structure which restrains it. If the restraining structure is capable of resisting the lateral thrust, the expanding steel structure may be even stronger, at least for a period of time.

If steel beams are restrained, as by a masonry structure, and the temperature of the fire is sustained in the 1,000°F range, the expansion of the steel may cause the displacement of the masonry, resulting in a partial or total collapse. A one-story mercantile building in Wheaton, MD was constructed of concrete block walls and a steel joist roof. The steel joists were restrained by the walls. The ceiling structure was combustible. A fire broke out in the concealed space above the ceiling, and employees, after smelling smoke, searched for the fire for nearly an hour before the flames were seen from outside. The long-burning, but not excessively hot, fire apparently provided ideal conditions for the steel joists to elongate causing severe damage to the supporting walls. By contrast, in hot, fast fires, failure temperatures are reached rapidly and the lateral thrust against the wall is minimized.

A relatively small, but hot, confined fire did this severe damage to steel bar joists. In hot, fast fires, failure temperatures are reached rapidly.

In a one-story and basement noncombustible building, a heavy girder was supporting a one-wythe thick brick wall, which added to the decor of the reception lobby. There was a hot fire in the basement. The steel beam attempted to elongate but could not because of restraint. It took the only other course to absorb the increased length; it twisted, dropping the wall completely across the lobby. Fortunately, no one was in its path.

This overturning can be anticipated when unprotected steel girders are used to support wooden floors in combustible buildings.

Steel Fails

When steel is raised to temperatures above 1,000°F, it starts to lose strength rapidly.

Temperatures above 1,000°F are quickly reached in fires. Recognizing this, the standard fire test, ASTM E 119, used to test building materials and assemblies, reaches 1,000°F in 5 minutes. Temperatures developed in Bureau of Standards tests of typical basement room fires reached 1,500°F in 5 minutes, reflecting today's plastic fire loads, rather than any special significance of basements.

In standard tests of the "fireproofing" of steel columns, the test is ended when a temperature of 1,200°F is reached at one point or 1,000°F is reached on the average in the column.

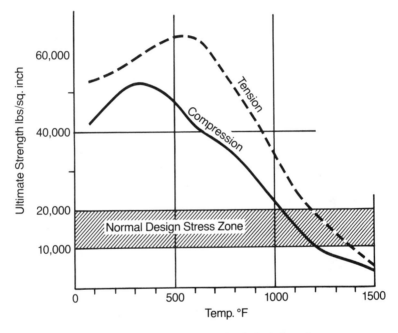

Effect of temperature on structural steel strength.

How fast the fire temperature will be achieved in the steel is another matter. A principle variable is the weight or mass of the steel unit. For instance, a bar joist will absorb heat very rapidly, giving the rule of thumb that, in a fire equivalent to the standard test fire, the bar joists will fail in about 7 minutes. In one observed fire in a one-story and basement bar-joist building, the bar-joist supported floor was collapsing before the people could get out of the first floor. On the other hand, very heavy steel sections might survive an hour of the standard fire test without failure.

Failure of steel bar joists dropped this floor and opened a hole in it before occupants escaped the building.

The amount of ventilation may also be significant. It appears that if a fire is well ventilated, the temperature in the steel may be less than if the fire is confined. This may help to explain observed differences in the effect of fire on steel in different fires.

The heavier the steel is loaded, the faster it will fail.

Unprotected Steel

In almost all building codes, steel at some distance above the floor is not required to be protected, possibly because it is believed that flame impingement is not likely at this height. The concept is probably valid if there can be an ironclad guarantee that there will never be a significant fire load in the area below the steel. Such a situation is unlikely; the situation which existed at Chicago's McCormick Place in 1967 is far more probable.*

The McCormick Place Fire

McCormick Place was a huge exhibition hall, built in Chicago in 1960. It burned January 16, 1967, with a loss of $154,000,000.[4]

The main exhibit area provided a clear area of 320,000 square feet. The roof was supported on 18 steel trusses, 60 feet on centers. Each truss had a 210-foot central span and a 67-foot cantilever on each side from the column to the exterior wall. The trusses were 16 feet deep at the column, tapering to 10 feet at the center of the building; they were made up of heavy steel members.

The columns were trusses themselves. In accordance with the previously mentioned provisions common to many building codes, the columns were fire protected up to a height of 20 feet above the floor. The roof trusses were unprotected.

The exterior walls consisted of freestanding steel columns supporting precast

---

*The mistake is akin to what the logician calls the fallacy of "arguing from the particular to the general." The fallacy is not unknown in other code areas.

concrete sandwich wall panels. The sandwich consisted of concrete interior surfacing, foamed plastic insulation, and exterior sculptured concrete.

At the time of the fire, preparations were being completed for the exhibit of the National Housewares Manufacturers Association. Exhibit booths crowded the floor, some two stories high. The fire loading has been estimated at 80,000 to 120,000 Btu per square foot. In addition, the combustible material, both by nature and arrangement, was fast-burning, providing a high rate of heat release. Fortunately, the fire started in the early morning hours; the possible life loss would have been staggering had the fire occurred during the day.

The fire was discovered early, but soon spread beyond the capabilities of the maintenance employees. There was some delay in the alarm (about 6 minutes). A second alarm was transmitted on arrival. The first-due engine company penetrated the hall a few feet when they heard a great roar as the flames approached. At this time, their water supply failed. It developed later that the hydrants in the area were the responsibility of the exhibit hall management, but the hydrants were out of service. In retrospect, at least one close observer thought this was a blessing. Had the water been available, it would have made little or no difference in the loss, but it is possible that a number of fire fighters would have been in serious danger advancing lines into a collapsing building.

Within a half hour after the start of the fire, portions of the massive roof trusses began to fail. The failing trusses pushed out a number of the wall panels, making close approach hazardous. Several of the panels collapsed.

The New McCormick Place

After the McCormick Place fire, there were a number of studies and reports. The following excerpt from the report by the American Iron and Steel Institute is pertinent:

> For the size and occupancy classification of the building, the construction of McCormick Place apparently met the fire protection requirements of the Chicago Building Code and most other widely recognized building codes. Most contemporary codes* would classify the building as an assembly occupancy. However, at the time of the fire, it more closely resembled a mercantile occupancy such as a department store, discount house or shopping center. In this type of occupancy different fire protection measures would have been required by most modern building codes.
>
> If viewed as a mercantile occupancy, modern building codes would

---

*Building codes do not attempt to relate the fire potential of the building, in terms of fire load, to requirements for fire resistance or fire suppression systems, except in the crudest, most general way. "Warehouses shall have 4-hour fire resistance, mercantile buildings shall have 2-hour fire resistance," are the sort of requirements which are typical. The warehouse may contain anchors, and the store may have foamed plastic stacked 40 feet high. Be on the alert for high fire loads in buildings not rated for such occupancies; a potential disaster is in the making.

have required a building of this size to be equipped with an automatic sprinkler system throughout. Emergency heat and smoke venting and draft curtains at the roof level would also have been required but additional or supplementary fireproofing of the structural members would not have been required.

Automatic sprinklers installed throughout the exhibit areas, draft curtains and emergency roof vents probably could have limited the spread of this fire and enabled the fire department, with a properly operating water supply, to bring it under control with a minimum of damage.[5]

In the new McCormick Place, structural steel is protected with directly applied fireproofing delivering 1-hour fire resistance. The entire building, except electrical enclosures and enclosed stair towers, is sprinklered. Sprinkler systems protecting the high fire load exhibit area are hydraulically designed to deliver 0.30 gpm per square foot over areas as large as 6,000 square feet. Ninety-eight percent of the floor area can be reached with 100 feet of hose from standpipes. Elaborate provisions have been made for smoke venting. Supervisory control has been provided for water supply. Training programs for employees and prefire planning with the Chicago Fire Department are provided.

The new McCormick Place has had the benefit of professional fire protection engineering from the start, not as an afterthought but as an integral part of the design procedure. It is now apparent that a building which can house 50,000 people and is the equivalent of a tinder-dry forest simultaneously cannot be protected by adherence to codes which did not contemplate such a building, nor by neglect of equipment once installed, nor by assuming that employees will take proper emergency action. At least it is recognized in Chicago. How apparent is it to your local armory, hotel ballroom, exhibit hall? Is your fire department satisfied that the extent of their responsibility is to see to it that all "rubbish" is removed?

Some Test Experience

After the fire disaster in Chicago's McCormick Place, a number of studies were undertaken. One study at Underwriters Laboratories Inc. examined the validity of the argument that sprinklers would have controlled the fire.[6] Examine the data developed, not at this time relative to sprinklers but relative to the temperatures developed at quite ordinary fires.

The fire tests were conducted in a building 30 feet high. (One should note the relationship of this height to the height above the floor at which it is presumed that steel needed no protection.) The fuels used were lightweight, readily combustible materials typical of those used for exhibition hall displays. Plywood, tempered hardboard, cardboard cartons, and packing materials made up the fire load of 20 pounds per square foot average. The materials had a high surface-to-mass ratio and thus would produce a fast, hot fire. The display material was set up over a 20- by 30-square-foot area; thus, the total weight of combustibles was about 12,000 pounds, which would yield almost 100,000,000 Btu if completely consumed. This was a rugged test, but not at all unrealistic. (Make some calculations for yourself at the next trade show or exhibit you attend.)

The first of the test fires is of great interest. The plan was to keep the sprinklers turned off for 6 minutes after ignition, corresponding to the reported gap between ignition and time of discovery of the McCormick Place fire. The report gives no indication that even the experienced engineers running the test had any qualms about the plan of holding back the water for 6 minutes. At 5 minutes and 45 seconds after the start of the fire, temperatures of 1,500°F were being recorded at the ceiling. The sprinklers were turned on at this time to avoid damage to the test facility.

Steel beams were monitored as to temperature by thermocouples peened into the surface. A bar joist at the ceiling reached 1,540°F at 5-plus minutes. An I beam reached 1,355°F at 5-plus minutes. Temperatures at the ceiling level exceeded temperatures in the booth by a slight margin.

The fire load used in the test cited is not excessive either in quantity or rate of heat release. There are many unprotected steel buildings with such a great fire load that collapse will have started prior to the arrival of the fire department, even with prompt alarm and normal response.

Hazards of Unusual Fire Loads

Even when average fire loads are low, an unprotected steel building may be endangered by a highly concentrated fire load in one area. The occupants of the buildings are not always as enamored of the big open space as the designer. While unprotected steel buildings are usually only one-high story,* it is not uncommon to find a mezzanine built into the building to provide office or storage space. Often the mezzanine is built entirely of wood. Where problems of noise, temperature, pilfering, or privacy exist, "buildings" may be built within the building to provide office space, controlled environment, privacy for secret processes, or security for valuable merchandise. Such structures are often combustible. The structure represents a hazard not only to the building but to that which it is designed to protect.**

The "experimental hall" at a scientific research facility is often built of

---

*The term "one-high story" is used to designate buildings of greater than usual height from floor to ceiling. A description such as "one equals five" describes a one-story building of a height equal to five ordinary stories.

**A hydrogen fusion test device, representing millions of dollars of research time, was classified "secret." It was necessary for uncleared workmen to do work in the area. Concerned only about security, the security authorities ordered a "house" built around the device. Concerned only about cost, the business manager ordered it built of a quite combustible "building board" made of pressed paper. Concerned about making the machine run, the scientists in charge recognized no hazard. There was something of an awakening when the author, who was inspecting the project, asked the question: "How much of this machine could stand a temperature of 1,000°F for 5 minutes?" For a few who couldn't fathom the nature of the problem, a simpler question was asked: "Suppose the Boy Scouts, collecting paper, asked to store it up against your device? What makes the paper any different because it has been macerated and pressed out into sheets?"

lightweight steel construction to provide the greatest free floor area at the least cost. The steel is usually unprotected; the fire load is expected to be minimal. In recent years, however, the practice of purchasing ordinary house trailers to use as offices and to house telemetry equipment inside such halls has grown. These provide privacy, some modicum of comfort, and a bit of status. From the fire protection point of view, however, they provide a fine potential source of heat. Grouped closely together, the trailers expose one another, increasing the possibility of a serious loss.

Prefabricated metal buildings are sometimes used in place of the trailers. Often these are of sandwich construction with an insulator placed between the inner and outer walls. The nature of the insulation should be determined; many insulations present severe fire problems.

At Kent Island, MD, at the eastern shore end of the Chesapeake Bay Bridge, the U.S. Army Corps of Engineers has a 14-acre undivided unprotected steel-trussed building, housing a nine-acre working model of Chesapeake Bay, our nation's largest inland sea. It is well worth a visit. Tidal and current flows are duplicated with meticulous accuracy. It took several years of work to get the model calibrated. The entire operation is menaced by the combustible house trailers and flammable gases stored in the building. Enough heat would be generated in what might otherwise be an ordinary trailer fire to do substantial damage to the steel. Water used for fire fighting would surely destroy the calibration of tidal flows. The structure has a huge water supply and it would be simple to sprinkler the necessary trailers, and remove the flammable storage.*

At one time, those operating a scientific facility felt immune to the potential for loss by fire. "We are so important they will give us a new one," was a common attitude. Today, it is much more likely that a scientific project which is destroyed will simply be dropped.

It is worthwhile to try to get the management of such projects to see the consequences of hazards of which they are not aware. Scientists, in the author's experience, tend to look on fire as a chemical phenomenon. Steel does not burn so it is no problem. The management of one facility finally enforced a rule: "Trailers are to be sprinklered or moved outside."

Following the medical term "triage," in which casualties are divided into three groups, we can "triage" the heat given off by a fire:

- Heat going off the space, leaving the building. To the heat conservationist, this is a tragedy, to the fire fighter, it is a blessing, though often not recognized. How many times have you seen heavy caliber streams, ladder pipes particularly, used to shove the heat back into the building "where it belongs"? Even when exposures are to be protected it is almost always better to absorb the heat at the exposure.

---

*The Corps of Engineers was given an opportunity to comment on this paragraph. They are confident that fire hazards are adequately controlled and stated categorically that "water used in fire fighting will not destroy the callibration of the model."

- Heat that is keeping the fire going. This is of secondary importance. A chair which is half-burned is all destroyed. This is often regarded as the principal target by those whose only thought is to "hit the fire."

- Heat that is being absorbed by something which will be ignited or damaged by the fire. This is the most important of the three divisions of heat.

Cooling Steel with Hose Streams

The heat being absorbed by unprotected structural steel or by the metal deck roof is very probably the most important heat to be removed by the fire suppression water. This may require violation of another unthinking slogan: "Never throw water into smoke." It is certainly justified to throw water where you cannot see the target if you know the target is unprotected steel. If the owner had adequately sprinklered the steel, it would not be necessary for us to "improvise a sprinkler system." (Learn to use the right words!)

This matter of throwing water on steel has long been confused. Many fire texts speak of causing steel to collapse by cooling it. This is simply not true. The author maintains the following is the case:

"If steel is elongated due to heat, and cooled with water, the steel will contract to its original shape. If the steel has started to fail and is cooled, the steel will remain in the shape it has assumed. Literally, it will be frozen."[7]

The author expressed this view in the first edition of this book, and it was questioned by readers who had the notion that hose streams on exposed steel invariably lead to collapse. Private correspondence with steel industry officials

These two identical steel beams were subjected to the same fire. One was protected by hose streams; the other (arrow) couldn't be reached, overturned, and dropped the roof.

substantiated his view as expressed. This correspondence sparked interest by other publishers when brought to their attention. They subsequently reviewed and modified their texts on the subject of hose streams and steel.

The steel will neither crack nor explode when struck with water. If steel is not too badly fire damaged, it can be straightened by heating and jacking, a process known as "flame straightening."

Planning for the cooling of steel should provide for a safe location for personnel. If a safe location is not available, this should be pointed out to the owners of the building so that they may be forewarned as to what the consequences of a fire will probably be.

The distortion of the steel due to fire may alter the structure of the building so that it will not be able to carry the superimposed loads or unexpected stresses, particularly torsional, that may be created due to buckling. However, structural steel that has been exposed to fire and water will not have significantly different properties than steel unaffected by exposure to fire. Unfortunately, the erroneous impression that it can be hazardous to put water on unprotected steel is rather widespread in the fire service. This mistaken belief can lead to unnecessarily ineffective fire fighting operations.

It is often difficult, even for technically trained people, to understand how fast steel can be destroyed. One of the HPR (highly protected risk) insurance companies had great difficulty in convincing policyholders of the necessity for sprinklering steel industrial buildings in which the only fuel was the hydraulic oil in extrusion presses. They set up a fire using a pan of oil, which exposed a steel beam, loaded with barrels of gravel to simulate typical loading, and the beam failed. The demonstration convinced the doubters.

Noncombustible Buildings

Building codes generally classify steel buildings as "unprotected noncombustible." A further classification is "protected noncombustible." In this case, major columns and beams are protected to some degree. This is often found in conjunction with sprinkler protection. The concept is to hold the building, and therefore, the sprinkler system together, until the sprinklers can turn down the heat production of the fire.

Noncombustible buildings can be and have been destroyed by fire in the contents.

There is another problem with the code definition of "noncombustible building." Determine what is permitted locally with respect to the roof. Is a combustible metal deck roof of the type described later in this chapter permitted? If it is, the building is most assuredly subject to destruction independently of the contents.

In studying the possible fire in an unprotected steel building, the first consideration should be the basic type of construction. Column, girder, and beam construction is used for both single-story and multistory construction. If a masonry bearing wall is substituted for some of the exterior columns, the building is "wall-bearing." Column and girder buildings are characterized by relatively short spans, so the failure of one column may affect only one portion of the building. As with all such generalizations, watch for the exceptions. The author knows of a church that is built of steel columns supporting steel roof rafters. The roofing is wood. The curtain walls are composite brick and block. Should there be a severe roof fire, the heat absorbed by the roof steel would cause it to elongate, pushing the columns out of alignment. The moving columns might well bring down the walls.

Just recently, the author studied a church under construction in Louisiana of a type unique to him. It is code classified noncombustible. The framing is steel. There is a huge fire load in the form of combustible fiberboard sheathing and a wooden balcony. The structure is to be brick-veneered. The roof is a conventional (therefore combustible) built-up metal deck.

Here bar joists are supported on rolled steel girders which in turn are supported on unprotected round steel columns. The failure of one column may affect only one portion of the building. Correctly identifying the components of unprotected steel construction is essential in anticipating structural stability in fire conditions.

The interior of the structure and the underside of the wood-floored balcony are "protected" by ⅝-inch fire-rated gypsum board. (In Chapter 5, Garden Apartments, there is a discussion of the widely held fallacy of extending to gypsum board itself, fire ratings achieved in the laboratory on specific test structures.*) It is unlikely, though, that the voids under a steel-framed balcony would be divided into limited areas, such as is the case with a floor on firestopped wood joists 16 inches on center. (The firestopped joists limit the air supply to the unexposed side of the floor hazards.) In addition, there is the further

---

The Fire Resistance Directory of Underwriters Laboratories Inc. states under "Design Information Section-General": "Fire Resistance Ratings apply only to assemblies in their entirety. Individual components are not assigned a fire resistance rating and are not intended to be interchanged between assemblies but rather are designated for use in a specific design in order that the rating may be achieved."

fallacy of assuming that fires will only burn upwards. Fuel in a church balcony might well include hymn books and cushions possibly made of foamed plastic. A fire in these fuels would readily burn down into the void.

A fire involving the wooden balcony or the metal deck roof could well cause the steel framing to move and thus cause one of the brick-veneered walls to fall. A well-advanced fire in such a structure should be treated with caution.

Two items are of the utmost importance. In any truss or space frame, every part of the truss is important to its stability, so the failure of any element may lead to the failure of the entire truss. Because of the wide spans and the interdependence of trusses one with another, the failure of a truss may have serious consequences far from the point where the initial failure occurs.

Here lightweight truss beams are supported on a lightweight truss girder. Because of the interdependence of trusses, the failure of one truss may affect the stability of the structure at a point far removed from the initial failure.

How long an unprotected steel structure will retain its stability in the face of a fire delivering a high Btu per minute attack is impossible to determine in advance due to the many variables involved. The more weight of steel, the greater the heat absorption and distribution capacity provided, but more massive steel does not guarantee longer survival time for the building. The rapid heat absorption of a lighter member may have serious consequences when the lighter member stresses the heavier member in a manner not provided for in the design.

If heat can escape, the steel may sustain its load indefinitely. Relatively small fire loads in confined areas can do extensive damage. In a New York subway tunnel fire, a wooden platform was the major portion of a fire load which caused several columns supporting famed 42nd Street to fail. This variation in heat escape can lead to erroneous opinions on how serious the problem of unprotected steel actually is.

It is probably futile, therefore, except in the most general way, to attempt to estimate the time during which a given unprotected steel structure containing a

A small, but hot and confined fire caused this beam to tear off the girder as it attempted to elongate. The beam was supporting a brick wall which collapsed.

given fire load will remain substantially unaffected by the fire, but it is usually less than most people estimate.

The distortion of the steel due to fire may alter the structure of the building so that it will not be able to carry the superimposed loads or unexpected stresses, particularly torsional, that may be created due to buckling. However, structural steel that has been exposed to fire and water will not have significantly different properties than steel unaffected by exposure to fire. Unfortunately, the erroneous impression that it can be hazardous to put water on unprotected steel continues to be found in the fire service. This mistaken belief can lead to unnecessarily ineffective fire fighting.

Steel Highway Structures

Overpasses on highways provide a good example of unprotected steel which is vulnerable to the occasional gasoline truck fire which does enormous damage, possibly accompanied by huge consequential losses to those whose vital traffic is interrupted. For at least some interchanges, an analysis might prove that protection is justified, but any effective action is most unlikely.

Fire departments should preplan for water supply and anticipate a flammable liquid fire. Water supply must be carefully thought out. If water is used to cool steel and mixes with gasoline, it may provide a flowing fire. Will the fire flow harmlessly, or will it flow into nearby buildings? In the latter case perhaps, the use of water is not indicated due to the greater loss possibility. *

A bridge on the Trans Canada Highway in Nova Scotia was melted when

---

*Over 50 years ago, when prefire planning was literally unknown, the New York Fire Department developed very specific preplans for fires on the East River bridges, recognizing the vulnerability of the unprotected steel. At a time when a daytime phone alarm from an orphanage would draw only an engine and truck, the bridge assignments of several units were required to be filled out, if assigned units were unavailable.

struck by a fuel truck that exploded. The bridge was out of service for 69 days and the repair cost $350,000.

A Pittsburgh bridge passed over a plumbing warehouse. A 10-alarm fire in the warehouse severely damaged the bridge; estimates were that it will be out of service for at least a year. The total loss may reach $10,000,000 taking into account the losses suffered by the 18,000 who had used the bridge daily.

The fire plans for buildings which expose highway structures should take into account protection of the structure. It could be the case, where life safety is not a factor, that the highway structure is far more valuable than the building, and cooling the steel might be more important than extinguishing the fire. Tactics not normally acceptable, such as driving the fire down into the building, may be necessary.

Calculated Risk

The lack of fire protection for unprotected steel may be the result of a "calculated risk." We must always closely examine the means by which this "darling" of management is arrived at. Perhaps the calculations went like this:

> The fire load in this noncombustible building is minimal. Unpackaged noncombustible heavy metal sections are stored on pallets. The combustible load averages about 2 pounds per square foot. There is no apparent source of ignition. No special protection is recommended. Fine.

On the other hand, suppose the calculation went like this:

> This noncombustible building is of low value and is slated for demolition in the 5-year plan. In the meantime, it is used for the reserve storage of foamed plastic shipping containers. The value is low and the loss of the containers would not hamper production. The cost of any special protection could not be recovered during the life of the building.

Fine for the auditor: "in case of fire — let it go." Since that is what the owner thinks, the fire officer should be guided accordingly. A high fire load, in an unprotected steel building, particularly of truss roof construction, can bring the building down unbelievably fast. In such a building, there is often literally no time for routine "inside-attack-first" tactics. Any serious fire in such a building should be fought with the safety of men uppermost.

Automobile parking garages are an example of an engineering calculation which demonstrated that protection of steel is not necessary.

Steel and reinforced concrete compete directly. The necessity of protection of steel is an economic disadvantage particularly since it is a specific item in the budget. The "fireproofing" of concrete is buried in the overall cost.

If steel parking garages were required to be protected in accordance with building codes, they would be at a disadvantage relative to concrete. The

American Iron and Steel Institute sponsored tests which demonstrated that an automobile fire has no serious effect on the steel of a parking garage.[8] As a result, most codes permit unprotected steel if half of the garage is open to the air. (This permits the escape of heat.) This engineering calculation could not be applied to the parking garage noted in a Texas city in which the floors are supported on bar joists, rather than the heavy steel tested.

An engineering calculation, which would have to be marked "failed," was made for an "air rights" parking garage in Maryland. Since the steel garage is built over a railroad track which carries flammable liquid tank cars, it was decided rightly that the steel columns of the garage should be protected. This was done literally — the columns and steel directly at the track were protected. If the anticipated fire occurs, the fire will roll out beyond the protected elements and destroy the building. The same sort of poor engineering was seen on a Canadian car-passenger ferry. What had been passenger cabins was converted to car space. Overhead beams are protected to the extent that they are under enclosed space above. The part under the weather deck is not protected. It is rather like camping out in winter and piling all the blankets on one leg. It will not work.

In considering "calculated risks," keep in mind the computer programmers' "GIGO" or "garbage in — garbage out." No calculation is any better than the information used.

"Calculated risk" is often a synonym for "forget it."

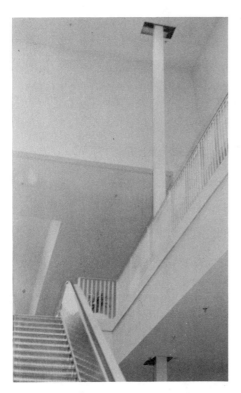

These unprotected steel columns, prepainted to match the walls, were being slipped into a fire resistive building to support a failing roof. An alert fire fighter spotted them. This sort of thing can compromise the integrity of an otherwise sound fire resisting structure.

Protecting Steel from Fire, by Design

Steel can be protected from fire in one of several ways:

- Automatic sprinkler protection is provided with the dual function of suppressing the fire and cooling the steel.

- Automatic sprinkler protection is provided to cool the steel (of a roof structure), while high expansion foam is provided to suppress the fire.

- The steel is "fireproofed" by one of several systems to provide protection, thus qualifying for a fire resistance rating if the particular system of protection selected had previously passed a standard fire test (NFPA 251, *Standard Methods of Fire Tests of Building Construction and Materials*, ASTM E 119).*

- The steel is located out of range of the heat that the fire will develop.

Automatic Sprinkler Protection

Automatic sprinklers can provide protection to steel unprotected by "fireproofing," provided that the water supply is adequate in quantity and distribution. Otherwise, it is quite possible that the early failure of the steel structure will cause sprinkler piping to break, resulting in loss of the sprinkler system to the fire department. If this is a possibility, the fire department should have the best possible knowledge of the entire sprinkler system, particularly the exact areas served by each valve, so that only the valve(s) supplying the disabled system(s) will be shut down.

Sprinkler piping is attached to steel buildings following the practice developed in heavy timber buildings. If the building fails, the piping can be disrupted. Piping is often located above the ceiling with main laterals hanging from trusses. The void space can be quite roomy and is a tempting space to store lightweight material, as for instance, the whole stock of plastic plates and cups over the winter. In addition, if the roof is a combustible metal deck roof, as is probable in the usual mercantile occupancy, a roof fire can destroy the trusses and drop the sprinkler system. The sprinklers operating below the ceiling will not affect the roof fire.

Fire loads that produce rapid, intense fires, such as high-piled combustible

---

*It is now well accepted that there is no such thing as a "fireproof" building, but the term has survived as the designation of the systems by which steel is protected. A subcontractor may contract to "fireproof" the steel: the material is known as "fireproofing." Steel which has been fireproofed is known as "protected steel"; however, with the usual perversity of the building industry with respect to language, sheet steel which has been coated with a weather protective coating is also known as "protected."

pallets, high-piled rolled paper, merchandise packed in combustible plastic (this includes almost everything), and rubber tires (these are often stored on racks right up to the steel overhead), can produce temperatures in the steel structure which will cause sufficient distortion to rupture sprinkler piping in a few minutes.

Standards for the protection of structural steel by sprinklers over high-heat-release fire loads have been prepared by the Factory Mutual System. They require one sprinkler for each 50 square feet of space (a most unusual sprinkler density) and a minimum water pressure of 25 pounds per square inch on the most remote sprinkler, thus providing a discharge density of 0.6 gallons of water per square foot per minute. Directly applied fireproofing, sufficient to develop 1-hour fire resistance, may be necessary in addition to the sprinklers, if the sprinkler system is to survive long enough to suppress the fire. These are based on actual measurements made at Factory Mutual Research Center.[9]

Do not be lulled into a false sense of security by the mere fact that an unprotected steel building is sprinklered. Compare the sprinkler system as installed with the current standard, NFPA 13, *Installation of Sprinkler Systems*. Study the fire load. Examine the possibility of rapid fire spread over the ceiling. Note well any local concentrations of fast-burning fuel. Look for places where the sprinklers might be ineffective because of obstructions to their discharge or because fire in a particular type of material may be difficult to extinguish (rubber tires are a good example). Check on the volume and pressure of water available. Make plans to supply the sprinklers immediately with water at the highest pressure the system will stand to provide the maximum water discharge density from sprinklers. Supplement the sprinkler protection for the steel with hose lines from pumpers at hydrants which have been preselected so as not to rob the sprinklers. Stand clear of any possible collapse zone.

The fire tests conducted at Underwriters Laboratories Inc. after the McCormick Place fire,[10] discussed earlier in this chapter, showed that sprinklers can be effective in protecting exposed steel from the type of fast-burning fire which can be expected in exhibition hall contents. Sprinkler discharges of 0.20 gallons per minute per square foot were adequate. The successful results were based specifically upon combustible loadings of 15 to 20 pounds per square foot, and the report of the tests cautions against uncritically applying the successful results of these tests to materials of faster burning rate than those tested. A caution can be added against applying the results to buildings in which the fire loads, in pounds of combustibles per square foot, are substantially higher.

Denver's new exhibition hall was built to take advantage of the provision allowing unprotected steel above a certain height. The completed building, which has a 640-foot by 240-foot clear span made possible by unprotected steel space frames, has been described as a "modern Crystal Palace." We can only hope that the author of that term is not a prophet. The original Crystal Palace in England was destroyed by fire. The American Crystal Palace, erected in New York in 1853, was also of steel and glass construction. It was destroyed in a spectacular fire in 1858. As a result of dogged effort on the part of the Denver Fire Department, the city's exhibition hall is now fully sprinklered.

Aircraft hangars are usually built of unprotected steel and are equipped with deluge sprinkler systems in the high bays and conventional sprinklers in the shop and office areas. Draft curtains are provided to confine the heat and to prevent sprinklers from operating wastefully far from the fire. (This can be a severe problem if the doors are open.) Proper drains should be provided to avoid a floating gasoline fire. The fire department should be trained to manipulate the valves properly, so that the system can do its job of cooling the steel.

Lightweight steel sections make up the space frames which provide the columns and roof of the Denver Exhibit Hall. Due to strong fire department efforts, the hall is sprinklered.

Some aircraft hangars are provided with fog-foam systems to protect the steel and also suppress flammable liquid fire on the floor. Fire fighting operations should be well thought out to prevent hose streams from washing away the foam.

Fully enclosed rooms, or "buildings within the building," present a special problem. It is often assumed that the plywood or other combustible roof will burn away in the event of a fire, and thus, there is no need to sprinkler within the structure. This is not so. Repeated demonstrations conducted at the Atomic Energy Commission Fire Loss Management Courses at the Naval Structural Fire School in Norfolk, VA, demonstrated that even the thinnest plywood would resist burning through, while the water from sprinklers played ineffectively on the outside. There are plastic panels, listed by Underwriters Laboratories and approved by Factory Mutual System, which can be installed in the ceiling of a room within a building to melt out when exposed to fire so the sprinklers can operate unimpeded. *

---

*These panels are quite light in weight. At one location, the opening and closing of doors caused the panels to "float" out of position. A "practical" handyman solved the problem; he laid plywood atop all the panels.

High Expansion Foam in Combination With Sprinklers

A recent, most promising approach to the protection of high-piled stock in un-protected steel or combustible buildings is the combination of high expansion foam and automatic sprinklers.

The Industrial Risk Insurers (formerly Factory Insurance Association), a leading insurer of well-protected industrial plants, asserts that fires in material such as rolled paper, foamed plastic, foamed rubber, or tires or similar fast-burning high-heat-release material can present serious danger of collapse in as short a time as 5 or 6 minutes in a non-fire resistive or combustible-roofed building.

One system described combines sprinkler protection for the roof with massive high expansion foam installations, so that the entire stock in the building is submerged in foam in a few minutes. The foam system, in one installation, consists of six foam generators, each capable of generating 54,600 cubic feet of foam per minute. When the stock is submerged in foam, five of the six units shut off automatically, and the sixth operates to maintain the foam level.[11] Such monster units are not to be compared with the "shaving lather dispensers" with which some fire departments attempt to use high expansion foam.

If the fire department is to operate efficiently where a combination sprinkler-high expansion system is installed, there must be a complete understanding of the system's operation and capability on the part of the fire ground commander.

Fireproofing of Steel

"Fireproofing" of steel is accomplished by several methods:

- The steel is encased in a structure. Concrete, terra cotta, metal lath and plaster, brick, and gypsum board are representative of materials used. This method is called "encasement."

- Fireproofing is applied directly to the steel, usually by the spray method. Asbestos fiber and an intumescent coating (it swells and chars when exposed to flame) containing noncombustible fibers are two materials used. This method is called "direct application." (When the encasement and direct application methods are used, the members are said to be individually protected.)

- A suspended ceiling is used as part of a fire resistive floor (or roof) and ceiling assembly. This is often called "membrane fireproofing."*

---

*Individual protection (or "fireproofing") and membrane protection (also "fireproofing") are antonyms.

- A newly developed method, first used on the United States Steel Building in Pittsburgh, provides protection for exterior box columns by filling them with water.

Fireproofing is applied to meet the standards required by the local building code. If, for instance, the code requires 3-hour fire resistance on columns, the designer will select a system which has been approved by the local building department. As a supplement to the building code, the building department will indicate which systems tested at which laboratories are acceptable.

The building department may require modifications. Recall from Chapter 6, Principles of Fire Resistance, the procedure that a manufacturer develops a fire resistance system which is tested at a laboratory to determine whether or not it meets the standard. To shave costs, for instance, some screws may be 1½ inches while others need be only 1¼ inches. The building commissioner of a major city refused to accept such precise details, arguing that they were impossible to enforce and required 1½-inch screws throughout.

The efficiency of fireproofing depends entirely on the competence and integrity of the building department in inspecting the installation and on the fire department in seeing that the integrity is not compromised. This is particularly important in the case of spray-on fireproofing and membrane-floor and ceiling assemblies.

A high-rise hotel, which was involved in a large loss of life fire, was found, when it was built some years before, to have very poor fire protection of the structural steel. It was so poor that it was initially downrated to "unprotected steel" by the insurance engineer. This raised the insurance premiums by a very large sum annually. The insurance organization later reversed itself.

Encasement Method. Terra cotta tile was used early for encasement of steel. One of the errors made was to leave the bottom web of beams unprotected. This error was corrected by the development of "skewback" tiles, which were shaped to fit around the steel. Typical of what can happen to such fireproofing was the removal of the terra cotta from the bottom of a beam in an early 20th-century building of the old Bureau of Standards, then used by the Washington Technical Institute. The purpose was to mount a folding door to divide a space into two classrooms. Such an alteration in a building with a high fire load could lead to a serious collapse. Fireproofing that is easily removed is forever at the mercy of the person making alterations who has no knowledge at all of why this "useless material" is "in the way." In altering a Washington, DC department store for the construction of a subway entrance, the contractor removed the protection from a major column. About a hundred cylinders of propane gas were stored adjacent to the column. Look in on "rehabilitation" work on buildings; you may pick up vital information.

Concrete became quite popular as a protective covering for steel, particularly where concrete floors were being used because of the simplification of work. The wood falsework required for placing concrete, however, provides a high fuel load and has been involved in a number of serious construction fires.

Concrete has the advantage of being the most permanent fireproofing, being difficult to remove, though it is not unusual to see the concrete knocked off beams and columns by a motor truck.

The disadvantage of concrete is its weight. In the ceaseless effort to reduce the deadweight of a building, and thus its cost, the fireproofing is a tempting target. Encasement systems of gypsum board, or wire lath and plaster were developed to save weight.[12]

Directly Applied Fireproofing. The search for lighter fireproofing led to sprayed-on cement coatings. Sprayed concrete is called "Gunite," another trade name that is increasingly used as a generic term. It can spall badly when exposed to fire.[13] Many combinations of asbestos fibers and other materials have been developed. While some of these materials can pass laboratory tests, some serious questions are raised about their reliability in the field.

Writing in *Fire Journal* for May 1965, Robert Levy, superintendent of the Bureau of Building Inspection for San Francisco, pointed out that when sprayed-on assemblies are tested, the steel is cleansed with carbon tetrachloride or some other solvent. In the field this is impracticable. Being sure that the material is applied at the specified thickness and density is another problem. [14]

The sprayed-on material is easily knocked off by other trades intent on doing their own work. The author was taken by an insurance executive to the 30th floor of an almost completed San Francisco skyscraper. It was found that the fireproofing had been stripped from all the columns by the plasterers.

There was a serious fire in the Time and Life Building, a modern high-rise office building in New York. The steel members of this building were protected by sprayed-on fibers, but it was observed that they were washed off in several places by hose streams.[15] The loss of fireproofing either prior to or during the fire can have serious consequences.*

Asbestos fiber fireproofing is under fire from another angle entirely; there is a serious health hazard seen in its use. Health regulations have practically eliminated the use of asbestos fireproofing. Newspaper accounts report the removal of asbestos fireproofing from buildings. The accounts are silent as to what was done to replace the asbestos. There are a number of compounds now available which do not contain asbestos.

Another type of sprayed-on fireproofing makes use of the intumescent principle developed for surface coatings to reduce flame spread. Binders are incorporated into the intumescent material, and when it is sprayed onto a steel member it dries to a hard tough coating about $\frac{3}{16}$ inches thick. When heated to 300°F, it swells up to a cellular mass 2 to 3 inches thick. A $\frac{3}{16}$-inch coating provides a 1-hour fire resistance on a steel column.

---

*One of the author's fire science students reported finding sprayed fireproofing generously applied to all exposed steel surfaces in a building including the sprinkler system, sprinklers and all.

Membrane Fireproofing.   Membrane fireproofing is accomplished by the use of a suspended ceiling that provides a "membrane" under the entire area to be protected. The membrane may be of plaster on wire lath, but in recent years, the use of a suspended acoustical tile ceiling as the membrane has become popular. The steel "protected" by the membrane is left bare.

In rating such a fire protection system, Underwriters Laboratories Inc. tests and lists a "floor and ceiling assembly" or a "roof and ceiling assembly" (further discussion will refer to floors only; for convenience, roof may be substituted wherever appropriate). The whole assembly, i.e., the ceiling, hangers, electrical fixtures, floor joists, left-in-place formwork for the concrete floor (usually corrugated steel), air ducts and diffusers, and the concrete floor,* are tested as one unit as it would be built. Therefore, it is incorrect to speak of a "fire-rated ceiling"; the ceiling is but one part of the total assembly. Such floors are sometimes referred to as "Q" floors.

No present test procedure evaluates the protection of columns passing through the void space between the ceiling and the floor above. The UL *Building Materials List* states: "All ratings are based on the assumption that the stability of members supporting the floor or roof is not impaired by the effect of fire on the supports."[16]

To the author's knowledge, no test procedure evaluates a floor and ceiling assembly as a possible path for horizontal spread of fire to other occupancies on the same floor.

It is the author's opinion that the use of suspended ceiling or membrane fireproofing for the protection of floors, and more recently of columns passing through the plenum space above the ceiling, represents a serious menace to the safety of fire fighters who are, in the main, completely unaware of the dangers. The considerations which lead to this opinion are:

- Long years of observation indicate that building tradesmen sometimes are prone to disregard detailed instructions particularly when they do not understand them or find them inconvenient. (One of the author's fire science students, a fire inspector, found aluminum wire being used to support the ceiling channels. "What's the matter with aluminum?" asked the mechanic.)

- Laboratory listing is contingent upon the field assembly being accomplished exactly as performed in the laboratory. (Where is the building inspector with sufficient expertise and time to follow and examine every item?)

- There is no guarantee of a supply of replacement units such as light fixtures identical to those tested (probably unnecessary as it is most unlikely they would be called for).

---

*Steel channels may be inserted into the concrete floor for electrical distribution.

- The ceiling system is at the mercy of every owner, operator, tenant, tradesman, and mechanic who has "reason" to remove tiles. Access to utilities and additional storage space for items such as rugs or ladders are only two of the reasons. Where the plenum space is part of the air-conditioning system, smart employees soon learn that displacement of a tile improves airflow at their location.

- There are no legal provisions, in any code of which the author is aware, requiring that the membrane protection be maintained. Even if there were, the difficulties of enforcement are insuperable. Replacement acoustical tile, which appears on the surface to be identical to the original tile, may be variously combustible, rated for fire hazard only, rated as part of a fire resistance system but not the system installed in the building in question, or may be the proper tile, the only tile which meets the specification of the listing. All electrical fixtures, air duct openings, and other penetrations of the ceiling must be rated as part of the ceiling system. What procedure ensures or even requires that original listed units be replaced with the same units?*

  The term "fire-rated," when applied to tiles, is often confused with tiles which are a part of a fire resistance system and tiles which merely meet flame spread requirements. A steel bar joist floor with concrete topping with flame-spread-rated tiles below may appear to be fire resistive. In fact, it is only noncombustible construction.

- Owners, managers, and occupants, even of buildings where fire protection equipment is well maintained, are usually totally unaware of the significance of absolute integrity of the ceiling system.**

- The laboratory fire tests are necessarily conducted under a slight negative pressure to remove smoke and fumes. Fire generates positive pressure, and lay-in ceiling tiles may be easily displaced by fire pressures. When a tile is displaced by a mechanic, it is, by and large, just dropped back down into place without any restraint to upward motion.

---

*The fact that tiles are obviously missing does not necessarily mean that a fire resistance system has been violated. The building may not be required or intended to be fire resistive. It may be simply of noncombustible construction. In such a case, ceiling tiles are at the option of the owner and need merely meet flame spread requirements, if any. The building may be of concrete construction and the ceiling again is at the option of the owner.

**A common practice is to remove tiles from the storeroom (who needs a fancy ceiling in the storeroom?) to replace tiles in public locations. Tiles are damaged by water leaks. Tiles are removed by tradesmen. Holes are cut through tiles. Displays are hung from the metal grid (floor-ceiling systems are not tested with superimposed loads). The list of ways in which such a floor-ceiling assembly can be rendered useless as a fire resistance system is endless.

- The addition of insulation, particularly in a roof and ceiling assembly, can significantly affect the fire performance of the assembly. The insulation will cause heat to be retained in the channels supporting the tiles, causing them to fail earlier than expected.

- The tests do not embrace the question of extension of fire horizontally through the fire floor. In effect there is a cockloft between the ceiling and floor. In one incident of which the author has knowledge, fire, starting in one room, traveled across a hallway above the ceiling and came down through joints in the tile ceiling of another room to ignite books on a top bookshelf. Only alert fire fighting prevented the full involvement of the floor. Some code provisions are made for firestopping, but the use of the plenum space for various services makes it probable that the firestopping will conform to our definition of "legal firestopping" (see Chapter 3, Wood Construction).

- In some buildings, the use of deep, long-span trusses to provide clear floor areas creates plenum spaces several feet in height, high enough for a man to walk upright. "Waste space" is anathema to any conscientious building manager. What is more logical than to provide access to this space and use it for storage?* Whether the membrane fireproofing is provided by suspended tile or plaster on metal lath is immaterial, the fire load has been placed adjacent to vulnerable unprotected steel, in a location in which fire fighting will be almost impossible.

- Most important of all, is the fact that some codes now permit the omission of fire protection for columns passing through the plenum space, in the face of the clear warning contained in *Fire Protection Through Modern Building Codes* that protective materials should extend to the full-column height.[17] There should be no discontinuities, i.e., either by cut-offs at the ceiling lines, or at other constructions that may abut the column. A report of laboratory tests justifying the omission of column fireproofing in the plenum space carries this vital caution qualifying the protection of a column by a floor-ceiling assembly, "the column supports only a *single* floor or roof."[18] It is very probable that the removal of one tile in a high fire load area, near an unprotected column, is the most serious hazard, since this would concentrate the heat output of the fire at a *single* vulnerable point.

---

*Engineers closely associated with hospital construction call this void "interstitial space," borrowing the term from anatomy.

The membrane ceiling is supposed to protect the bare column in the plenum space. The ceiling of this storeroom was robbed to provide replacement tiles elsewhere. A high fire load and an unprotected steel column can be equal to disaster.

In some cases, the column protection in the plenum consists of gypsum board assemblies. True to the widespread erroneous belief that the taping of joints and nail-setting are merely cosmetic, the nail heads and joints are left untaped. Exposed to fire, such incomplete assemblies will fail rapidly.

It is not sufficient to say that there has been no bad experience with rated ceilings. The potential for serious problems exists. Contrary to the maxim, experience is not the best teacher, but it can certainly be the most expensive. There are dangers, and the fire department must be prepared to present its point of view to code authorities, train its own personnel, and properly educate the owners and occupants of membrane fireproofed buildings.

There is a further discussion of this subject later in this chapter in the section on prefire planning.

This column of a building under construction is "fireproofed" where it passes through the plenum space above the membrane ceiling; however, a common error was made. Closer examination by the author revealed that the nails were not set nor the corners (arrow) covered properly. These are not cosmetic; they are part of the fire resistance system.

Fireproofing with Water.   We have long "fireproofed" our internal combustion engines with water. The heat of combustion is removed from the cylinder wall by water, thus preventing distortion. The water releases its heat to the atmosphere and recycles to the engine block.

The United States Steel Company wanted to make its new headquarters in Pittsburgh a showcase for steel, particularly its "Corten" alloy. Corten weathers to a dark brown color, the color provided by an iron oxide coating, which, unlike

Cutting this opening for a conveyor destroyed the membrane protection provided by the floor and ceiling assembly.

ordinary rust, does not penetrate beyond the surface. The material does not need to be painted. The advantages of such an alloy are lost if it must be buried under fireproofing.

The columns are hollow boxes. Water with antifreeze added has been placed in the columns. Columns are cross connected with piping so that heated water can circulate by gravity. Though not formally rated, it is probably quite valid to consider these columns as having almost infinite fire resistance. The author has been told that the builder felt that the location of the columns on the exterior by itself was sufficient to assure that no fire would cause them to reach critical temperatures. The building department, however, insisted on the water. Since then, several other buildings have been designed similarly, and it will be interesting to observe whether a new fire resistance technique with wide application has evolved more or less by accident.

The structural supports of tanks containing flammable liquids present serious problems of possible collapse during a fire due to run-off water carrying floating fire. It would be simple to design such supports as piping systems and provide for water circulation inside the pipes. The interface of water and flammable liquid would thus be avoided.

Steel Out of Range of Heat

The architectural interest in exposed steel elements and the constant search for methods to reduce weight and cost led to design concepts in which important steel elements are located out of reach of the fire.

A new 54-story building has been built in New York for United States Steel. Major parts of the exterior will be formed by 70-inch deep exposed steel spandrel beams. The beams are fireproofed on the interior surface and 14-gage steel covers are provided for the top and bottom flanges for flame shields and architectural cladding. The exterior web is left exposed.

A furnace was especially constructed at Underwriters Laboratories Inc. to submit an assembly of this type to the standard fire test as defined in NFPA 251, *Standard Methods of Fire Tests of Building Construction and Material* (ASTM E 119). The maximum temperature measured in the exposed web was 640°F.

The floors of the Bank of Ireland in Dublin are suspended on steel cables from beams cantilevered out from the top of the central core. An analysis made by a prominent British fire researcher showed that they would not be overheated by a credible fire.

The cables suspending a dormitory at a university in the northwest might be another story. They are right against the building and it would appear that a credible room fire might locally overheat a cable with possibly serious consequences. Cold drawn steel cables at best can resist temperatures up to only 800°F. It would be necessary to heat only a small section of cable to induce failure.

The Insulated Metal Deck Roof Fire Problem *

In August 1953, the General Motors Corporation transmission plant at Livonia, MI burned. The loss of $32,000,000 was the largest industrial fire loss to that date. The metal deck roof was the principal contributing factor to the destruction of the plant.

Metal deck roofs consist of metal sheets laid over steel joists, spaced quite closely. The joints are crimped together making a seal which is not gas-tight. Usually insulation is added, often a low-density fiberboard, though plastics are making inroads. Insulation which has absorbed moisture is useless, so it must be protected from moisture driven through the roof by capillary attraction, as a result of heating the building for comfort. An adhesive is necessary also to secure the insulation to the roof deck and to prevent loss of the roof covering in a wind-

---

*The concept of an "approved roof," the generally accepted term for roofing which meets UL standards for roofing that is resistant to the propagation of fire from building to building, by creating or receiving flying brands, is not pertinent to this discussion. See "Roof Covering Materials" of the current UL *Building Materials List* for a description of the factors which enter into a listing of a roofing material.

storm. A bituminous coating serves as the adhesive and sometimes as a moisture-stopping vapor barrier. On top of the insulation go successive layers of bituminous material and roofing felt. From the successive layers grows the term, "built-up" roof.

When a fire occurs below this type of roof, the metal deck heats up. The heat is conducted through the deck to the bituminous adhesive and the adhesive liquefies and then vaporizes. The gas cannot escape through the roofing; instead, it forces itself down through the joints in the deck. When the gas mixes with the air below, it burns after pilot ignition by the fire below. This gas fire, which the author has observed to be as much as 6 feet deep, rolls along under the roof, heating additional roof areas which generate more gaseous fuel; thus, the fire is self-sustaining and independent of the original fire. The roofing insulation and some of the roofing felt may burn, showing fire on top of the roof, but this is of little consequence; the problem is at the underside of the roof.

In 1956, the Atomic Energy Commission plant at Paducah, KY, suffered a $2,100,000 fire loss in a metal deck roof fire which was a classic example. The fire started in an explosion in process equipment, and in less than 10 minutes the fire had spread to the roof. The insulation was glass fiber so essentially only the vapor seal was involved, and there was practically no combustible material in the plant, so the burning tar which fell from the roof found no fuel to feed upon. The 70,000-square-foot roof was completely destroyed and as it collapsed, it pulled down the walls. The building was destroyed by a 1-hour fire, but the equipment it housed was back in operation within 24 hours except for the unit where the fire started, proof that the combustible adhesive-vapor barrier was the principal culprit.

Faced with the possibility of alarming losses in industrial buildings they insured, the Factory Mutual Insurance Companies conducted extensive experiments on metal deck roofs. From these experiments came two FM designations for metal deck roofs. They are: Class I metal deck roof construction that is noncombustible or has combustibility reduced to an acceptable degree and Class II construction encompassing all other typical roof deck constructions.[*][19] Even Class I roof construction, if combustible to a degree, may not always be acceptable. For example, a fire on February 7, 1966, in a large concentrator building at the Wabush Mines complex in Labrador, badly damaged a metal deck roof of Class II construction 90 feet above the floor. Investigators theorized that, in this particular instance, if the roof had been of Class I construction, damage still would have been severe.[20] Failure of unprotected steel supports undoubtedly would have let the roof sag, exposing to the fire asphaltic materials in the built-up covering over the roof insulation.

Tests also have conclusively shown that well-supplied automatic sprinklers will

---

[*]The term "Class I" may be taken to indicate "completely satisfactory under all circumstances" by persons not fully familiar with this subject. If a Class I roof has any combustibles at all it may not be satisfactory for use over a high-value installation, such as a computer area, and total noncombustibility should be sought. Intumescent coatings can be used to convert a Class II roof to a Class I roof.

control extension of fire under metal deck roofs. After the AEC plant fire in Paducah reported above, the AEC (which had upgraded its standards for new roofs only days before the fire, based on an evaluation of the General Motors fire) studied the problem of protection for literally thousands of acres of metal deck roofs covering some of the most vital production plants in the country. Many proposals were examined. The final decision was to sprinkler the plants. Over $20,000,000 was spent on sprinklers and water supply. When a similar fire to the first Paducah fire occurred in a sprinklered plant at Paducah, the fire was confined to the area of origin. Over 2,000 sprinklers opened, causing a water flow of millions of gallons.*

A bowling alley in a Maryland community was destroyed by fire. The fire started in the concession stand and extended upward to the metal deck roof. The roof burned very rapidly and collapsed. There was essentially no fire on the bowling alley floor level except for the concession stand.

Careful reading of fire reports indicates that the metal deck roof is a factor in many losses where its significance is not understood by the fire suppression force. Later in this chapter, there is a discussion on fighting metal deck roof fires. Suffice it to say here that it is impossible to fight a metal deck roof fire except by getting water onto the underside of the roof deck.

In May 1977, the Beverly Hills Supper Club in Kentucky burned with the loss of 165 lives. The author has had occasion to make an extensive study of the circumstances of that fire and is convinced that a most significant factor in the huge loss of life was the tremendous amount of black, choking smoke developed by the fire in the metal deck roof. It is true that the roof was cut up by parapeted walls (the building had been added to many times, but undoubtedly there were penetrations through the walls). A tongue of flame passing through a small hole, impinging on the metal deck of the next section would be sufficient to extend the fire. (Factory Mutual tests on metal deck roof assemblies have shown that it takes only 800°F for 5 minutes for heat impinging on the surface of steel decking to start a self-sustaining roof fire. This fire is independent of the original fire.)

It is not at all clear that if the supper club building had been sprinklered according to the usual practice that the loss of life would have been substantially reduced. Under typical practice, the void would be regarded as "noncombustible" and left unsprinklered. Since it was above the ceiling insulation, a separate dry-pipe system would have been required to prevent freezing. Of course, had code requirements for exits and fire resistance been met, the loss of any life would have been doubtful.

It appears that in most jurisdictions, an ordinary combustible metal deck roof would be permitted on a "noncombustible" building. Sprinklers below the ceiling in such a building will not protect against a metal deck roof fire moving through the void. However, they would most likely prevent a contents fire from reaching the metal deck roof.

---

*This fire and others led many to the conclusion that sprinklers in such plants should be of higher temperature ratings, to prevent them from opening far from the fire scene.

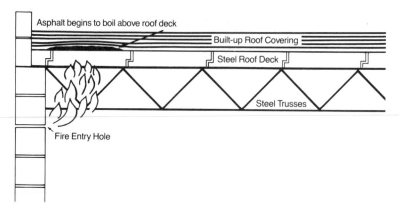

Fire under a small part of a steel roof deck heats the deck to the melting point of asphalt, and the asphalt layer in the built-up roof covering above the deck begins to boil. (Adapted from sketches by Chris Brannigan.)

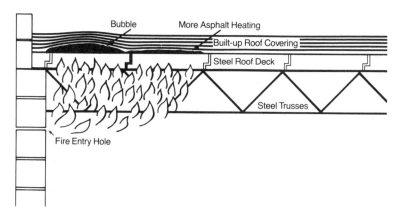

Pressure from boiling asphalt forms a trapped bubble and forces volatile gases down through a joint in the roof deck and the gases ignite. The spreading fire, which includes asphalt dripping through the opened joint, heats up more asphalt above the deck. The gas bubble above the roof deck continues to spread forcing more gas and hot asphalt down through adjoining joints in the deck. This cycle repeats itself until the whole roof can become involved. (Adapted from sketches by Chris Brannigan.)

High-Piled Stock

In setting the thesis for this book, the subject of "contents" is set aside as not germane to a discussion of buildings; but note must be taken of a new type of building, the one-story warehouse, 100 feet or more in height, in which the contents are stored on unprotected steel racks. Typically, such buildings are of noncombustible construction. Most likely the warehouse is fully automated. Computers keep stock records, and a signal from an operator at a keyboard can send the traveling elevator to the correct location. The elevator platform will then ascend to the proper level, and the desired refrigerator, or pallets of radios in polystyrene packaging, or automobile parts or whatever, is pushed onto the elevator. In some cases, the stock moves automatically all the way to the truck which will carry it to its destination.

A rack storage warehouse represents, in effect, a multistoried building without the resistance to the spread of fire provided by even the poorest floor. Sprinklers located at the roof line only are completely inadequate to control such a hazard; sprinklers must be provided at intermediate levels as well. Water supplies must be strong, reliable, and hydraulically calculated to provide adequate quantity and distribution. Massive high expansion foam systems show much promise in protecting these huge risks, but the foam production rate must be adequate to fill the building in a very few minutes, and provision must be made to confine the foam within the building.

High rack storage demands special protection.

Projected rack storage buildings are upwards of 2,000,000 square feet in area. Delivery of a single 2½-inch stream within the building may require relay pumping. Sprinkler systems should be backed up to the utmost. Hydrants used for pumpers supplying handlines should be preselected to avoid "stealing" water from the sprinklers. Handline operations, even inside, may be futile, and master stream appliances may be necessary.

An estimated $14,000,000 loss occurred in a rack storage warehouse in Kernersville, NC, in March 1981. Sprinklers were not provided in the racks in a section 15 feet, 9 inches high. Though the fire had been burning only 10 minutes when the fire department arrived, an interior attack was impossible. About 15 minutes later, the roof started to collapse.

There was no fire department connection to the sprinkler system. (NFPA 13, *Installation of Sprinkler Systems*, makes this connection optional, a serious error in the author's opinion.)

There is an ongoing technical controversy on the validity of venting sprinklered warehouses. One insurer of such properties recommends that the fire department leave the warehouse closed up and let the sprinklers control the fire. This is discussed further in Chapter 10, Smoke and Fire Containment.

Some conditions may be unbelievable. Warehouses are not designed for ventilation, and even where roof vents are provided, they may be totally inadequate to deal with the volume of smoke, particularly from the plastics so common in products or packing material.*

The only hope for successfully combating a fire in a rack storage building rests in the closest cooperation between the owner, the operator (often different from the owner), the insurer, and the fire department, not only of the city involved but also of those who might be called for mutual aid. It is not at all unlikely that the building is situated in more than one governmental jurisdiction. The problem of who is in command should be settled before the fire.

Every possible scrap of important information should be sought out ahead of time. The location and operation of all pertinent features of the water supply should be determined. Make contact with the engineers of the insurance company carrying the risk. A study should be made of fire walls and any deficiencies so that preassignments can be made to guard any weak spots. All the information should be reduced to writing and provision made to have it available to the fire ground commander and the company officers, even if the responding units are not the regular assignment. An employee brigade should be encouraged but should not be permitted in any way to delay calling the fire department immediately. The chief on night duty might swing by the building, chat with the watchman, check the water pressure, note that the rubbish is properly stored away from the building, and otherwise demonstrate that the fire department is concerned.

One argument might be, "That's an awful lot of trouble for one building," but warehouses now in the design stage represent, with contents, possible $100,000,000 losses. A lot of effort would be expended to manage a fire where only a fraction of this sum is at risk.**

The fact that such tremendous risks are developing seems to have caught American industry unaware. Writing in the January 1970 *Fire Journal*, Charles S. Morgan, then NFPA general manager, gave a glimpse of the results of an incomplete survey of industry, which attempted to determine at what level of management the decision to take such risks was made. The results were not reassuring. The possibility of a disastrous fire is covered by the fact that insurance is purchased. The determinable annual cost of insurance makes it easy to factor it into the cost of operation, and thus the problem is disposed of by the "problem

---

*Industrial Risk Insurers (formerly, The Factory Insurance Association) has produced an excellent movie, *High Piled Stock*. It is well worth seeing, particularly to see the effect of the "flues" between the stacks, and to gain an appreciation of the hazards of collapse. The film was made, however, with all the roof vents in the test building fully open. Had the vents not been open it would have been impossible to make the movie due to total smoke obscuration.

**Press reports of a fire in a K Mart warehouse in Bucks County, PA, in June 1982, spoke of a loss of almost $100,000,000 and 300 jobs.

solvers" whose management "genius" is principally financial.

Since the insurance system is in truth a privately managed taxing system that portions the cost of fire mismanagement out to all, the body politic may someday awaken and limit the extent to which all of us contribute to make good huge losses which are totally unnecessary in the light of available technology.

There are standards for high-rise rack storage based on some tests and other limited information. It is a fact that we simply cannot expect the success rate from automatic sprinklers that we have come to expect in simpler structures. Extreme care must be given to the safety of personnel. Never get into a position where stock racks can collapse behind you. Be on the alert for fire flashing over the entire ceiling of the building and prepare to evacuate promptly. These buildings are built to minimize handling costs and maximize profits; fire protection is almost always an afterthought.

In an article in *Fire Journal*,[21] tests on fires in storage of plastic containers are described. The full report should be studied by those with responsibility for this type of storage, but one item is of note. The article points out that it is advantageous for the piles of plastic containers to collapse since this aids in control of the fire. It recommends that if loads are tied or banded, that materials which will burn easily be used to *create instability of the pile*.

A similar hazard exists in many lumber yards or similar occupancies. Metal shelves supported on "columns" of lightweight steel, with many punched-out holes for flexibility and lightness, are used for storage of stock. These are structures within the building and extemely vulnerable to collapse.

Fire Walls

NOTE: The material in this section is principally drawn from *Fire Walls in Modern Industrial Buildings*, published by the Factory Mutual System (undated, approximately 1969).

The structural steel frame provides the stability for an unprotected steel industrial building. If a fire wall encases some of the steel frame, or is fastened to it, the fire wall can be destroyed by a fire which distorts the steel frame.

Steel members which sag due to fire will try to carry their loads as suspension members. This causes large horizontal forces; if these are transmitted to the fire wall, it can be destroyed.

The formula for the horizontal force is:

$$H = \frac{wL^2}{8S}$$

in which H is the horizontal force in pounds, w is the deadweight of a roof strip 1-foot wide and as long as the spacing between trusses or beams, L is the truss or beam span in feet, and S is the sag in feet which can be assumed to be between $0.07\,L$ and $0.09\,L$.

A wall is rarely designed to resist such a horizontal force. The wall anchor may not be strong enough to transmit it, and the anchor may fail.

A basic concept of the fire wall is the freestanding masonry wall as used in large wooden buildings. It is completely independent of the building, unsupported by the framing, and carries none of the load. Such walls are rarely found in steel-framed buildings.

We should be aware of the various types of fire walls which might be found in order to make an assessment of the reliability of the wall in the event of fire:

Tied Fire Walls.   These are walls which are fastened to, and which usually encase, members of the steel building frame. The pull of collapsing steel on one side must be resisted by the strength of the unheated steel on the other side. Should there be heavy fire on both sides of the wall, and not necessarily in the immediate vicinity of the wall, it will almost certainly be in danger of collapse. If there are only a few bays of steel tied to a wall on one side and none on the other, it will have very little strength to resist a horizontal pull from the other side, and is a "one-way wall."

Freestanding Walls.   These walls are not connected to the building at all. The steel frame on each side must be quite close, however, so that the unexposed steel can resist the lateral push on the wall by the heated exposed steel.

Partition Walls.   These walls may not be designed as fire walls but may serve very effectively as defensible barriers, if their deficiencies are recognized.

Combination Walls.   The author recalls observing a one-high-story steel-framed building under construction in which the fire wall parapet of brick was completed, even though there was no wall inside the building below. A steel girder had been placed with an upward camber, and the weight of the parapet, after it was built, flattened out the girder. A concrete block fire wall was then built up to meet the girder. The girder was then protected with directly applied fireproofing. Heavy unprotected steel girders were connected to the "fireproofed" girder. Any elongation of the unprotected steel would displace the girder. Such a wall should be watched for signs of failure of the protection on the steel which might precipitate the collapse of the parapet.

Prefire Planning

It is beneficial to note here that, since this is not a treatise on fire fighting, only such factors of prefire planning that relate to steel are considered. In a discussion of prefire planning, it is difficult to separate strictly fire fighting procedures from the necessity of adequate fire protection requirements for steel structures as well as proper maintenance of the protection as it is provided, whether it is static (fireproofing) or dynamic (sprinklers). The fire department must first develop a competence in the many facets of the overall problem, and while it is impossible

Masonry walls are not always the fire barriers they first appear to be. There are deficiencies here. The openings in the masonry walls will pass fire. The roof is a conventional metal deck roof and will burn independently of the original fire. In all probability, fire will pass over the wall.

for all officers to be equally competent and informed, one person or a small group should be designated to familiarize himself or themselves with the problems. Local building codes, as they are now, as they were in the past, and most importantly, the manner and degree of enforcement, are prime areas of concern to enlightened fire departments. It is a general, if not universal situation, that building officials are less than enthusiastic about fire department interference in "their business." Incredibly, in one metropolitan-type county, fire inspectors are not permitted to inspect a building until the certificate of occupancy is issued. It is doubtful, to say the least, that the air pollution people are so cavalierly excluded.

Where fire fighting plans indicate much water will be needed to protect steel, it might be beneficial to tell this to all who may suffer a loss in the event of fire. The point to present is: "You have created this problem. There are several alternative solutions, but only one of them is up to us; the others are up to you."

Preplanning considerations divide themselves between unprotected and protected steel buildings. The chief consideration in unprotected steel buildings must be the safety of personnel. The possibility of sudden collapse of large areas must govern all actions. The pell-mell rush to get a handline into the building must be replaced with studied action based on the awareness that the building can collapse in as little as 5 minutes, and that the minutes start to tick off when the fire starts, not when the fire department arrives.

A fatal collapse should serve as a grim warning of the dangers of buildings in which clear spans are achieved by lightweight trusses.

Four fire fighters lost their lives in the collapse of a Wichita, KS auto showroom which occurred about 10 minutes after fire fighters arrived. Being an auto showroom with repair shops requiring large open areas, the garage, described as "half-block square," was of lightweight steel truss construction. Fire was visible to units turning out, and a second alarm was transmitted immediately. Obviously, a large amount of heat was being produced, and we can conclude with little doubt that much of the heat was being rapidly absorbed by the steel, causing it to collapse soon after the fire department arrived and stretched a handline into the building.

Note that lightweight steel trusses were involved in the Wichita garage fire. Not only can such trusses heat up rapidly due to their high surface-to-mass ratios, but by the very nature of their design and installation they are susceptible to collapse over a wide area. Every connection is vital to the stability of the truss. The necessity for stability of the roof against wind loads and the inherent instability of long, thin trusses makes it necessary to tie the trusses together. By means of these ties, torsional loads generated as one truss fails can cause structural failure of trusses not affected by the heat.

Before the truss fails by collapse, it may expand against a masonry wall and cause the collapse of the wall.

By constant training, the attention of all personnel should be directed to the paramount necessity of cooling all steel. The quantity of water needed at any point is not excessive. Water supplies for spray systems designed to protect tanks and steel supports from flammable liquid spill fires are calculated on a requirement of 0.25 gallons per minute per square foot. It is most important to cool all the steel that is within reach of hose streams, and to give special attention to columns. The stream should be kept moving; overcooling in one spot obviously means undercooling, possibly disastrously, someplace else. It is important that the stream reach the steel. When a long reach is required, a solid stream tip might be better than a fog tip, even on its most "solid setting." A fog stream may absorb a greater amount of heat in penetrating the flame zone. Converting all the water to steam as it passes through the fire might not be the best use of water, particularly when the fire is uncontrollable.

The use of water in the manner urged here might bring criticism of "unnecessary damage." The damage potential was created by the person who placed the combustible loading in a thermoplastic building. If the building can be saved, it will probably be possible to save some of the contents; if the building collapses, the contents are lost too. In this day of close attention to sources of municipal revenue, it is not amiss to note that it is the building, rarely the contents, on which the tax base rests. To the argument that "we never throw water except when we see fire," the reply can be offered: "If the owner had sprinklered the building, as he should have, water would be discharged on the steel, whether or not the sprinklers can 'see' fire."

A Washington, DC junkyard consisted of several identical unprotected steel buildings joined together without walls. A fire in waste paper exposed two identical steel girders above the fire. One was successfully cooled with a deckpipe. The other, shielded by the first one, heated, overturned, and dumped the roof.

Fighting a Metal Deck Roof Fire

The author first became cognizant of the metal deck roof problem in 1946. A fire occurred in a 250-foot (between fire walls) section of a 1,500-foot by 100-foot warehouse at the Marine Corps Depot of Supplies, Norfolk, VA. The warehouse was loaded with field supplies. The fire was in the center of the section, involving stock on both sides of the center aisle. The author was supervising handlines inside the building on the burning stock. Blue flames extended down six feet or more from the underside of the roof, accompanied by dripping, flaming tar. When the heat and tar became hazardous, we would direct the streams upward, and, as we then thought, "wipe out" the fire. We made a "good stop," cutting the fire off right in the piles of stock 20 feet high. The pictures of the building however, told a different story. The roof was gone from fire wall to fire wall and it appeared as if the fire department had played no effective part in stopping the fire, but that rather the fire walls had done the job. After much discussion and study, we came to the conclusion that the roof had burned over our heads independently of the main body of fire in the contents.

It seemed that a fire involving a metal deck roof could be controlled only by the continuous application of water to the underside of the roof. By cooling the steel, we would prevent the tar from generating gas. Literally, the fire would be extinguished by removing the fuel.

The chance to try the plan came in about 15 months. In a similar warehouse section, the Marines had built a plywood office, sick bay, and theater, while the balance of the area was loaded with combustible stock. Like the first fire, this fire was well advanced upon arrival, and an immediate second alarm was sounded. Looking up along the underside of the roof, from the door at the end opposite the fire, the author observed the characteristic blue waves of fire several feet deep advancing rapidly towards him. A hose wagon with deck pipe was ordered into the building, taking a position in the center aisle opposite the door. The orders were to direct a heavy stream (1½-inch tip) at the underside of the roof, keep it moving, and ignore the burning contents. The operation was successful. The warehouse looked as if a knife had cut through the roof and walls; the building was lost only in the area where the huge body of initial fire had done its work before our arrival.

A prefire plan for a metal deck-roofed building should require taking a safe position where all parts of the underside of the roof could be reached and the use of heavy stream equipment to provide a stream of sufficient reach. Don't hesitate to "throw water into smoke." The target will probably be invisible, but it is there.

A ladder pipe is useless on a fire under a roof deck. The roof is a structure designed to keep out rain, and the fire is always many feet beyond the point at which the roof is opened up. The fire that can be seen on top of the roof is in the fiberboard or plastic insulation or the upper layers of the built-up roofing and is unimportant. Stop the progressive heating of the "frying pan" from underneath and the roof fire will die out or decrease to easily manageable proportions.

It is important to note here that the metal deck roof is not found exclusively on steel-framed buildings. Many are built on masonry buildings but the problem is the same. In December 1970, in Kensington, MD, fire fighters responded to a serious fire in the contents of a one-story building housing a janitor supply business. There was heavy smoke, and the ceiling was not visible through the smoke. One of the fire fighters felt hot tar dripping on his neck. Later he related:

"I immediately thought 'metal deck roof,' and though we couldn't see the overhead, we directed a big line up to the ceiling area to cool everything. When the smoke cleared away we were able to see 'icicles' of chilled tar hanging from the joints in the steel."

The metal deck roof fire is not always readily recognizable, particularly where a ceiling is installed. An unsprinklered shopping mall in Winter Park, FL, was constructed of one-story store sections with metal deck roofs on either side of a high-bay central mall of precast concrete. A fire in a store extended to the metal deck roof. The steel of the lower section elongated and moved the supports of the high-bay section. A large part of the T-beam roof fell, fortunately without causing injury. An alert fire officer pulled the ceiling of a walkway in the one-story section and saw the fire moving under the roof. He placed a monitor nozzle in position and cut off the fire.

The metal deck roof fire cannot be cut off by ventilation. It can move very fast. Tests show that only huge vents would be at all effective. Working on a roof over a metal deck roof fire would be extremely hazardous. An article discussing collapse of wood trusses makes the point that wood trusses collapse while steel trusses sag. This is not necessarily a blessing. The sag may cause fire fighters on the roof to slide downhill on a slick liquid tar surface.

Preplanning Your "McCormick Place"

With the advantage of hindsight, and a study of the McCormick Place fire, consider a possible preplan of a fire in an unprotected steel building being used as an exhibit hall. For our purposes, the building is a typical National Guard armory, i.e., brick-bearing walls, unprotected steel trusses supporting the roof. There is a balcony running down both sides which is used for sports events but not for exhibits. Recommendations for automatic sprinklers have not been carried out, no money is available, the state will not permit the city to "dictate" to it, and finally, most of the time the fire load is low.

The first move might be to set up a bruising battle for either sprinkler protection or denial of the facility to high fire load exhibits. The battle is set up, not to be won, but with the full expectation that it will be lost. In losing, however, the fire department gets the "right" to some concessions. Without being too specific, get in writing an agreement that the fire department can take whatever steps it deems necessary to protect life and property during exhibits. Having beaten back the financial threat, the managers of the armory can only be magnanimous and yield a point to the fire department. When they learn the proposed precautions, they may be quite upset, but the fire department has the written agreement, and in a public battle, it is not the fire department who is "willing to risk the lives of our citizens."

Consider the worst possible case. An exhibition, such as a housewares show, a home show, or a furniture show, is scheduled. Calculations of the combustibility of typical exhibits together with the floor area develop a fire load in the 20 pounds-per-square-foot range; the surface-to-mass ratio of the material and its arrangement will provide a fast fire with early high heat release at a rapidly increasing rate. The same fuel load in a more restricted area might give off heat more slowly due to restricted airflow, but in the exhibit hall we can assume an unlimited supply of oxygen, so the air supply will not be a limiting factor.

The fact that a hot, fast fire is anticipated indicates a more severe test of the structure than a slower fire. The radiant energy of a fire is proportional to the fourth power of the absolute temperature (actual thermometer reading in degrees fahrenheit plus 459); thus, a 2,000°F absolute fire of 20 minutes' duration could have a radiant energy potential 16 times that of a 1,000°F absolute fire of the same duration. Because of the large void above the bottom chord of the trusses, a huge smoke storage area is provided, and for quite some time there might be little smoke to protect the bottom chords of the trusses from radiant heat. If the bottom chord fails, the truss fails.

If the facts are set down truthfully and the logical conclusions drawn, it becomes painfully apparent that instant notification of a fire in an exhibition hall, even by a fire fighter equipped with a walkie-talkie radio, is not fast enough to provide substantial assurance that the fire department can cope with the fire.

Only one plan is left. Get set up ahead of time. Provide a temporary "sprinkler system"; not by piping and sprinklers but by prepositioned hose streams. In a building of ordinary ceiling height, we are accustomed to the operation of one to three sprinklers on the average fire. On a high bay fire, expect the operation of a greater number of sprinklers even under the best conditions. In Underwriters Laboratories Inc. tests, simulating an exhibition hall occupancy, the first sprinkler opened 2 minutes after the start of the fire, and a total of 12 sprinklers opened over an exhibit booth in a section with a 30-foot-high ceiling. Since the sprinklers were on 100-square-foot spacing, an area of 1,200 square feet was wet down to control the booth fire. If the building was sprinklered, accept the fact that the minimum area which would receive water damage in the event of fire would be approximately 1,200 square feet. Since we are attempting to provide an improvised "sprinkler system," we are "entitled" to plan to wet down a similar sized area.

Hose lines would be prepositioned in the balcony feeding fixed nozzle holders. Portable deluge guns could be used, or simple nozzle holders could be fabricated in the fire department shops. The point is that there is a nozzle covering every part of the hall. If there is a standpipe in the hall, it would be used to supply the lines; if not, large-size supply lines would be laid as temporary fire mains and "thiefs" used to take off the supply to the nozzles. A fire department pumper would be hooked up to a hydrant and to the supply lines. Lines would be charged and then pressurized. Of course, good hose, tight fittings, and good gaskets are a must; leaking connections would be intolerable.

One or more fire fighters would be assigned to the fire watch along with a pump operator. If fire breaks out, the fire watch, equipped with a radio, sounds the alarm and orders the operator to provide pressure. He moves to the nozzle which covers the fire area. Depending on local planning, 60 to 90 seconds might be allowed for exhibitors or the maintenance staff to bring the fire under control, or perhaps for it to burn out. If, after the predetermined time, the fire is not obviously under control, water is started into the overhead at a point directly over the fire.

On first glance such a plan appears ridiculous. Men and equipment cannot be tied up to fight a fire before it happens. The precautions would upset the patrons of the show: "We never did anything like that before. Who will pay for the time? We have a fire, we call you, you come. That's the way it's always been."*

The author can recall lecturing to Navy personnel, during World War II, to the effect that "there is no law saying you must have a fire before you open a can of foam. Foam can be used to suppress a flammable liquid hazard before the fire starts." A revolutionary idea then, but now the foaming of runways in anticipation of a crash landing is commonplace.

There is nothing the human intellect resists more than a new idea. Step by step the logic leads inexorably to the only practical solution; prepare for the fire before it happens. The only other solution is to trust luck, and most often this will work. Fire is an improbable occurrence in any event, and chance is on the side of those who do nothing.

If the building is sprinklered, study the probable water supply requirements, particularly against the possibility of the opening of ten to twenty sprinklers. It might well be advisable to make a standby hookup to the sprinkler system siamese connection so that the supply could be augmented immediately upon the start of a fire.

Of course the foregoing depends on the fire department being aware of what is going to occur in the community. The author recalls learning of a camper show which was being held in an armory somewhere in his area. The fire department was called. It knew of no such show when in fact the armory, in its response area, with typical unprotected steel girders, was loaded with highly flammable trailers. There were several hundred people at the show and there was only one narrow door open. Fire fighters should be eternally curious. Just about everything that happens has some impact on the fire protection situation. Evaluate it and take action to contain any problem.

Fire Resistive Steel Buildings

If a building is rated as fire resistive, two considerations are vital. They are:

- Is the legally required level of fire resistance adequate for the fire load as it exists in the building?

---

*Since many tests have shown that a number of sprinklers will undoubtedly open on any fire in a high-bay area, perhaps the engineering of protection should be rethought. Instead of overhead piping, perhaps the concept of an automatic system similar to the manual system described here should be investigated.

- Has the protection of steel been completely provided and is it maintained?

If the answers to these two questions are negative, is there any legal relief, or is it up to the fire department simply to do the best it can in the event of a fire? If the fire department estimate of the situation indicates disaster unless heroic countermeasures are taken, or inevitable disaster, who, if anyone, is notified?*

All personnel making inspections should be made aware of the requirements for fire resistance as applied to specific buildings and the manner in which fire protection of steel can be degraded by the innocent actions of persons completely unfamiliar with its necessity. Sprayed-on fireproofing and membrane protection are the two methods most likely to be degraded.

Sprayed-on Protection.   Sprayed-on protection is often removed by mechanics accomplishing other work.

It may also be removed to dispose of the asbestos hazard. The fire department should insist on its being replaced with an adequate safe material. It is also vulnerable to being washed away by hose streams. Caution should be exercised when returning to areas from which the fire department was driven by the fire.

Membrane Ceilings.   There are so many ways a suspended acoustical tile ceiling can be tampered with that a prefire plan for any building in which one is installed should draw special attention to the possibility that collapse might occur in the event of severe fire exposure. This is particularly a possibility where steel columns are unprotected in the plenum space. The scenario for severe damage to a column could go like this: An interior stockroom has a characteristically high fire load. A ceiling tile has been removed for use elsewhere, or the ceiling fails. The column is heated to failure along a limited length and the column drops a foot or two. The effect would be staggering. Fire fighting plans might well subordinate the customary small line, low water-use tactics, which fire departments customarily employ in fire resistive buildings, to heavy caliber stream attacks on the fire to absorb the maximum Btu output before the heat is absorbed by the steel.

Where heavy stream tactics are deemed necessary to prevent collapse, it would be beneficial to explain them to the building management in advance, because if the tactics are successful, the water damage may be severe. The necessity of using the tactics employed might not be apparent after the fire, and the fire department may find itself ridiculed by the architect, the builder, the building department, and the steel industry for even thinking that a collapse might have occurred. It's good to get the point across beforehand.

---

*The record of the NFPA Annual Meeting in New York City in 1969 shows that Chief John O'Hagan of the New York City Fire Department took serious exception to the fact that the Port of New York Authority (whose representative had just presented a glowing account of the fire protection precautions taken during the construction of the World Trade Center) had built the world's biggest "lumberyard" in the form of plywood formwork and structures, and stated that the city fire department could not assume responsibility for damage to structural steel.[22]

The fire department should be aware of the extent to which membrane fireproofing is permitted under the local code. Discussions should be held with the building department and local installers to determine the extent of their knowledge of the necessity for strict adherence to the details of the installation as specified by the manufacturer's listing.

A particular study should be made of the extent to which the protection of columns is omitted when passing through ceiling spaces.

All must be aware of the difference between fire hazard (flame spread) listings of tile and the fire-rated tile as part of the fire resistance system of the floor and ceiling assembly. Similar-appearing tile is often used to accomplish desired results in these two quite different categories.

Not all suspended ceilings are a part of the fire resistance system. The ceiling may be installed for any reason, such as to reduce the room volume for better air conditioning, to hide ducts and pipes, to conceal an old ceiling for acoustical treatment, or whatever. At one fire it was found that the fire resistive membrane was provided by suitable lath and plaster, that the visible tile ceiling was only acoustical, but that to install the tile ceiling, many holes had been punched in the fire resistive ceiling to accommodate the support wires.

Through the inspection procedure, owners and occupants should be put on notice that the "ceiling" (as they view it), including light fixtures and air diffusers, is an integral part of the fire protection of the building (if in fact it is), and that the fire department or building department must be consulted before any changes are made. Inspecting and preplanning personnel must be cognizant of the hazard of tampering with the ceiling and an appropriate item should be provided on checklists.

Particular note should be taken of high fire load areas, where fast, high heat output is possible, or where inaccessibility may provide a lower but long-sustained Btu output. Plans should provide for fast, high-heat absorption capacity lines. The fast, hot fire may be more serious, but the long, slow fire may exceed the fire load against which the assembly was tested. Fire load is discussed elsewhere, but a rule of thumb is 10 pounds of combustibles per square foot equals a 1-hour standard test fire.

In studying the building, estimate the consequences of a ceiling failure. Is the ceiling protecting only the steel from overheating or, as is possible in ordinary construction, is it protecting a combustible roof supported on unprotected steel? In this case, the spread of fire to the roof will undoubtedly distort the steel. The distortion of the steel may cause the collapse of masonry bearing walls.

Be alert for the possible passage of fire laterally through the ceiling structure.

It has been noted previously that the term "interstitial space" is used to describe in steel-framed buildings what we called in ordinary and wood construction buildings "inherent voids." Such spaces can represent several feet of height between the floor above and the suspended ceiling below, particularly where long-span trusses are used. (Long-span trusses must be several feet in depth.) Building operators should be warned against any attempt to use such space, and, if necessary, adequate legislation should be sought.

Steel Culverts

Large culverts of corrugated steel, many feet in diameter, are used singly or in groups to carry streams under highways in place of what would previously have been sizable bridges. These culverts are coated with bituminous material and often paved with asphalt. In dry weather, there may be little or no water flow. Grass fires or fires set within the culvert can cause an expensive loss. Never hit such a fire with fog streams from both ends at once. Decide on the best approach, usually from upwind, and hit the fire from one end only, to avoid injury.

The possibility of a brush fire involving a culvert can be eliminated by removing foliage and laying down stone. Point out the potential for loss to the highway department.

References

[1]Huntington, W. C., Chapter 7, *Building Construction*, 3rd ed., Wiley, New York, 1963, pp. 373-443.

[2]"Connection Cited in Hyatt Collapse," *Engineering News-Record*, March 4, 1982, p. 10.

[3]Tryon, G., ed., *Fire Protection Handbook*, National Fire Protection Association, 13th ed., 1969, p. 8-38.

[4]Juillerat, E. and Gaudet, R., "Chicago's McCormick Place Fire," *Fire Journal*, Vol. 61, No. 3, May 1967, p. 15.

[5]"The McCormick Place Fire, Chicago, Illinois," Feb. 15, 1967, American Iron and Steel Institute, New York.

[6]Webb, W. E., "Effectiveness of Automatic Sprinkler Systems in Exhibition Halls," *Fire Technology*, Vol. 4, No. 2, May 1968, p. 115.

[7]Brannigan, Francis L., *Building Construction for the Fire Service*, 1st ed., National Fire Protection Association, 1971, p. 249.

[8]"Automobile Burn Out Test in an Open Air Parking Structure," Gage-Babcock Associates, Westchester, IL, 1972.

[9]"Protection of Steel," Loss Prevention Data Sheet 2-78, November 1969, Factory Mutual System, Norwood, MA.

[10]Bono, J., *New Criteria for Fire Endurance Tests*, Underwriters Laboratories Inc., Chicago, 1969, pp. 5 and 6.

[11]*FIA Sentinel*, March 1967, Factory Insurance Association, 85 Woodland Street, Hartford, CT.

[12]See Ref. 3, Fig 8-7G for illustrations of various encasement systems.

[13]See Ref. 3, p. 8-115, Part I.

[14]Levy, R., "The Building Official Looks at Fire Tests," *Fire Journal*, Vol. 59, No. 3, May 1965, p. 52.

[15]"With the Society," (A digest of a talk by C. R. Anderson and W. R. Powers), *Fire Technology*, Vol. 5, No. 3, August 1969, p. 242.

[16]"Floor or Roof and Ceiling Construction and Beam Protection," *Building Materials List*, Underwriters Laboratories Inc., Chicago, published annually with bimonthly supplements.

[17]*Fire Protection Through Modern Building Codes*, 5th ed., American Iron and Steel Institute, Washington, DC, p. 126.

[18]Sauer, G., "Suspended Ceilings as a Fire Protection for Columns," *Building Standards*, Part I, p. 6, July-August 1970, National Gypsum Company.

[19]Factory Mutual Engineering Corporation, "Building Construction and Materials," *Handbook of Industrial Loss Prevention*, 2nd ed., McGraw-Hill, New York, 1967, pp. 5-5 to 5-8.

[20]Grant, D., "The Wabush Mines Fire," *Fire Technology*, Vol. 4, No. 3, August 1968, p. 267.

[21]Lenton, T., "SPI Fire Tests of Stored Plastics in Warehouses," *Fire Journal*, Vol. 73, No. 3, May 1979, p. 30.

[22]"The Seventy-Third NFPA Annual Meeting, May 12-16, 1969," *Fire Journal*, Vol. 63, No. 5, September 1969, p. 25. See Ref. 3, Table 8-1B, p. 8-8.

CHAPTER 8

Concrete Construction

Concrete is a cementitious material which results from the chemical reaction of portland cement and water to which inert materials called aggregates are added. To produce good concrete, the water and cement should be in proper proportion; the aggregate must be of the proper type, size, and amount; and the mixing must be thorough but not overlong. Further, the concrete must be transported from the mixer to the form and set in the form in the proper manner to prevent formation of air pockets or separation of aggregate, and the concrete must be properly "cured." Shortly after it is mixed, concrete "sets" into a solid mass, but it takes several days for concrete to cure to its full strength. *

High early strength concrete will come to full strength in a shorter time. While curing, concrete generates heat (heat of hydration). During the first part of the curing time, the concrete must be protected from freezing. Low ambient temperatures retard the cure of concrete, and freezing is harmful.

Good concrete is a product of carefully controlled materials properly handled. Because there are so many possibilities for poor work, high factors of safety are used in concrete design.

Concrete is very weak in tensile strength and has poor shear resistance. Its compressive strength is good, particularly when the cost of resisting a given compressive load with concrete is compared to the cost of steel to resist the same load.

Concrete Construction

Until the post-World War II era, concrete construction was regarded as suitable only for massive low-rise utilitarian structures in which aesthetics played little part. High-rise buildings were universally built of steel frame, often "fireproofed" with concrete and with cast-in-place concrete floors. Precast con-

---

*Concrete continues to cure indefinitely. Construction specifications set a date by which the concrete must reach its required compressive strength: for instance, concrete required to reach design strength in 28 days is sometimes spoken of as "28-day concrete."

crete was known principally in the form of "cinder block."*

Concrete warehouses and factory buildings built before World War II are readily recognized. The panel walls, usually brick or brick and block composite, were built on each floor, between the exterior columns. The outlines of the building structure showed clearly that the columns and floor slabs stood out as if drawn in an elevation view. It was a simpler age, and buildings were of an easily identified, named type.

Today, there is a bewildering variety of mix-and-match elements. Steel-framed buildings now can have cast-in-place concrete floors made with forms that are removable or which can be left in place, or precast-cored units. The floors may or may not contribute to the stability of the building. Precast concrete and prefabricated metal wall panels are common.

Precast and cast-in-place units are combined in every conceivable way. An example of the new innovations in construction is the Housing and Urban Development Building in Washington, DC. Its walls are built of 13-ton precast units, each 10 feet wide and one story high, literally, a massive set of blocks. The walls are supported on cast-in-place spandrel beams. The spandrel beams rest on two-pronged pilotis (columns supporting a building above an open ground level). Precast concrete double tees extend 30 feet from the exterior walls to a poured-in-place continuous beam girder supported by cast-in-place reinforced concrete columns 30 feet on centers.

Some concrete building framing appears similar to timber framing with its columns, beams, and joists. Regardless of material, the principles of construction remain the same.

---

*"Cinder blocks" (so called) are concrete blocks in which cinders were used for the aggregate to produce a light, relatively inexpensive block. "Concrete blocks" (so called) are concrete blocks in which other materials are used for aggregate. The surface is usually smoother than that of cinder blocks and they are heavier.

"Underwriters blocks" are concrete blocks produced under Underwriters Laboratories' classification and follow-up service. The manufacturer is authorized to issue a certificate giving the number of units and type delivered to the job. Such blocks are used in assemblies required to meet fire resistance standards.

In some cases, the architect may wish to emphasize the massiveness of concrete construction. Far from considering form marks to be defects, the architect orders them accentuated, as in the Morris Mechanic Theater in Baltimore, MD. In other cases, tons of brick veneer and brick and concrete block composite veneer walls are used to conceal the concrete construction. This does not refer to masonry curtain walls which serve a purpose, but to purely decorative veneer. A report describes the loss of several stories of brick veneer from a reinforced concrete building.*

Fire Department Problems

The problems of the fire department with concrete construction can be divided into three distinct subjects:

- Collapse during construction with no fire.

- Fire when under construction.

- Fire in completed, occupied buildings.

Definitions

There are two main divisions in concrete construction: cast-in-place and precast. Cast-in-place concrete may be plain concrete, reinforced concrete or posttensioned concrete. Precast concrete may be plain concrete, reinforced concrete, or pretensioned concrete.

Concrete

	Conventional		Pre Stressed	
	Plain	Reinforced	Pre – Tensioned	Post – Tensioned
Cast In Place	X	X		X
Pre Cast	X	X	X	

---

*High early strength concrete tends to shrink. The brick veneer does not. If proper provision is not made for the contraction, the brick veneer will crack and may fall off.

The definitions needed to understand the foregoing are found in this section in which the principal terms necessary to understand concrete construction are given.[1]

> *Aggregates*   Materials mixed with cement to make concrete. Aggregates are divided into fine aggregates and coarse aggregates. Fine aggregate is usually sand; coarse aggregate may be one of a great variety of materials, depending upon availability and the characteristics desired in the finished product. Concrete often is described by the name of the aggregate used. Crushed stone, gravel, cinders, "breeze" (a coke by-product no longer used), expanded slag, shale, slate, clay, vermiculite and perlite (natural materials expanded by heating, thus, light in weight), natural lightweight stones, such as lava and pumice, and, recently, fly ash reclaimed from boiler plants have all been used as aggregates.*

> *Cast-in-place concrete*   Concrete which is moulded in the location in which it is expected to remain.

> *Spalling*   The loss of surface material when concrete (or stone) is subjected to heat. It is due to the expansion of moisture in the concrete. Some concrete and some aggregates are more subject than others to spalling. Concrete made with ground-up fire brick is very resistant to spalling and is used at fire testing facilities.**

> *Explosive spalling*   Spalling which occurs violently, throwing out bits of concrete like projectiles.

> *Plain concrete*   Concrete without reinforcement or with only light reinforcement to resist temperature changes. A sidewalk slab or a concrete base in the ground for a child's swing are examples of plain concrete. Some early concrete bridges, such as the Taft Bridge which carries Connecticut Ave. over Rock Creek Park in Washington, DC, were built of plain concrete.

---

*Special heavy concretes are used to provide gamma radiation shielding. Neutron radiation energy is absorbed by hydrogen atoms, so if neutron shielding is required, hydrogen-rich aggregates, such as magnetite, are used.

**The making of concrete is something of an art. During World War II, the author was officer in charge of a Navy fire fighting school. The scarcity of steel dictated that a 15-foot demonstration tank be built of concrete. An experienced concrete foreman, judging the mix by feel, produced a tank which spalled only slightly in hundreds of fires. Another foreman, following his instructions, produced another tank which spalled violently.

Reinforced concrete A composite material of steel and concrete. Steel provides the tensile strength that concrete alone lacks. Steel may also be used to provide compressive strength. The term "reinforced concrete" is a misnomer in that it often leads to the belief that the concrete has a certain strength and that the steel reinforces that strength. This assumption leads to failure to recognize the seriousness of the situation that exists when there is a failure of the bond between the concrete and the steel. Reinforced concrete is a composite, and both elements are equally important. Fire fighters who would walk gingerly across a concrete floor from which chunks of concrete had fallen show little concern about an overhead in which the steel has separated from the concrete. *When the bond between the two equally important elements is broken, the structural strength ceases to exist, and all that remains is deadweight.*

The reinforcing rods in concrete columns are in themselves columns helping to support the load. The next rods will be connected to the protruding ends.

In order to accomplish the designed function, the concrete and steel must remain in intimate contact, to assure unified action. In addition, and most important, exposed steel reinforcement lacks fire resistance, as does any unprotected steel.

Reinforcing bars or rods Steel rods or bars that are

usually "deformed"; that is, there is a raised pattern in the surface of the steel which aids in the transmission of stress from the concrete to the steel. The workers who install rods are called "rodmen" or "rodbenders."

Temperature rods   Thin rods installed near the surface, usually at right angles to the main reinforcing rods to help the concrete resist cracking due to temperature changes. Failure of the bond between the temperature rods and the concrete does not materially affect the strength of the structure.

Chairs   Small devices designed to keep the rods up off the surface of the form, so concrete will flow underneath the rods.

Precast concrete   Concrete that has been cast at a location other than the place where it is to remain. The precasting may be done at a plant miles from the construction site, on the construction site, but remote from the ultimate installation location, or immediately adjacent to the point of use. The common "cinder block," or concrete block is precast, plain concrete. Other precast concrete units are reinforced, or tensioned.

We can't see the reinforcing rods, but this precast concrete stairway could not stand without the rods to provide the tensile strength which concrete lacks. Under load, the stairway performs as a beam.

Tilt up   The term applied to casting wall panels on the ground along the outside perimeter of the building. When the panels are cured, they are tilted up into place and attached.

Lift slab A type of construction in which the columns are erected first to their full height. The first floor slab is cast on the ground. Each floor slab and then the roof slab is cast, one directly on top of the other. A bond breaker is used between each slab so there will be no adhesion. Starting with the roof, the slabs are successively lifted or jacked into position. Steel deck roofs have been assembled on the ground and lifted into position in a similar manner.

Casting The process of placing the fluid concrete into molds, generally called "forms," in which the concrete is permitted to harden to the shape of the form. Cast-in-place concrete is concrete cast at the place of final use. Continuous casting, as its name implies, is a process for casting or pouring concrete without interruption from start to finish. This avoids the problem of joining new concrete to concrete already poured. If such a joint is not properly made, so called "cold joints," which are planes of weakness, result. In particular, "laitance," a milky white substance, must be removed from the surface of the concrete already set.

Slipforming A technique by which forms are moved, usually upward, as the concrete is poured, so that continuous casting may be accomplished.

Continuous slipforming Pouring concrete continuously as forms move upward so that continuous casting may be accomplished.

Monolithic construction When all the concrete in a building is properly bonded together, the building then is a monolith (from the Greek "one stone").

Pretensioning A process for putting concrete under compression. It is more fully explained in the section on prestressed concrete.

Posttensioning Also a process for putting concrete under compression. It too is explained under prestressed concrete.

Composite construction While reinforced concrete is a composite material (see above), steel and concrete also can be combined in a variety of other ways. In many steel-framed buildings, the concrete floors simply transmit the live loads to the columns, but in composite construction, the concrete floor is well-tied to the beams and assists in resisting the shear load. During construction, such beams can be recognized by the studs set in rows along the top flange. In a concrete building, it may be found advantageous to use some structural steel; this is also spoken of as composite construction.

Sometimes the top floor, or penthouse,* of a concrete building will be of lightweight steel construction. The extra story is literally "free." In one case, the top floor of a concrete high-rise in Florida is of wood.

This concrete high-rise will have a lightweight steel top floor. It saves on construction costs but complicates fire control by introducing different problems.

The ceaseless search for less costly building has brought the use of corrugated steel for forms for floor slabs. The steel becomes a "left-in-place" form and the steel and concrete, reacting together under the load, provide a composite floor.

Chemical reactions   Many substances are not suitable for close contact with concrete. Aluminum, for instance, in the form of electrical conduit, will react with concrete and cause decomposition. Some natural stone cannot be used for aggregate because of undesirable reaction.

---

*A penthouse is any structure above the roof line. A penthouse apartment literally is an apartment built on the roof.

Footings Thick concrete pads, usually heavily reinforced, that transfer the loads of piers or columns to the earth.

Drop panels Thicker sections of floor at columns to assist in resisting the natural tendency of the floor to shear off at the column.

Mushroom caps Tapered extensions at the tops of columns which assist in transfer of loads from floor to column. Columns in heavy-duty concrete construction may show both mushroom caps and drop panels.

The mushroom cap and drop panel are characteristic of heavy-duty concrete construction, designed for strength, not appearance.

Composite and combination columns Steel (other than reinforcing rods) and concrete are combined into one unit. As in all composite construction, the different elements must react together, and any failure of the bond between them means that the column is distressed. Composite columns were formerly used on the lower floor(s) of a heavy-duty building to avoid surrendering floor space to large diameter concrete columns made solely of reinforced concrete.

Lally columns Steel pipes are filled with concrete to increase their load-carrying capacity. They are popularly called "Lally columns," a trademark name. In the 19th century, cast-iron Lally columns were usually drilled to provide a vent for steam generated in the concrete. Steel pipe concrete columns seen today are not so drilled. Exposed to heat, they may explode as their predecessors did.

Concrete Structural Elements

Columns

Since a chief virtue of concrete is its high compressive strength (compared to cost), concrete is used for columns, even in structures otherwise built of steel, as for instance, highway bridges. Short columns that are very wide in proportion to length are called piers and may be of plain concrete; otherwise, columns are of reinforced concrete.

The reinforcing bars serve not only to create a composite material but also to carry some of the compressive load. In early concrete construction, it was the practice to enlarge the diameter of the columns successively on lower floors to cope with the increasing loads.* At the lowest level, the size of the columns might be so great as to interfere with the use of the building. In some cases, steel columns were used within the concrete to take advantage of the far greater compressive strength of steel. In modern buildings, columns increasing in size would be unsatisfactory in that the rentable area would vary from floor to floor. This is overcome by increasing the size of reinforcing steel as the loads increase so the

Development of better steel reinforcement makes it possible for concrete columns to have the same dimensions on all floors. This provides an economic advantage over previous conditions which required larger concrete columns on lower floors to carry the building load.

---

*If this sounds backwards, remember that buildings are designed from the top down, though built from the bottom up.

outside dimensions of columns are identical on all floors. At one building examined, the columns had so much steel in them that a special additive was mixed into the concrete to permit it to flow into the relatively small spaces between the steel.

The reinforcing rods are long and of relatively thin diameter; however, they are columns in their own right. Sleeves, reinforced overlaps, or welding are used to connect the ends of rods and to transfer the load. The rods in a column are tied together with lateral reinforcement called ties or hoops. The ties or hoops serve to cut the long slender column up into a number of relatively short columns one on top of the other. Recall from Chapter 2, Principles of Construction, the effect of lengthening a column section by the loss of bracing. The same effect will occur if column reinforcement bracing is lost. Together with the concrete to which they are bonded, the rods form a composite structure. If the unity of the composite fails, such as by loss of concrete either under normal circumstances or as a result of fire, the steel rods may act as Euler's law columns (see Chapter 2), and buckle under load. Buckling will cause the rods to protrude from the column.

Protruding rods are a serious sign of possible collapse and should bring about the immediate evacuation of the area, for a column failure may have catastrophic results. Fortunately, column failures have been rare.

In reinforcing round columns, the ties connecting the outer rods are often formed in a helix. Such reinforcement is erroneously called spiral reinforcement.*

Beams and Girders

Concrete is strong in compression, weak in shear, and is not credited with any tensile strength at all. As was learned in Chapter 2, Principles of Construction, when a beam is loaded, it deflects. This deflection brings about compression in the top of the beam and tension in the bottom of the beam. Reinforcing bars are usually placed in the bottom of the concrete beam to provide the necessary tensile strength. Precast beams often have the word "Top" cast into the top of the beam to ensure that the beam is installed right side up. In a cantilever beam, however, the tension is in the top of the beam. In a continuous beam, there is tension in the top of the beam in the area over the tops of the columns and tension in the bottom of the beams between columns. In all cases, reinforcing must be properly designed. Vertical reinforcing bars in concrete beams designed to prevent cracking under shear stresses are called "stirrups." Some of the tensile reinforcing bars may be "bent up" to accomplish the same purpose.

When we consider the foregoing, we realize that concrete below the neutral plane serves very little purpose. Sufficient concrete is needed to bond with the

---

*A spiral is a line which starts from a point and moves away from it in a continuous circular path. A helix is a line inscribed around a cylinder.

reinforcing rods to deal with the relatively slight compressive stresses, and to protect the steel from fire; but the rest can be minimized. This leads to the characteristic "tee beam" design in which the neutral plane coincides with the bottom of the wide, thin floor slab. "Double tees" are floor slab and beam combinations with two beams, and floor-beam combinations with four beams are manufactured.

As in columns, where space requirements are paramount, steel may be substituted for some of the concrete to handle compressive loads. For any unit area, steel has 15 times the compressive strength of concrete. Steel is, however, more costly.

Concrete Floors

Concrete was first used for the leveling of brick and tile arch floors. From this, it was but a short step to the development of concrete arches both segmental and flat, sprung between unprotected wrought iron or steel beams. Early concrete buildings were built of identifiable individual beams supporting a floor slab, all poured at the same time. The appearance of the underside is reminiscent of wood joists, and this method is still used at times. In so-called waffle concrete, closely spaced beams are set at right angles to one another, and the lower side resembles a waffle.

Many concrete floors are designed as continuous beams, that is, the entire floor is one big beam, just as a wooden floor conceivably might be made of one huge sheet of thick plywood. In heavy construction, mushroom caps and drop panels are used at columns to cope with high shear stresses. In lighter construction, the floor may be just a flat plate, with no projections below the floor line. This gives a smooth surface, easily finished, and it is popular for offices and apartment buildings.

When precast tee beam units are used, additional concrete is cast in place on top of the units, which are left rough on top when cast initially to ensure a bond, and the entire unit becomes an integral beam and floor element.

Precast concrete floor units often have cylindrical openings cast lengthwise through the units (cored out) to remove unnecessary weight, and to provide channels through which building services can be run. Concrete floors may be simply load-bearing, or may be an integral part of the structural stability of the building.

Concrete floors may be built in buildings of any structural type. Building codes in some cities require that the street floor of buildings of ordinary construction be of "fireproof" construction, so a concrete floor or, more likely, a concrete topping over a wood floor, may be found in a building of ordinary construction.

The technique of building a cast-in-place concrete floor into ordinary wall-bearing masonry buildings sometimes provides a real hazard during construction and perhaps for the life of the building. At times, in construction, the wall is carried up a story or more beyond the point where the floor is to be located. A "slot" is left in the wall at the point where the floor is to be cast.[2] The concept is that concrete will fill the slot, thus uniting the floor and wall. If a windstorm occurs during the time that the slot is unfilled (usually a few bricks set on end, unmortared, are relied on for temporary support), a collapse may result. A ladder or hose line over the top of the wall may also cause collapse. For all practical purposes, the wall is "undercut." Unless the concrete is carefully pushed up into the gap when the floor is poured, the wall will always be weak at this point. The finished floor may give no evidence that the continuity of the wall is interrupted by a void which creates a plane of weakness.

Concrete floors in steel buildings may be precast, or cast in place. If cast in place, they may be cast integrally with concrete fireproofing for the steel. They may be only load-bearing, or they may provide structural stability by being designed to resist shear loads such as wind. In the latter case, studs are welded to the steel beams and the reinforcing bars are attached to the studs.

Concrete floors in cast-in-place concrete-framed buildings are cast integrally with columns, providing a monolithic rigid-framed building. Concrete floors that have been cast in place may or may not be designed to assist in providing structural stability. Concrete floors in precast pinned concrete buildings may be only load-bearing and may not contribute to the structural stability.

Note that when concrete floors are cast onto corrugated steel, so-called "left-in-place" forms, the steel actually provides the necessary tensile strength. If the bond fails, the floor section may fail.

Precast units can be connected so as to develop a monolithic unit in a building. Reinforcing bars are designed to protrude from the finished precast slabs or other units. When the slabs are laid down, a space is left between them. The protruding bars of one slab extend past the ends of the protruding bars of the other slab. They overlap without touching. An 18-inch space, for instance, may be left between slabs with bars usually hooked, protruding 12 inches. A form is made, and concrete is poured to level. The sections, thus joined by a wet joint, become monolithic.

Precast columns are provided with "haunches" (a sort of corbel or shelf) cast into their sides on which beams are set. The connectors are then welded together. This, of course, is pinned, not monolithic construction.

Prestressed Concrete

Up to this point, conventional reinforced concrete, as it has been known since the invention of the reinforcing bar in 1885, has been discussed. In recent years, "prestressed concrete" has been developed. Prestressing places engineered stresses in architectural and structural concrete to offset the stresses which occur in the concrete when it is placed under load.

Consider a stack of books side by side. As a "beam," such a stack will fail of its own weight without any superimposed load due to the lack of shear resistance between the books. Drill a hole through the stack of books laterally, pass a wire through the books, and tighten the wire against the end books. The stack of books would be compressed by putting tensile stress in the wire and compression in the books. The "beam" could be placed across two chairs and stood on. The beam was prestressed sufficiently to counteract the stresses placed in it by the load.

Special high-strength cables, similar to those used for suspension bridges, or alloy steel bars are used. They are technically tendons but are known as "strands" or "cables" by those working with them. The high-tensile-strength wire, ordinarily used for prestressing, is relatively more sensitive to high temperatures than is ordinary steel. There is virtually complete loss of prestress at 800°F.[3] The protection of the tendons from overheating by fire is of paramount importance, and several potentials for disastrous collapses due to fire will be noted later in this chapter.

There are two methods of prestressing. They are pretensioning and posttensioning. The prefix "pre" and the descriptor "post" refer to whether the concrete is poured before or after the tension is applied.

In pretensioning, high-tensile-strength steel strands are stretched between abutments. Concrete is then placed in the forms which are built around the strands. As the concrete sets, it bonds to the tensioned steel. When the concrete reaches a specified strength, the tensioned strands are released from the abutments. This prestresses the concrete, putting it under compression, thus creating a built-in resistance to loads which produce tensile stresses. Pretensioning is done in a plant and the completed units are shipped to the job site.

In posttensioning, high-tensile-steel strand wires or bars are encased in tubing (plastic and paper are common) or wrapped to prevent any adhesion between steel and concrete, positioned in the forms, and then the concrete is placed. After the concrete is set and reaches a specified strength, the steel is stretched* and anchored at the ends of the unit.

Sometimes the posttensioning is performed in increments, as the load of the building increases as stories above are placed. Some bridge girders are tensioned enough to make shipment possible, then posttensioned after being placed. After tensioning is completed, some designs call for "pressure grouting" in which a cement paste is forced into the space between the tendon and the concrete to provide a bond.

Some tensioned floors consist of small precast units through which tension cables are run, then tensioned. Posttensioning is most often done on the job site.

At times, the terms "precast" and "pretensioned" are erroneously interchanged. Concrete can be cast in place or precast. Concrete can be conventionally reinforced or tensioned. Cast-in-place concrete can be conventionally

---

*This process is known technically as stressing the tendons, but colloquially, in the trade, as "jacking the cables." The necessity for being conversant with both the technical and colloquial terms was forcibly brought home to the author when he went to photograph the operation. The question, "Are you going to tension the tendons today?" brought the reply, "I don't know about that, I'm too busy jacking these cables." Funny under the circumstances, but tragic if a fire officer doesn't understand the warning, "They haven't jacked those cables yet."

Washington's famed Watergate Apartments are of posttensioned concrete. This provides lightweight and thus, thin floors. One resident is supposed to have said there was no need to bug the place as the floors transmitted sound readily.

reinforced or it can be tensioned. Precast concrete can be conventionally reinforced or tensioned.

Reinforced Masonry

In the discussion of ordinary construction in Chapter 4, Ordinary Construction, it was learned that there is a built-in limit to the economic height of an ordinary brick, bearing-wall building because of the requirement that the walls increase in thickness as the building's height increases. The limit is about six stories.

In recent years, it has become possible to build brick, bearing-wall buildings 20 or more stories in height, with no wall thicker than 12 inches. Concrete block can be used in a similar fashion.

In one method of construction, two wythes of brick are built, the width of one brick is left between them. Reinforcing rods are placed vertically in the cavity. After the brick masonry sets, concrete is poured into the void. Floors in one such building the author knows of are of precast-tensioned concrete double tee slabs, but other systems can be used.

Reinforced masonry construction was first developed to resist earthquakes. It is unsuitable for buildings in which large clear spans are required, but it can be used for buildings where interior walls are not undesirable, because the interior walls are equally load-bearing with the exterior. Apartment houses and motels with their repetitive box construction are well adapted to this method. Office buildings require more flexibility. The high level of soundproofing is another benefit of this type of construction.

The 17-story reinforced-masonry Travelodge motel at Lake Buena Vista in Disneyworld, FL is the tallest wall-bearing concrete block building. It is built of special block using epoxy mortar.

Conventional wall-breaching techniques would be ineffective and possibly dangerous if used on a reinforced masonry wall.

Eleven-inch reinforced masonry walls (4 inches of brick, 3 inches of reinforced concrete, and 4 inches of brick) have achieved a 4-hour fire resistance rating.

In low-rise buildings (up to about 70 feet), some recent designs eliminate the reinforced concrete in the wall. High-strength bricks, special mortar, and the liberal use of masonry reinforcement trusses (see Chapter 4, Ordinary Construction) between courses and to tie cross walls to the exterior walls, together with precast concrete floors integrated into the building by the use of wet joints, produce a masonry, wall-bearing building several stories high with no wall thicker than 8 inches.

It is interesting to note that this type of construction has become popular in some resorts. Some recently built seaside motels consist of individual concrete boxes. When these are serviced by outside open-air stairways and balconies, the epitome of life safety in a multiple-unit occupancy is being approached, and the occupant is safe from all folly but his own.

Collapse during Construction

From time to time, some concrete structures under construction collapse. If workers are trapped, the fire department is often involved in rescue operations. The following suggestions are offered as aids in planning a department's role in rescue operations when a building collapses.

While the authority of the senior fire officer present at a fire is clearly spelled out in the organic law of almost every community, the authority to act or to direct others at emergencies other than fires is rarely defined. The fire officer should be well informed on the legal position of the fire department. Many codes assign the power to order the removal of a dangerous structure to the building commissioner (or similar official), thus making it quite clear that the fire department has no right to demolish a structure just because it represents a hazard in the view of the senior fire official. It would be good if every community had a clear plan determining who is in charge at emergencies other than fire.

This may be touted as the Computer Age but it is also the age of the lawsuit. After the dust settles in a building collapse, there will be a series of legal actions as the owner, architect, general contractor, subcontractors, and the victims attempt to determine who will be financially responsible. Collapses do not just happen — they are caused — and some of the people at the scene may well know, or at least fear, that the blame may be imputed to them. It would be expecting too much of human nature for a person in such a position to guide all his actions by the principles of justice. In other words, somebody may be trying to "cover up." It may not be your duty to prevent this, but it certainly is not part of your duty to assist. Be wary of such suggestions as: "Let's knock down the rest of this before it falls." A study of the uncollapsed portion of a structure may show why the collapse occurred.

(A summary report of the collapse of a reinforced concrete condominium under construction in Cocoa Beach, FL in 1981 in which 11 died is contained in Appendix B. Readers will find it an interesting account of the difficulties facing fire fighters when they are called to the scene of a major building collapse.)

Problems of Falsework

Falsework is the temporary structure erected to support concrete work in the course of construction. It is composed of shores, formwork and lateral bracing. Formwork is the mold that shapes the concrete. Shores are members that support the formwork. Lateral bracing is usually diagonal members that resist lateral loads on the falsework.

Falsework can represent 60 percent of the cost of a concrete structure.

Concrete falsework is a temporary structure, designed without the extra strength calculated into a building to compensate for deterioration. It is usually built at the lowest possible cost out of low quality material by people rarely versed in the basic engineering and mechanical principles involved. With this structure, the builder hopes to contain a fluid that can provide a head pressure of up to 150 pounds per square foot (psf) for each foot of height. Large fluid loads are placed at the tops of slender columns, and the loads are then vibrated. The structure must often absorb the impact load of the sudden stopping of motor-powered buggies carrying heavy loads of concrete. It is not surprising that falsework failures occur; what is surprising is that they are relatively rare.[4]

Falsework for walls or columns must have adequate strength to resist the tremendous head pressure of the heavy fluid concrete. Depending upon the weight of the concrete, the head can vary from about 75 pounds per square foot per foot of height to as much as 150 pounds per square foot or more. As concrete sets, the pressure is reduced due to internal friction. If a wall is designed for several successive pours, depending upon the set of the concrete to reduce head pressure, and an overpour occurs, collapse may result. (The set of concrete is temperature dependent; cold weather slows it. Concrete poured at 50°F will develop one third more pressure than at 70°F.)[5]

When falsework bursts, it usually goes out at the bottom. Workers may be trapped, particularly between falsework and the sides of the excavation. There will probably be no reinforcing steel to interfere with rescue, as it will be held inside the forms. Rescue involves uncovering faces, supplying oxygen, digging the victims out, and avoiding further injury. If it is necessary to retard the setting of concrete, as much sugar as can be obtained should be worked into the concrete; concrete containing sugar will set slowly, if at all. Bear in mind that the concrete can transmit pressure; avoid standing in concrete which might transmit the weight of your body to the victim.

Concrete requires time to cure; the length of time depends upon the type of cement used and the temperature during the curing period. After a proportion of the design strength is reached, falsework is removed for use elsewhere. Some "shores" are set in place in order to help carry the load of the still-curing concrete. This is called "reshoring." Evidence of reshoring is an indication that the concrete is not yet up to full strength; it may be unable to handle large superimposed loads, such as debris removed from a collapse, excess numbers of people, or additional apparatus or construction equipment. Avoid adding to loads that already exist, such as equipment and piles of construction material.

Most falsework collapses involve falsework supporting floors. Very often the collapse is of framework supporting a high bay floor, one in which the falsework is double-tiered, i.e., two sets of normal floor-high shores used one on top of the other. Such structures have all the instability characteristics of long, narrow columns; proper crossbracing is vital as is proper footing. Often the formwork is resting on earth, sometimes on unconsolidated backfill. The planks on which the shores rest are called "mudsills," but if water liquefies the earth and turns it to mud, the bearing may be inadequate. The operation can pass all such pitfalls successfully, but the repeated impact load of vibrating the concrete to provide a dense smooth surface may be the straw that breaks the camel's back and a collapse occurs.

Most of the victims will be on top of the falsework. Use bolt cutters to cut reinforcing rods to manageable lengths. Thicker rods may require torch cutting. The rods are tied together in a matting. Further harm may be done by attempting to remove too-large sections.

When concrete is being poured, almost invariably there are carpenters and helpers under the structure, bracing and wedging any shores that appear to be distressed. Determine from the carpenter foreman whether all of his workers are accounted for. Missing workers may be under the falsework.

In the previous edition, it was stated, "Reinforced concrete which has set hard to the touch usually has developed enough strength to be self-supporting, though it may not be capable of handling superimposed loads." This was based on information from experienced concrete designers. It was proven not to be true in the Skyline Towers collapse in Arlington, VA, in 1973. In this collapse, the shoring was removed from the topmost floor. The floor collapsed, and the collapse was progressive, ending at the ground level. (The removal of the shoring by laborers is no different than the removal by fire. The result would be identical.) A consequence of this collapse is to blunt the distinction in collapse potential between conventional reinforced concrete and posttensioned concrete. In short, any concrete falsework failure presents the likelihood for catastrophic collapse, though the potential is even more pronounced in the case of posttensioned concrete. Few concrete buildings are designed to withstand the collapse of one floor onto another. Progressive collapse is almost a certainty.

Techniques of Posttensioning

The construction of posttensioned, prestressed concrete presents catastrophic fire collapse potential and it is vital that a fire department be fully aware of any posttensioned structures being built in its area and be able to recognize the clues which indicate posttensioning. (The term "structures" here is intended to include bridges and other structures not usually considered in prefire planning.)

When posttensioned prestressed concrete is poured, samples are taken in standard concrete sample cylinders and are sent to a testing laboratory. When the concrete tests up to a designated fraction of its ultimate strength, hydraulic jacks are used to tension the tendons. Until this is accomplished, the entire weight of the concrete is on the falsework. There is no bond between the tendons and the concrete — quite the contrary, the tendon is contained in a sheath to prevent bonding. The weight of the concrete is transferred to the columns only when the tensioning is completed, several days after the concrete has been poured.

Here tensioning cables are laid out. The concrete floor will be poured over them. As the concrete hardens, the tendons will be successively tensioned (stretched) giving the concrete load-carrying ability. If they lose tension in a fire, the floor may fail.

In the meantime, the concrete is hard to the touch and can be walked on. The formwork for additional floors can be erected. A description of the construction of a posttensioned concrete building in Rio de Janeiro, Brazil, mentioned that at one point the top four floors had not developed enough strength to be posttensioned. Until then, the four floors were supported solely by pipe shoring, which was also left in place in the six floors below them. It is the author's opinion that any failure of the shoring on any of the floors would have precipitated a collapse of the four untensioned floors; the collapse of these floors would, without doubt, bring down the the entire structure.

The advantage of this construction is that thinner floors are possible. In the building described above, conventional construction would have yielded only 33 floors. The posttensioned structure has 40 floors. The additional value is about $20,000,000.

A "cured" concrete bridge awaited tensioning in California. Apparently, failure of the formwork caused the total collapse of the bridge, killing a man driving underneath it. An adjacent bridge in the same stage of construction was immediately and successfully tensioned. Fire department operations at this collapse are described in an article in *Fire Command!*[6], supplemented by a letter in the "Voice" column of that magazine by this author.

Collapse Potential of Reinforced Masonry Structures Under Construction

Since large numbers of masonry units are used in construction, it is easy for workers to overload a portion of a floor with brick or block. This happened in Pittsburgh, PA, and the excess load caused the partial collapse of several stories of precast floors. Since the broken-off "stubs" of the floors were keyed into the wall, they could not be removed and the gap was bridged by cast-in-place reinforced concrete floors.

Collapse of Precast Concrete Structures under Construction

Precast concrete buildings under construction are very unstable until all connections are completed. Temporary bracing, such as tormentors (adjustable poles set diagonally) or cables, holds units in place. In some cases, wooden temporary shoring is used. The completed structure may be of pinned construction, with bolt-and-nut or welded connections, or may be made monolithic by the pouring of "wet joints" to join precast units together. A collapse of an almost completed parking garage was laid to washers that were oversize and thus passed over the nuts that were supposed to hold them.

A Florida shopping mall collapsed just before opening. A haunch supporting a main girder was not provided as designed. Workers installed a haunch using $\frac{3}{8}$-inch lag bolts.

Fire Problems of Concrete Buildings under Construction

Concrete buildings under construction can present serious fire problems. Fire in falsework can result in major collapse with potential for massive life loss.

Numerous fire causes are present: welding and cutting, plumbers' torches, temporary electrical lines, and the ever-possible arson. Fuel, such as wooden falsework, flammable liquids, and plastics for insulation, is available. Glass-fiber formwork is also combustible. This is often a surprise to those handling it.*

The most dangerous hazard is heating. Sometimes workers light some scrap wood in a steel barrel and some kerosene heaters are used, but most often, heating is supplied by the use of liquefied petroleum gas (LPG). In order to confine the heat, the building is sheathed in plastic. Inside the heated area, LPG bottles are scattered at random together with the open-flame salamanders in which the gas is burned. The stage is set for a tremendous disaster. Most codes for the use of LPG require storage of the gas away from any open flames, for good reason. On Oct. 10, 1963, 74 people died in an LPG explosion at the In-

---

*Glass fiber is actually glass-fiber reinforced polyester resin plastic. The plastic will burn leaving the glass fibers. Be very careful in overhauling this material; the broken fibers can cause severe skin irritation.

Heat is often necessary during construction to prevent freezing of the concrete but creating a fire hazard. Note the LPG tanks and salamanders in this construction enclosure. If the tank leaks, a disastrous explosion is likely.

dianapolis Coliseum. A gas-fired cooker was in use under the concrete stands. A leaking cylinder exploded when the gas reached the flame of the heater. According to a rescuer, one victim was found in the center of the arena with a huge piece of concrete on him. Do not be deceived by the fact that the enclosure is of plastic. It may vent and preserve the building, but all in the area will die.

In one west coast city, the practice is to store the gas bottles at ground level and pipe (usually with plastic tubing) the gas to the salamanders. An excess flow valve should be installed. Such a valve senses a sudden increased flow, as from a broken line, and shuts off the gas.

The situation is not necessarily improved when noncombustible falsework is used. Aluminum falsework in combination with fiberglass forms is a popular combination. The aluminum is subject to failure by melting. Steel tube shores are better but can be caused to fail by high fire loads as from gas cylinders or lumber piles.

The fire hazard represented by the massive wooden formwork for the construction of the hyperbolic parabaloid dome of the new St. Mary's Cathedral in San Francisco was recognized, and temporary automatic sprinkler protection was provided for the lower portion of the formwork. The author was informed that there were some questions, particularly as to whether in such an open structure, a fire would not move upward ahead of the opening sprinklers.

Hazards of Posttensioned Concrete

Posttensioned concrete presents an even greater hazard of catastrophic collapse during a fire than conventional reinforced concrete. As has been noted previously, the entire weight of a posttensioned concrete floor (or floors) or beam rests on the falsework until the tensioning lifts the load onto the columns. A falsework fire, therefore, could cause the sudden collapse of an entire concrete floor slab, or the entire structure, onto fire fighters. Under no circumstances should personnel be in the collapse zone (or close enough to be hit by missiles) during or after a serious fire in formwork supporting posttensioned concrete.

The ends of tendons are left exposed for varying periods of time after tensioning. In the case of floor slabs, several feet of excess tendon may protrude from the socket until it is cut off and the socket "dry packed." In the case of beams, the anchors may be left exposed, possibly for further tensioning, or just as a normal construction delay. In either case, this represents a serious threat to the stability of the building during a fire, as the tendons will lose their stress at about 800°F.

The substantial fuel loads provided at construction sites, or by exposing buildings, may well provide sufficient heat to cause the tendons to fail. Failure of a tendon or tendons will cause the collapse of that part of the structure supported by the failing tendon.

One high-rise motel the author knows of consists of 13 floors of repetitive cast-in-place concrete framing rising above a "transfer floor" (a floor where the column loads are rearranged to provide different column spacing, in this case, in the garage area), which incorporates several huge posttensioned beams. When heavy loads are to be carried on tensioned beams, the entire prestress cannot be applied at once. The prestressing takes place in increments as the building load increases. The tension anchors were left exposed for months on the side faces of the third-floor level and in pits within the building to facilitate successive tensioning. The ends within the building were particularly exposed to an "ordinary" construction fire. A loss of tension in this major beam would probably have caused total collapse of the motel.

A Texas fire department learned of the hazards of posttensioning during a seminar conducted by the author. Looking over their area, they found several posttensioned buildings under construction. In one case, wooden falsework potentially exposed partially tensioned tendons to severe heating. The prefire plan called for apparatus to proceed to a fire in the falsework through a passage under the partially tensioned building. They changed the plan.

Just a few years ago, a posttensioned building under construction in a midwest city suffered a falsework fire. The fire department operated from the adjacent completed section, extinguished visible fire and notified the contractor that he had a problem. A second fire occurred. As the fire department was setting up, the entire 18-story building collapsed. No fire fighters (or anyone else) were injured.[*]

If a concrete building is to be built in your area, learn whether it is to be posttensioned, and if so, plan accordingly. You will likely find that the construction superintendent will first be incredulous when you speak of total collapse, but press him for facts. Much posttensioning is done. It is only a question of time before a disaster of the type discussed here occurs. The hazard will then be recognized. This is one opportunity to control a disaster before it occurs. Insist

---

[*] The chief told the author that, between the fires, he reread the relevant material in the first edition of this book. Press reports indicated that the owner blamed "water thrown by the fire department" for the collapse.

on "fireproofing" of tendon anchors immediately after tensioning is completed and temporary protection for incrementally tensioned tendons. Instruct all personnel in the hazards of to-be-tensioned concrete. If your alarm office functions as it should, as a critical information center, crank "posttensioned construction underway" into the system as a hazard item. Learn to recognize the signs of posttensioned concrete construction, the protruding cables, the coils of tendons, the characteristic anchors.

In a fire, it is vital that any exposed steel connected to stressed tendons be cooled immediately and for the duration of the fire.

Excavation Hazards

In Chapter 2, Principles of Construction, the hazard of unprotected steel used to brace excavations was discussed. This steel bracing complicates the building of the structure, as it must be coordinated with a permanent structure of the building. While this bracing is often spoken of as excavation bracing, in fact it is literally holding up the neighborhood. The standing structures exert an axial load on the earth. Depending on its shear resistance (solid rock is best), the earth tries to shear sideways into the excavation, thus dropping the buildings supported on the earth. The loss of excavation bracing can be extremely serious.

Where soil conditions permit, holes are drilled horizontally into the earth. Anchors are placed in the bottom of the holes to hold steel cables similar to the tendons used for posttensioned concrete. The cables hold the excavation sheeting in place. The cables are then tensioned. The ends of the tensioned cables stick out into the excavation where they could act as heat collectors. A falsework fire could heat these tendons. If the temperature past the anchor reaches 800°F, the tendons will fail and a collapse will probably result.

This problem is especially serious in the deep excavations of ten stories or more, so common today for parking garages under buildings. The hazard goes unrecognized and will remain so until a spectacular and tragic collapse occurs. Even if the fire department recognized that cooling unprotected structural steel or tendon ends is an absolute must in a below-grade formwork fire, it is often impossible to cool the steel due to obstructions.

Precast Buildings

When precast concrete units are being erected, temporary bracing or support is provided until the full space-frame relationships of the building are developed. Columns are usually braced by telescoping tubular steel braces, but wood can be used. Wooden falsework is often used to support horizontal elements until connections can be made. The loss or upset of any such temporary support may precipitate a general collapse.

This pinned precast concrete building is temporarily braced with cables which may fail in a fire causing the building to collapse.

Cantilevered Platforms

Cranes are often used to deliver construction materials to buildings under construction. This necessitates a temporary platform extending out from the floor. A common method of securing such a platform is to brace it into place by the use of wooden shores delivering a compressive load to the floor above. A fire involving such a platform would destroy the shores and drop the platform to the street. Keep all personnel and apparatus well clear of the collapse zone.

Tower Cranes

The high tower crane, which is the symbol of modern construction, can literally climb up the building. As it climbs, it is supported on the building. The weight of the crane may be distributed over several floors by falsework. A fire involving this falsework will bring down the crane. Falsework on an otherwise apparently completed floor should be investigated. It may be supporting a patch of a hole that was left in the building to handle material, or it may be supporting a heavy load such as the crane or supplies of building materials.

Formwork for concrete placement on the 23rd to 25th floors of a high-rise under construction burned. Flaming debris fell to the street, and the tower crane operator was trapped in his cab above the 28th floor. Fire fighters protected him with a heavy caliber stream from a nearby roof until an engine company fought its way through the fire to save him.

Fire Fighting

There is no point in taking undue risks with personnel in a concrete building under construction. Rest assured that all the concrete even remotely involved in the fire will be torn down. Since it is impossible to put a limit to the effect of the loss of even a small portion of the falsework, command officers should err on the side of caution. Heavy caliber streams from safe locations deserve serious consideration. The falsework structure, at best, has a minimum of redundancy and thus cannot stand the loss of material caused by a fire. In the case of the collapse of falsework for a bridge, it was stated that the collapse of *one* 4 by 4 caused the general collapse.

In all buildings under construction, move very deliberately. Floor openings may be inadequately guarded, and side rails may be nonexistent or of poor quality.

Fire Problems in Finished Buildings

When concrete construction was first developed, it was regarded by some as the answer to all fire protection problems; it was truly "fireproof." After a series of disastrous fires, it became evident that concrete, like any other noncombustible material, can be destroyed by fire, given sufficient fire load and fire duration. Unfortunately, many people, including some who should know better, confuse noncombustibility with firesafety.

Concrete is inherently noncombustible. It may have been fabricated to meet a fire resistance standard.

Concrete itself carries no guarantee of life safety. The reinforced concrete Joelma Building in São Paulo, Brazil burned on February 1, 1974 with a loss of

Floor to ceiling glass windows make exterior fire extension a real possibility. Glass fails quickly in a fire exposing the contents which provide additional fuel.

179 lives. The structure escaped with relatively minor damage. Factors other than the structure itself, such as interior finish, and the inability of the occupants to reach a nontoxic environment speedily may be far more important.

Concrete construction can be built to deliver any required level of fire resistance. Unlike steel, which is fireproofed by protecting it with an insulator, the fireproofing of concrete is integral. Some of the concrete, particularly that below the reinforcing rods, is not necessary for structural strength but is necessary only for fireproofing.

In evaluating the potential for fire in a concrete building, the first determination that should be made is whether the building was even intended to be fire resistive. If it was not, then the concrete is merely noncombustible and there is no real inference of its integrity in a fire. A noncombustible (code definition) concrete building can have a lightweight, totally unsafe (for ventilation) wood, gusset plate truss roof.

While the only true test of any fire resistive building is a severe fire, there are signs of possible trouble which can be determined as part of a prefire plan.

Deteriorated concrete, spalling which exposes reinforcing rods, and cracks in concrete which can admit corrosive moisture to the reinforcing rods are some signs that indicate the building is in distress. Reinforced concrete is a composite material in which the concrete provides compressive strength and the steel provides the tensile strength. Any failure of the bond between these two materials means that the composite has failed, and you do not have what was intended. You may have only useless and possibly dangerous deadweight.

This came to attention tragically in a collapse of a Miami office building. A slab had been added to the roof to surface a parking area. The building collapsed with the loss of seven lives. This prompted a unique program in which the building department makes inspections of the structural integrity of existing buildings. It was found that some of the buildings were beyond economical repair and were required to be demolished. One common cause of failure was the use of sea sand with a high salt content in the concrete. The salt corroded the reinforcing rods. It has been necessary to demolish a number of concrete buildings which were found to be unsafe and beyond economical repair.

When a concrete structure is in trouble, it is often repaired with steel. Very often no consideration is given to providing equivalent protection for the steel. Just about nothing done to a building after it is completed, except for installation of automatic sprinklers, benefits the fire suppression effort. Fire fighters should be eternally curious. A curious fire fighter-student saw steel columns being brought into a 3-hour rated fire resistive shopping mall. They were prepainted an exact match to the wall color in the mall. The roof was failing and they were provided to shore up the roof. Logically, if they were structurally necessary, as is obviously the case, they should be protected to the same requirement as the structure. The building department, to whom the case was reported, took several years to get the columns fireproofed.

It appears that if steel is designed into the structure initially, the proper degree of fireproofing will be specified. If it is an afterthought or part of a repair job, it

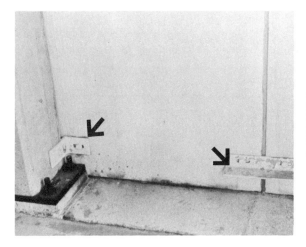

Precast wall panels often are held on with unprotected steel connectors (arrows). If these fail in a contents fire, the panels may fall on fire fighters, equipment, or other structures.

will usually be unprotected.

Steel connectors for precast units are often unprotected.

An alteration of an old Florida hotel showed a practice which may be common. In altering the building, it was decided to convert the high-ceilinged ballroom into two floors. A steel bar-joist floor and ceiling assembly was used because of its relatively light weight. Altered buildings in particular may not be consistent in their construction.

Concrete structures represent a massive heat sink compared to other structures. This is beneficial in that every Btu that goes into heating the concrete is one less to keep the fire going. However, heat fed back from the concrete can seriously affect the fire suppression forces. This problem should be recognized. It might be wise to delay overhauling until the structure has cooled down to minimize the debilitating effects of heat on personnel.

In any fire resistive building, the existing fire load both for the building as a whole, and in high fire load areas should be compared with the fire resistance rating of the structure. (See the discussion of fire load in Chapter 6, Principles of Fire Resistance.) Where high fire loads exist and where for any reason fire suppression efforts will be inadequate, severe structural damage is a distinct possibility.

One of the fire protection advantages of concrete construction is that there are no inherent voids. Voids may be created in the finishing of the building, but they are not a necessary part of concrete buildings.

Beams projecting below the ceiling, and waffle slab, may be quite satisfactory for industrial construction, but they are usually considered aesthetically unattractive. It appears to be a simple matter, and indeed it is, to nail or glue up some furring strips and improve the appearance and the acoustics by installing an acoustical tile ceiling. Unfortunately, this is often accomplished after the building is finished, and combustible tile, of high fire hazard rating, is used.

The combustibility is not intended; it just happens that combustible tile is cheaper than tile of low fire hazard rating. The interconnected voids above the furred-out tile make it possible for the tile to burn on both sides simultaneously. When inspecting a concrete building, take particular note if air supply diffusers are hung below the floor above, with the finished outlet in place, showing that it is intended that a hung ceiling be installed in the future. Warn owners against the dangers of combustible tiles and concealed spaces.

When suspended ceilings are installed as part of initial construction, it is more likely that they will have satisfactory fire hazard characteristics. But it is necessary for fire officers to be aware of the local code requirements. The tile usually will be only as safe as the law requires.

Combustible tile need not be suspended to create a serious hazard. Tile applied with a flammable contact adhesive to a noncombustible surface was a key factor in a fatal hospital fire.

Combustible voids can be created in a variety of ways. A heavy wooden decorative suspended ceiling is located in a concrete restaurant. The sprinklers are below the ceiling. Fire could burn unchecked in the void. Nothing is known of the fire resistance of the ceiling hangers, which are attached to anchors in the concrete. A fire might cause the collapse of the false ceiling. Wall paneling and screens for unsightly areas may also create combustible voids.

While voids are not an inherent part of concrete construction, they can be very useful. A church was built without voids. When the church was to be air conditioned, the only solution was to run the ducts on the roof. This created costly maintenance problems. In a modern office building, with its huge communications requirements, as much as one-third of the height from floor to ceiling may be in the ceiling or under floor voids.

Even when the void is noncombustible, it may still present problems. Combustible thermal insulation, combustible electrical insulation, and combustible plastic service piping may be concealed in the hung ceiling. Do not fail to examine the ceiling void promptly when seeking a hidden source of smoke.

Since the fire resistance of the building is achieved by the concrete construction, the hung ceiling, though it may be identical in appearance to the ceiling used for membrane fireproofing of steel (see Chapter 7, Steel Construction), is not required for the structural integrity of the building. The owner is free to remove it or modify it as he sees fit, within code requirements for flame spread (fire hazard as distinguished from fire resistance).

The Integrity of Floors

If a building is rated fire resistive, it is presumed that the floor will be a barrier to the extension of fire. The use of the building requires, however, that the floor be penetrated for utilities, plumbing, heating, air conditioning, electrical service, and communications wiring. There are proper ways to do this so that the integrity of the floor is not compromised, but they are often not used. Some of the methods used are open to question. A typical method of protecting an opening in which a heat duct is located is to erect a wall of steel studs and gypsum board around the opening. This is accepted as joining the floor to the ceiling to make a fire resistive barrier. The author knows of no test which demonstrated that this structure will not fail, possibly due to expansion of the steel studs, which may be fitted tight to the floor and overhead.

"Poke throughs" are holes provided to draw utility services up to a floor from the void below. The penetration of the floor for poke throughs may well negate the fire resistance rating of the floor. An article in *Architectural Record* quotes John Degenkolb as explaining that even if utility services openings are properly protected in the initial construction, tenants may ask for additional services, and that the common method of providing them is to drill through the floor.[7] Where this is recognized as a hazard, the opening is often stuffed with mineral wool. Tests showed that this is ineffective. In addition, the article cites the case of a fire high up in a high-rise building in a California city. The water that was used spread downward to all floors via poke throughs.

In addition, any holes that violate the integrity of concrete floors may provide passage for deadly fire gases. It is not safe to assume that there cannot be an extension of fire from floor to floor in a building because the floors are of concrete. It would be good to be aware of the general local practices in providing utility services.

As penetrations of floors for services increase, the floor may in fact be unable to resist the passage of fire adequately, and a legal requirement may be made for a suspended ceiling, which, together with the floor, will develop the necessary fire resistance.[8] In this case, the owner is not free to modify the ceiling.

The discussion above points up a serious gap in the public management of the fire problem. The building official, who approves the building as built, rarely ever sees it again. The fire official who inspects the building is often uninformed of building code requirements or has no responsibility for enforcement. Someday, an entire building and its unsuspecting occupants will fall through the gap. .

If it is known that utilities were installed in this manner, plans should provide for immediate deployment of units to the floor above the fire floor to control extension, and units should be assigned for salvage work on the floor below.

Panels and Partitions

In many modern buildings, both steel-framed and concrete-framed, prefabricated wall panels of various design are hung on the exterior of the building. The resistance of the joint where the wall panel abuts the concrete floor slab is open to serious question. Expansion of the panel may open a gap along the wall and permit the passage of fire.

Precast concrete and imitation concrete panels are often sealed with foamed plastic at the point where they meet the floor. The plastic, even if inhibited from flame, will melt out in a fire and permit extension along the perimeter of the building.

When panels like these are being installed, the fire department should look into the question of perimeter integrity. Openings between panels and floors may provide unexpected channels for fire extension.

Concrete shrinks, and at times the shrinkage is sufficient to develop large cracks. Such cracks may permit the passage of fire from floor to floor, even though the concrete is structurally acceptable. Shrinkage, distortion, or settlement may put door frames out of alignment, thus preventing the closing of fire barrier doors. Metal expansion joints have been known to conduct enough heat to ignite combustibles in direct contact on the floor above. Combustible fiber expansion joints can provide a difficult fire problem.[9]

If extension to the floor above is a possibility due to the defects cited here, the prefire plan should give consideration to putting a few inches of water on the floor above if extension is a serious threat and the available forces cannot control it. Again, bear in mind that the fire fighting forces are doing only what a sprinkler system would do if it had been installed. Do not allow fear of "unnecessary water damage" to cause the loss of a building. It is appropriate to discuss the implications of the tactics made necessary by the defective construction with the occupants before any fire. Cut the ground out from the "If you had only told me that — I would have ...."

Based on personal observation, it is certainly unsafe to assume that masonry partitions provide a fire- or smoke-tight seal at the top against the floor above, particularly where a suspended ceiling is installed.

Even if the tight seal was designed, installation of utilities may have created penetrations through the barrier. The same problem occurs in the case of firestops.

Imitations

This age is many ages, and one of them is the age of the fake. Imitation concrete panels are used, particularly on the exterior of buildings. In one case, they are made of two sheets of asbestos board, with a combustible foamed plastic interior. Epoxy glue is painted on the surface which is then coated with small stones.

The top panels of the building are genuine precast concrete. At the base, concrete block panel walls are painted with epoxy adhesive and stoned. The entire building appears to be of identical finish. The fasteners that hold the panels on the building are held in the plastic. If the plastic burns or melts, the panels will drop off the building.

Most imitation concrete panels are made of a steel framework with asbestos millboard. Real concrete is applied in a thin coat, or stone may be epoxyed on.

The current craze for wooden interiors provides fake wooden beams in all sorts of buildings. They may be of solid wood or of wood boxes or easily ignited and totally combustible polyurethane. They are held up in all conceivable ways, none of which have any degree of fire resistance. Expect such jimcracks to come down; do not be under them when they do.

Concrete's Behavior in Fire

The concrete in fire resistive construction serves two purposes; it resists compressive stresses and it protects the tensile strength of steel from fire. The latter function is accomplished approximately in direct proportion to the thickness of the concrete cover. By sacrificing itself, the concrete buys time to extinguish the fire; however, this sacrifice may also cost strength. Any concrete subject to severe fire attack should be suspect and avoided. If holes must be drilled or cut through a concrete floor to reach fire below, personnel should be supported independently of the possibly damaged floor panels. (Lightweight scaffold flooring might be a useful addition to the fire department's equipment.)

Proposed impact loads (such as from sledges or jackhammers) to damaged concrete should be carefully evaluated. All areas where spalling has reached reinforcing steel should be shored by competent persons before any work is done in the area. Concrete floors, seriously damaged on the underside, may give no clue on the upper side to the distress below.

When a concrete building is finished, there are no clues to help the fire officer determine whether the floors are of conventional reinforced or tensioned construction. In cutting through conventional reinforced floors, reinforcing rods may be cut to the extent necessary, without seriously weakening an otherwise undamaged floor. Tensioned concrete is quite another story. If the stranded cables which signal tensioned concrete are found, they must not be cut. Fortunately, they cannot be cut with bolt cutters so officers in command of units equipped with cutting torches or power saws should be wary of any request for such equipment "to cut reinforcing rods." To quote one expert, "It's just like sawing off the tree limb you are sitting on."

Opening a hole over the fire would probably subject one or more tendons to temperatures which would cause failure. In short, cutting into posttensioned concrete floors in a fire is not safe.

Precast Concrete

A cast-in-place, monolithic, concrete building is very resistant to collapse. Because it has a rigid frame, the loss of a column does not necessarily cause collapse. The load will be redistributed, and other columns, if they are not then overloaded, will carry the load.

Precast concrete buildings are another story. The individual columns, floors, girders, wall panels (they may be load-bearing or non-load-bearing) are usually "pinned" together by connectors. The fire resistance of the completed structure is very dependent upon the protection afforded to the connectors. Connectors are cast into the precast element, and mating connectors are provided on the structures to which it is to be attached. Bolts and nuts may be used, or the connections may be welded. The recess provided for the joints is then "dry packed," i.e., packed with a stiff mortar. Often no protective covering is provided for the connectors; it may not even be required.*

The possible effect of an explosion in a precast, pinned building should be evaluated where possibilities of explosions exist. Such buildings have none of the redundancy which characterizes the rigid-framed monolithic concrete building.

Probably the most spectacular disaster to date involving precast concrete construction took place in England. It was the so-called Ronan Point collapse.[10] A 24-story apartment building was constructed of precast floor and bearing wall panels, literally, a giant concrete house of cards. A gas explosion high up in one corner blew out one wall-bearing panel. This caused the collapse of the floor above which fell to the floor below. The impact load was too much for the next lower exterior bearing wall panel and it buckled. Thus, the corner of the building progressively collapsed down to ground level.

The gas explosion provided an impact load for which the building was not designed. The immediate palliative measure was to eliminate gas service from buildings of this type. It was discovered later that there was a basic design defect in the connections involving a simple brass bolt. After the interim report of the investigating tribunal, local authorities were required to notify all residents of tall buildings whether their building was built by this system. Presumably, as responsible adults, they could decide for themselves whether they wished to assume the risk. There is much to be said for this novel approach. Later, a number of such buildings were ordered evacuated.

There is a special hazard in buildings built of tilt-slab construction. Note that when these are being built, the builder carefully braces the walls with tormentors (temporary and adjustable braces). This is required for stability until the roof is in

---

*If an ordinary or noncombustible building is permitted by the building code, a builder is free to build a structure which appears to be fire resistive but which in fact is not. Typically, in such a case, the connectors are unprotected. The problem should not be overestimated: the connectors must be inherently quite massive to resist wind and other forces, and the fire load usually must be quite severe to cause failure.

place and tied in, thus stabilizing the building. The roof may be of any type of construction. If the roof is being lost in the fire, beware of wall collapse as happened in a California warehouse. Even if the roof is of concrete tee beams, the potential exists, as the fire will spall the concrete protection from the bottom of the tees.

Often these buildings are not required to be fire resistive but merely noncombustible. This is of no merit to the fire fighter crushed under noncombustible concrete.

If the building is sprinklered, careful observations should be made of the density and volume of smoke. If heavy smoke is being generated in volume, then the sprinklers are not controlling the fire and the roof is vulnerable.

Bridges

The problem of highway bridges is discussed in Chapter 7, Steel Construction. The same consideration should be given to concrete bridges. The bridges are not designed to be fire resistive and the concrete structure may be no more fire resistive than an equivalent unprotected steel bridge.

In March 1969, the Chicago Fire Department fought a fire in a truckload of barrels of scrap magnesium on the Dan Ryan Expressway under the 91st Street bridge. The magnesium burned violently for some time but by the use of several heavy fog streams, damage to the bridge overhead was averted.

Orthotropic bridges, in which the entire bridge is constructed of one huge trapezoidal-shaped hollow box girder, present an interesting possibility. The sections of bridge are joined together with gaskets soluble in gasoline. Gasoline leaking from a tank truck can dissolve the gaskets and convert the bridge into an impromptu gasoline tank. The author was informed that this has already happened in a west coast city.

Some Fires in Concrete Buildings

Fire-weakened concrete can make secondary missiles when struck by heavy hose streams. The Mid Hudson Warehouse, in Jersey City, NJ, was a large, heavily reinforced concrete building. It was ignited by a massive exposure fire in May 1941. On the second day of the fire, the author watched the New York City fireboat *Firefighter* attack the upper floor fires with a stream from its 5-inch tipped bowgun. Very shortly, sizable parts of the building were falling to the street on the other side. The report of the fire notes that several structural members had "disappeared"; they were rubble in the street.

A six-story British department store was built of concrete with aluminum exterior panels. A fifth-floor storeroom fire extended to the sixth floor and did heavy damage due to the melting of the panels.

A shopping mall has one-story stores on either side of a one-high-story, wide central corridor. The stores are unprotected steel. The central corridor roof is of prestressed concrete tee beams supported on unprotected steel. Fire involved the built-up metal deck roofs of the store and deformed the unprotected steel columns. This dropped sections of the roof, fortunately, without injury. The metal deck roof fire was controlled by a heavy caliber stream set up inside the building.

A Christmas tree fire in a 16-story reinforced concrete high-rise Dallas apartment resulted in $340,000 in damage and a situation where there could have been a large loss of life except for competent preplanning by the fire department. Utility and vent pipes were punched through the ceilings. The gypsum and steel stud enclosures extended only to the suspended ceiling, not to the underside of the slab. Lighting fixtures and air-conditioning grills pierced the ceilings. Thus, smoke had an unobstructed path from the eighth floor of origin to all floors above. Note the similarity in concept to the "protection" of columns by the ceiling of a floor-ceiling assembly commented on in Chapter 7, Steel Construction.

The GSA-operated Military Records Center near St. Louis, MO was severely damaged in a fire in 1973. The building was of concrete construction six stories in height. The top floor, where the fire occurred, was over 200,000 square feet of undivided floor space. The incredible fire load was 21,800,000 military personnel files in cardboard boxes on metal shelves.

The fire destroyed the top floor, which was subsequently removed from the building. The roof elongated several feet almost all around the perimeter. The elongation caused top floor columns to fail at either the top or bottom. It should be noted that the portion of the roof which collapsed held together for about 22 hours after the start of the fire. The building did all that could be expected of it and more, considering the huge fuel load.

Computer installations in the Pentagon were destroyed in a fire which ate through the concrete floor of the concourse. The computer area had been provided with a suspended acoustical ceiling of low-density fiberboard. A 100-watt bulb set the ceiling on fire. The guard on duty had been provided with a CO_2 extinguisher, another example of the reverse thinking which worries about "damage" by the extinguishing agent more than damage by the fire.

A multistory concrete warehouse was filled with rubber tires. A fire destroyed the building.

References

[1]Huntington, W. C., *Building Construction*, 3rd ed., Wiley, New York, 1965, Chapters 8 and 9, for more complete descriptions of concrete construction and terms.

[2]McKaig, T., "Brick Wall, Louisville, Kentucky," *Building Failures*, McGraw-Hill, New York, 1962, p. 81.

[3]Woods, H., "Research on Fire Resistance of Prestressed Concrete," Bulletin 131, Portland Cement Association, 5420 Old Orchard Road, Skokie, IL, 1961, p. 59.

[4]Lew, H. S., "Safety During Construction of Concrete Buildings — A Status Report," *NBS Building Science Series 80*, January 1976, Center for Building Technology, National Bureau of Standards, Washington, DC.

[5]"Design of Wood Formwork for Concrete Structures," *Wood Construction Data* No. 3, National Lumber Manufacturers Association, Washington, DC, p. 1.

[6]"Rescue Operations," *Fire Command!*, October 1970, p. 27. (See also "Bridge Collapse," letter by Brannigan, F., *Fire Command!*, January 1971, p. 8.)

[7]"Makeshift Holes for Utilities Negate Floor Slab's Fire Resistance," *Architectural Record*, December 1968, p 147.

[8]Emerick, E. L., Jr., "Fire-Resistive Ceilings Can Protect Penetrated Floors," *Fire Journal*, Vol. 61, No. 6, November 1967, p. 9. (See also the discussion of membrane fireproofing in Chapter 7, Steel Construction, in this book.)

[9]Houvenaghel, Lt., "Expansion Joints: A Structural Advance Demanding New Fire Fighting Techniques," *Fire International*, No. 29, July 1970, pp. 81-86.

[10]"Systems Built Apartments Collapse," *Engineering News-Record*, May 23, 1968, p. 54.

CHAPTER 9

Flame Spread

Flame spread is a fire phenomena which has come to be recognized during the author's lifetime. It was not recognized as such two generations ago, and it is still imperfectly understood.

Flame spread, or rapid fire growth, which is the preferred term, often is determined by the interior finishes and furnishings. In addition to contributing to fire growth, the materials used in the finishes and furnishings may contribute heavily to the generation of smoke and toxic products, thereby threatening the life safety of both occupants and fire fighters.

While furnishings may contribute greatly to fire load and smoke generation, this chapter will concern itself chiefly with interior finishes with some mention of interior floor finishes. It should be noted that the NFPA *101®* , *Life Safety Code®* does not consider floor finishes to be interior finish. Rather, provisions for floor finishes are based on NFPA 253, *Critical Radiant Flux of Floor Covering Systems Using a Radiant Heat Energy Source.* As far as the fire fighter is concerned, however, they all become one problem.

At one time, lime plaster was almost the only finish for ceilings, though some ceilings were of embossed steel, usually called "tin ceilings," and some were of wooden boards called matchboarding. The plaster did not contribute to the fire, in fact, it absorbed heat and slowed the progress of the fire.

Today, interior finishes are many and varied. They include such materials as plastics, gypsum board, wood, plywood, fibrous ceiling tile, plaster, paper, paint, and fabrics. The fire properties of these materials have no relationship to fire resistance or fire resistance ratings; they can present the identified problems in fire resistive and non-fire resistive buildings. The conditions under which they become involved in a fire will influence their behavior. Despite extensive testing, it is not yet possible to predict the performance of materials under all conditions. The same material may be used in several different ways. If used in one way it may be subject to regulation, while not so subject if used in another way.

The Flame Spread Problem

The flame spread problem can exist on any surface — ceilings, walls and

Today, a wide variety of interior finishes may be found in buildings. Despite any fire resistivity of the basic construction, combustible materials such as this wood shingle ceiling, are subject to rapid and disastrous flame spread.

floors. The flame spread problem can also involve concealed building elements. Typical are combustible fiberboard sheathing, or sound-deadening board and the vapor seal on glass-fiber insulation.

There are now an almost unlimited host of materials used for interior surface finishes. Some of them are high-density fiberboard (called pegboard when punched with holes), fabric, paper, carpeting, plastics, etc. Some of them are recognizable for what they are; others are disguised as other materials, such as plastic imitation plaster, imitation wood sheathing and beams, and imitation glass mirrors.

The terrible 1944 circus fire in Hartford, CT, in which 168 lives were lost, was fueled by flames spreading rapidly on the paraffin impregnated canvas. (Parenthetically, it might be noted that the caged runway for the wild animal show was erected across the main exit "for only 10 minutes." So much for the "temporary" blocking of exits.) *

Flame spread restrictions in codes are typically strictest for corridors and less strict for rooms off the corridors. This has created a problem in the modern open-plan office. Where does the corridor begin and end? One fire department solved it simply. If there is no wall, there is no room; it is all a corridor. The next question, then, is about the flame spread on the dividers. Are they even regulatable, or are they furniture, presently almost universally unregulated?

Low-density fiberboard is used for sheathing and for soundproofing. It is concealed in walls. A common method of ignition is by a plumber's torch. The plumber sees only the gypsum board wall; the fire enters the wall along the copper tubing and ignites the wall. The fire often goes undetected till it erupts violently. A school under construction in New England had fiberboard on wood studs covered with gypsum board. The fire extended to the typical metal deck

---

*In 1961, 285 persons died in a circus tent fire in Rio de Janeiro.

roof. The damage was in the millions in this "noncombustible" building.

In a Des Moines, IA shopping center, sprinklers were omitted in one part of the concealed ceiling space. Combustible fiberboard formwork for the concrete roof topping had been left in place. The fire burned the combustible fiberboard. The estimated loss was $700,000.

The unexposed side of combustible surfaces can burn unobserved and protected from fire department streams. For this reason it is not sufficient to correct such a situation by flame retardant surface treatment of the exposed surface.

There are four ways in which interior finishes affect a fire:

- They contribute to the speed with which flashover occurs.

- They increase fire extension by surface flame spread.

- They add fuel to the fire.

- They generate smoke and toxic gases.

Hazards of Fiberboard

In the 1930s, low-density fiberboard, first made of wood fibers, then of sugar cane, came into use. It was produced in 4- by 8-foot sheets with a painted surface. It provided a cheap interior finish that could be quickly installed. When punched with holes, it acquired desirable acoustical properties.

Consider the situation in a building with deteriorated plaster ceilings. Replacing the plaster would have required removing the old plaster, placing the new plaster, and painting — three messy jobs. An alternative was for carpenters to come in over the weekend, nail furring strips to the ceiling and nail fiberboard tile

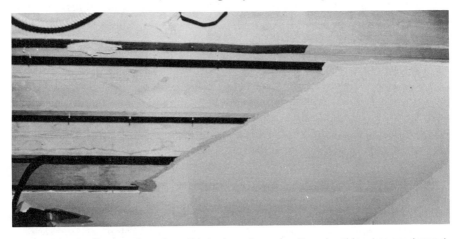

Low-density fiberboard, such as this broken piece of ceiling sheathing, has good sound-insulating qualities but is readily combustible. It is often concealed by gypsum board behind which it can burn unseen.

to the strips, producing a new attractive ceiling with acoustical benefits for Monday morning.

Fiberboard sales grew astronomically. The material was used both for initial construction and for rehabilitation. It was substituted for wood sheathing in frame construction. It was also widely used for interior finish in low-cost construction.

Fire officials slowly grew to realize that this brought a new problem, the spread of flame on interior surfaces. The industry vigorously fought any attempt at regulation. It developed a Department of Commerce approved "industry standard" fire test which amounted to a fire in a thimble full of alcohol. A light flame retardant coating enabled the board to pass this "test."

A very serious hospital fire occurred in the midwest. The hospital was lined with combustible fiberboard. The hospital sued the manufacturer. There was a sealed, but reportedly heavy, settlement in court. Thereafter, the opposition of the industry changed to cooperation in the development of fire standards.

Decorations and Contents

Decorations, usually hanging from the overhead and the walls, and contents, such as furniture or stock in trade, are very difficult to control. There are few regulations. They are difficult to write in the specific manner required by law, and difficult to enforce where regulations exist. Decorations often represent a serious problem because the hazard goes unrecognized even by people who consider themselves "fire-conscious."

One of history's most serious fires occurred in 1859 in a cathedral in South America. The overhead was decorated with fabric to resemble clouds. A fire broke out and about 2,000 persons were reported to have died. Closer to home are the fatal fires that have occurred when well-meaning adults create a "haunted house" for Halloween and use cotton sheeting, which has a high flame spread, in proximity to light bulbs or candles.

Combustible decorations are commonplace and difficult to control.

In the Cocoanut Grove Night Club in Boston, in 1942, nearly 500 persons lost their lives. The fire spread explosively along combustible decorations at the ceiling line*. A small counter-type restaurant in New York had plastic imitation wooden beams and plastic imitation fruit on the ceiling. The author photographed it for a lecture series. Not long thereafter, the building was involved in a multiple-alarm fire that started in the restaurant. Fire fighters who responded on the first alarm reported that flames were shooting across the avenue.

In a south Florida city, a two-story, brick and wood-joisted building has been "rehabilitated" into a typical tourist boutique. The ceiling is of rattan. The

---

*Recently, the author reviewed pictures of the Cocoanut Grove fire published in the *New York Times* of Nov. 30, 1942. One picture clearly shows the characteristic round blobs of mastic by which tile (undoubtedly combustible) was adhered to the ceiling. This fact apparently had not been noted previously.

Combustible decorations are commonplace and difficult to control. In this gift shop, the ceiling is of high flame spread rattan. The authority under which it was installed is in doubt.

building department blamed the fire chief. The author was curious but the chief never returned his telephone calls.

A scientist in a research laboratory was annoyed at the reverberations from the metal building. He designed pyramidal fabric bags that were filled with urethane scraps and hung them from the ceiling to suppress sound. He was quite surprised when they were ordered removed.

The author was giving a slide-illustrated seminar on combustible decorations in the northeast. For lunch, a group went to a popular local restaurant. The ceiling was draped in burlap. The fire marshal commented that there should be no problem since the place was sprinklered. He had no concept that fire could race over the burlap much faster than the sprinklers could operate. In addition, it is hard to imagine a greater panic-producing situation than fire falling from the ceiling on people, literally igniting hair and clothing.

In a popular west coast seafood restaurant, the ceiling is decorated with cork floats. It also is sprinklered, but again, the possibility of flame spreading faster than sprinklers can operate is unevaluated.

Contents have changed tremendously in the past few years. Fire loads and rates of heat release of the current contents of homes, stores, and offices are much greater than they were when most of our codes were written.

At one time, the only contents item that provided a real flame spread or rapid fire growth problem was the Christmas tree. While there are still some tragedies every Christmas season, such as the multiple loss of life at a San Francisco yacht club one year, the hazard is relatively under control. A new year-round contents hazard has arisen, however. The nature of furniture has changed. Plastics, both solid and foamed, have replaced much of the wood, cotton, wool, and other materials that were previously used. The hazard is often unrecognized.

The Consumers Product Safety Commission created a huge acrylic plastic lobby display to show what they were doing for the safety of people. This pile of "solidified gasoline" was located between the building's employees and the exits. A young attorney, schooled in fire protection, protested, and it was removed.

A Texas museum is very proud of a walk-through exhibit — a room full of figures at a rodeo. The larger-than-life-size figures are made of papier-mache. Visitors walk on pieces of foamed polyurethane covered in burlap. A possible second exit is blocked by bales of hay. Neither the creators nor the visitors have any concept of the potential hazard.

The author observed a serious fire in the Smithsonian Institution's National Museum of Science and Technology. The heat was unparalleled in the experience of fire fighters. The source of the heat was not understood until it was learned that huge sheets of clear plastic had hung from the overhead as part of the exhibit. They were completely consumed by the fire.

Testing

The National Bureau of Standards conducted tests of fires in typical residential basement rooms.[1] The "basement" could be dropped from the title as immaterial. The room could be a living room, a motel room, a fire station recreation room, a doctor's office, a reception lobby, an elevator lobby in a hotel, or wherever.

The purpose of the tests was to demonstrate the need for a revision of the widely accepted standard time-temperature curve used in many fire resistance tests. The significance of the tests, however, goes well beyond its stated objective in the author's opinion. For one thing, they lay to rest the comfortable concept that "15 minutes can elapse before the fire is serious." For another, the "advantages" of fire extinguishers in many occupancies appear to be open to question. There is such a short time between ignition and unbearable conditions in the extinguisher operating zone that the probability of a person being able to use an extinguisher effectively is greatly reduced. (The report of these tests is also referred to in Chapter 5, Garden Apartments, and in Chapter 11, High-Rise Construction. Some illustrations from the report are found in Chapter 5, including a typical room layout.)

The concept, espoused by some, that a high-rise fire can be allowed to burn unchecked for the 20 minutes or more it takes for fire suppression forces to react adequately also is certainly open to question. The full report is a gold mine of information. Do not be deterred by the dry technical writing. Obtain it and study it.

Consider Test No. 14 in the report: The test room was 3.3 by 4.9 meters in size (10.7 feet by 15.9 feet). The furniture consisted of a sofa, an upholstered love seat, an ottoman, two end tables, a coffee table, and two bookcases. The carpet was olefin pile with foam rubber backing. The walls were plywood. The ceiling tiles were wood fiber, 200 flame spread. The fire was started by matches in about a pound of newspaper on the couch.

The following times are abstracted from the report of Test No. 14:

Minutes

0	Ignition.
0:58	Back rest cushion of sofa burning.
1:33	Flames reaching almost to ceiling.
2:12	Wall paneling next to sofa burning.
2:25	Heavy flames out the doorway.
2:30	Olefin carpeting ignites.
2:38	Hot flaming zone extends down to 2 feet from floor.
3:55	Flames out doorway down to 5½ feet below ceiling.

Six minutes after ignition, the average gas temperature in Test No. 14 was reported as 700°C or about 1,300°F.

The rapid fire development shown in the tests should make fire officials wary of drawing the conclusion that a particular fire was incendiary "because it spread so fast."

Alterations

Fire departments should stay aware of any clues that alterations are planned and ensure that all pertinent codes are obeyed. The author has observed a number of instances in which interior finish materials that would not have been permitted under the code when the building was built were installed after occupancy.

A chain of low-priced steak houses decided to change the interior decor to have a rough, plastered look. Combustible paneling, finished to look like plaster, was used. It was done without consulting local code compliance officials. When the question was raised, the management provided certification from the manufacturer that the material met the local not-too-rigorous flame spread standards.

Rigid-foamed polyurethane has been used for interior finish in houses. There is at least one fatal house fire on record where such a practice was a contributing factor. When the extreme hazard was recognized, an attempt was made to reduce the hazard by plastering the interior.

A restaurant in a Texas university city is made into a cave by the use of foamed polyurethane. Reportedly, pieces broken off by students are easily ignited. A 1972 fire in a French nightclub, in which 144 persons died, was fueled by foamed plastic used as interior finish.[2]

Acoustical treatment is another potential source of trouble. Many acoustical installations are made without consideration of flame spread consequences of materials used.

Insulation

Insulation laid in ceilings must be kept free of light fixtures. The heat from the fixture can ignite the vapor seal. In one case, scores of sprinklers operated in vain below the ceiling while the fire spread freely over the insulation vapor seal. The actual weight of fuel which burned was minute; the damage was extensive. Again, this points out the fact that fire is not just a chemical phenomenon, not just a heat-balance equation, but that all factors that might affect the loss must be considered.

Foamed plastic applied to walls and ceilings for insulation has been involved in a number of disastrous fires.[3] When such insulation is installed, it should be protected from exposure to flame by a half-inch gypsum board covering. This only protects against ignition from a small source; in a well-developed fire, the plastic will be involved, probably almost explosively, as the gypsum fails or falls away.

Insulation on air ducts installed years ago was usually a hair felt with a high flame spread. The presently used aluminum-faced foil (not aluminum paper) glass-fiber insulation presents no flame spread problem.

Electrical insulation may be nominally self-extinguishing. The tests are conducted on wire not under load. When electrical wiring is operated at or above its rated capacity, the heat breaks down the insulation and flammable gases are emitted. The McCormick Place fire in Chicago possibly started in the flaming insulation of an overloaded extension cord. Large groups of electrical wires can provide mutual support for self-sustaining ignition.

A spectacular multiple-alarm fire occurred in the cable vault of a telephone building in New York City. The fuel was the insulation in the cables.

In another instance, workmen were foaming plastic insulation around wires to prevent cross contamination between two sections of the Browns Ferry nuclear reactor. They were using a candle to check for air leaks. The fire spread over the wiring. The actual amount of wire burned might have been worth $50,000. The loss to the taxpayers from the cost of coal over that of nuclear fuel was estimated at $300,000 per day for a year and a half.

Glass-fiber insulation was installed around an upstairs bathroom for sound-proofing in one of a series of connected expensive dwellings in a Texas suburb. The insulation excluded the piping from the house heat and it froze. A torch used to thaw pipes set fire to the paper vapor seal on the insulation and ten houses lost their roofs or upper floors.

Plastics

The first "structural" use of plastics of which the author became aware is a tunnel by which visitors pass through a shark pool at a Florida tourist attraction. The plastic tunnel supports tons of water. It is an interesting speculation to consider the possibility of flame spread on the surface of the plastic.

A feature of the Expo '67 exhibit at Montreal was the United States Pavilion. It was a geodesic dome covered in clear plastic. Architectural writers marveled at this beautiful lightweight futuristic building. After the building was turned over to the City of Montreal, an alterations worker accidentally set fire to the plastic with

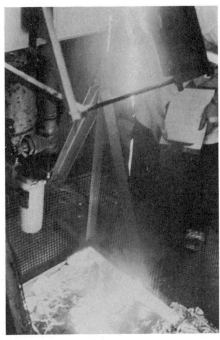

Plastics, in the many forms used for building construction, pose new and serious problems of flame spread and fire control. Note the dripping fire (upper right) as acrylic plastic burns in the National Bureau of Standards radiant heat panel tester.

a cutting torch. The structure was destroyed in 10 minutes.* Clear plastic has many inviting uses and not all designers understand the terrible hazard which is created.

This was certainly true at Summerland on the Isle of Man, part of the United Kingdom located in the inhospitable North Atlantic. In an attempt to create a Mediterranean type resort, the authorities built a huge entertainment complex called Summerland. A major portion of the walls and roof was acrylic plastic (polymethyl methacrylate), another large portion was asphalt-coated steel. The interior paneling was of pressed fiberboard. This created a combustible void wall. The stage was set for disaster.

A glass-fiber reinforced stand was set on fire outside the building. The fire extended to the asphalt-coated steel, then to the plastic. Fifty persons died in the fire.[4]

A recent newspaper article describes the rehabilitation of an old Washington, DC house by its architect owner. The floor of a child's bedroom is a sheet of clear plastic to allow light to penetrate below.

Even when the problem of combustibility is understood, other fire problems that are possible with structures incorporating large amounts of lightweight plastics are not self-apparent to architects.

---

*The author was told by an official of the plastics supplier that they were reluctant to supply the material and had impressed on the U.S. officials the necessity for strong precautions. He said that when the pavilion was turned over to the City of Montreal, they wrote to the city explaining the hazard. Some materials are probably too hazardous to use. Administrative controls are too unreliable.

The retired dean of Architecture at Columbia University describes a proposed pressurized membrane which would cover a complex of four to six government office buildings.[5] The structure would encompass 400,000 square feet. The purpose is to provide an ideal year-round environment independent of the atmospheric conditions. There is no consideration mentioned of the toxic atmosphere which would be created within the structure by the release of the toxic products of combustion locked within the contents and structures within the environmental envelope. These toxic products can be released with a single match.

Carpeting

The use of carpeting has changed in recent years. It is used now on walls and ceilings in addition to floors. It was not self-evident to designers or even to officials that the location where a material is installed may materially increase the rate at which a fire can grow.

A spectacular 1980 fire in the Hilton Hotel in Las Vegas spread from floor to floor outside the building because of flammable carpeting on the walls and ceiling of the elevator lobbies. The fire grew into one terrifying fire front extending many stories in height.[6]

Combustible Tile

The Hartford Hospital fire of 1961 developed in a soiled linen chute, rolled out, and attacked the ceiling. The ceiling was fire retardant painted combustible acoustical tile glued to a gypsum board ceiling. The fire roared down the corridor. Samples of the tile were tested and showed a flame spread not sufficient to give as intense a fire as was experienced. An entire 25-foot section was removed and sent to a testing laboratory. The flame spread was very high. The difference was due to the adhesive that attached the tile.[7]

The Clark County Fire Department Report on the MGM Grand Hotel fire in Las Vegas in 1980 notes that it required 12 tons of adhesive to adhere the tiles to the ceiling of the Casino. Because it was flammable, the adhesive added a large fuel load to the fire.[8]

Materials that are used can be confusing and it is not wise to make any assumptions. Specific knowledge is needed. For instance, imitation brick observed in a building materials display was made of vermiculite, a noncombustible mineral material. Imitation stone in the adjacent display was made of "glass-fiber reinforced polyester resin plastic." One common trade name, almost a generic name, for this material is "Fiberglas." The glass fibers are noncombustible but the plastic is another matter. Try a sample for yourself.

Prevention vs Suppression

About this point the objection might be raised: "Those problems are for the fire prevention division; this book is intended, you say, for fire suppression forces."

Laying aside the ridiculous concept on which many fire departments operate — that fire prevention and fire suppression are two disparate activities, there is a valid interest in the subject on the part of the fire suppression forces. Prefire planning should not start, as it often does, with "Engine Three will take hydrant A and advance a line up the stairway." Prefire planning properly begins with a scenario of the fire potential in the building with all its ramifications. The rate at which the fire will grow is certainly one of them. It makes a tremendous difference to know that the "fire resistive" building is lined with fast-burning materials. It may be, for instance, that the normal small line tactics surely will be ineffective and a waste of the time that is needed to bring heavy caliber streams into play.

Approaches to the Problem

There are several possible approaches to the problem of fast fire growth. One approach is to eliminate high flame spread materials. This is the path taken for U.S. merchant ships. Foreign ships may contain combustible trim and veneer. Reliance is placed on fire suppression from automatic sprinklers whose water supply might fail, particularly during an engine room fire. The rules for U.S. merchant ships severely limit the combustibility of construction materials and surface finishes. The rules for furnishings, however, apparently might not effectively rule out a large aggregation of combustible chairs such as those that provided the fuel for the huge fire that severely damaged the BOAC terminal at Kennedy Airport in New York.[9]

A second approach is to separate the material from the source of combustion. In 1946, the author was investigating a serious fire at a naval facility. An engineer working there related that the Public Works Department, when building an office in this warehouse, was recessing fluorescent light fixtures into the combustible tile ceiling. He pointed out to them that he had worked on the development of fluorescent fixtures and that a defective ballast can rise to 1,500°F. His advice to avoid contact with the ceiling had been ignored. He was happy to help the author locate the fixture which caused the fire. Thereafter, the *National Electrical Code*® (*NEC*®) was amended to include the following:

410-76 Fixture Mounting.

(a) Exposed Ballasts. Fixtures having exposed ballasts or transformers shall be so installed that such ballasts or transformers will not be in contact with combustible material.

(b) Combustible Low-Density Cellulose Fiberboard. Where a surface-mounted fixture containing a ballast is to be installed on combustible low-density cellulose fiberboard, it shall be approved for this condition or shall be spaced not less than 1½ inches (38 mm) from the surface of the fiberboard. Where such fixtures are partially or wholly recessed, the provisions of Sections 410-64 through 410-72 [for flush and recessed fixtures] shall apply.

Combustible low-density cellulose fiberboard includes sheets, panels, and tiles that have a density of 20 pounds per cubic foot (320.36 kg/cu m) or less, and that are formed of bonded plant fiber material but does not include solid or laminated wood, nor fiberboard that has a density in excess of 20 pounds per cubic foot (320.36 kg/cu m) or is a material that has been integrally treated with fire-retarding chemicals to the degree that the flame spread in any plane of the material will not exceed 25, determined in accordance with tests for surface burning characteristics of building materials.

This is a widely violated provision of the *NEC**. Unfortunately, while the code provides a technically correct definition of the combustible material, the problem is that most electricians do not recognize it except by its common, almost generic, name which happens to be that of a supplier of combustible fiberboard and many other building materials.

A metal-working shop in a northwest city had no fire problems, except one. There was a dip tank in one corner. Fumes from the dip tank rose to the overhead and coated the wooden roof with oil, creating a high flame spread hazard. A minor fire in the tank spread rapidly under the roof and destroyed the plant. The tank should have been isolated from the main part of the plant.

Some codes require a metal door sill separating more flammable carpeting, as in a motel room, from less flammable carpeting, as in a corridor, intending to prevent the spread of flame from a room fire to the corridor. The author has been told by an expert that this would be effective if the door is closed, but would have no effect if the door is open.

A third approach to the problem of the fast fire growth is to "flameproof" the material with a fire retardant coating. One difficulty with surface coatings applied after the hazard is discovered is that they do not protect the rear of the material. Fire can burn unimpeded in the void. The surface treatment of the ceiling tile in the Hartford Hospital fire, mentioned previously, was defeated by the adhesive.

Fire retardant surface coatings are effective only if applied as specified. The square footage per gallon is much less than that which is customary for paint. It is difficult to get painters to refrain from spreading the material too thin, not to mention the economic incentive to use as little as possible. One fire department requires that the requisite number of can labels be turned over to the inspector. This is not foolproof but it at least demonstrates the fire department's interest that the job be done right.

---

*The author observed this hazard in the display area of a building housing an antique automobile collection. He phoned the owner and explained the problem. Thereafter, a mechanic went to the building, lowered all the lights in the shop area 1½ inches (there was no ceiling there) and left the others as they were. A communications failure.

The ceiling of this corridor in a boarding facility was covered with low-density combustible tile reportedly painted with a fire retardant coating. The flame spread rating of the tiles, as installed and coated, was not known. The adhesive (the circular patterns on the ceiling surface) that affixed the tiles to the substrate contributed to the rate at which fire spread across the ceiling as the tiles were attached. In laboratory tests, the effect of the adhesive is a factor in establishing flame spread ratings.

Pressure-treated, fire retardant wood is often used where a wood surface is desired and flame spread requirements must be met.

Changes in the formulations of plastics can reduce their ignitability. Materials can also be formulated to be flame resistant. One manufacturer, for instance, offers a special hardboard that has a very low flame spread rating and zero fuel contributed and smoke density. Such a board would be an excellent substitute for the common hardboard and pegboard that have high flame spread characteristics.

Testing and Rating

Let's look at some tests and how they are used.

The Tunnel Test

The basis for regulation of flame spread is found in NFPA 255, *Method of Test of Surface Burning Characteristics of Building Materials*, commonly referred to as the "tunnel test." The test was developed by the late A.J. Steiner at Underwriters Laboratories Inc. and is also known as the ASTM E 84 and UL 723 test. The test is fully described in the NFPA's *Fire Protection Handbook*.[10] Essentially, a test sample 20 feet long and 2 feet wide forms the top of a tunnel or long box.

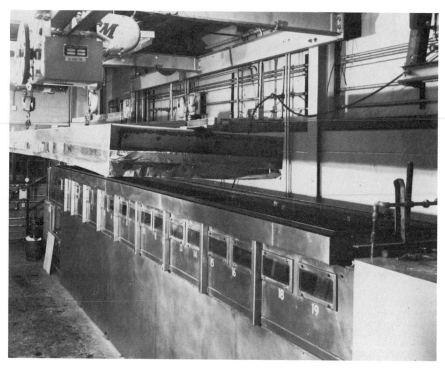

This is a Steiner Tunnel used for determining flame spread characteristics of materials and finishes. The sample is fastened under the cover. With the cover in place, a gas flame is lighted under the sample at one end and the rate at which flame spreads along the sample is observed through the window ports.

A gas fire is lighted at one end, and fire progresses along the underside of the top of the box or test panel.

There are two comparison points in evaluating test results. The flame spread over asbestos cement board is set at 0. The flame spread over red oak is set at 100, and tests have shown that fire can reach the end of a red oak test panel in 10 minutes. Other materials tested are rated on a scale determined by these points. A material on which the fire reaches the end of the tunnel in 5 minutes is given a flame spread rating of 200. If the fire has only reached half way down the tunnel when the 10 minutes have expired, the material is rated at 50.

In NFPA 101, the *Life Safety Code*, surface materials are classified as follows:

Class A Flame Spread	0 - 25
Class B " "	26 - 75
Class C " "	76 - 200

(In past years, Classes D and E were used for materials of very high flame spread. These have now been eliminated.)

Most building codes have requirements for ceiling and wall surfaces based on the tunnel test standard. A typical requirement might be for a Class A flame spread rating for corridors and exit ways and less restrictive requirements for offices. Sometimes higher flame spreads are permitted for sprinklered buildings, which is another reason for fire department's taking action when sprinklers are shut off.

When the tunnel test is run at UL, two other characteristics beside flame spread are measured. The other two are "fuel contributed" and "smoke developed." Together the three provide the "fire hazard" rating of the material being tested.

Fuel contributed is calculated by measuring the heat coming off the test and subtracting the calorific value of the gas burned in the test fire. The balance is the heat generated by the test sample. The standard is the same as for flame spread. The fuel contributed by asbestos cement board is taken as 0, that of red oak is taken as 100. The number assigned to the material tested is proportional to the two standards.

Smoke developed is calculated by measuring the obscuration as the smoke passes a photo electric cell placed in the stack from the test tunnel. Again, the loss of light from asbestos cement board is taken as 0 while the loss of light from red oak is taken as 100.

It appears that materials with smoke developed ratings of 300 or more can be expected to generate substantial amounts of smoke.

Note that only light obscuration is measured and not any of the other effects of smoke or gases, such as toxicity or irritation.

The NFPA *Life Safety Code* makes use of the smoke developed rating by including a requirement that surface finishes be classified at 450 or less on the smoke developed rating of NFPA 255. In the author's opinion, this cannot be regarded as a very restrictive requirement.

The flame spread testing standard is widely accepted. It is almost impossible to find literature for commercially installed ceiling tiles without finding reference to the flame spread rating. This is not wholly the case in the home market. Homeowners are frequently unaware of the flame spread testing standard or how it is applied to the tiles they buy.

In some cases, manufacturers have stressed low flame spread ratings, neglecting to point out that the material advertised received a very high "smoke developed" rating. All materials listed by UL are tested for all three characteristics.

A serious deficiency is the fact that a code may require a low flame spread rating for newly installed ceilings but usually does not require the removal of old combustible tile in a remodeling job. Left-in-place tile is an extreme hazard to occupants and to fire fighters; this caused the loss of two fire fighters' lives in Wyoming, MI in 1980.

There is confusion in the trade exemplified by the term "fire rated." Ceiling tiles that are part of a listed floor and ceiling assembly that is intended to provide fire resistance must be replaced by equivalent tiles observing the installation practices followed in the original installation. Tiles that merely meet flame spread requirements are not adequate for installation in required fire resistive assemblies; thus, the installation of a ceiling that meets flame spread requirements under a combustible floor does not impart any specific fire resistance to the floor itself.

A number of tradesmen and officials with whom the author has spoken seem to have no concept of the difference between installing a true fire-rated floor-ceiling assembly and simply putting up flame retardant tiles under an existing floor.

The Radiant Panel Test

The National Bureau of Standards developed the ASTM E 162 Radiant Panel Flame Spread Test. Samples for this test are only 6 by 18 inches. For some materials, results from this test can be correlated in a general way with the tunnel test but no direct relationship should be assumed. The radiant panel test has been used to develop information after serious fires, but would probably not be valid evidence in a criminal prosecution based on noncompliance with a code. This test is more fully described in the NFPA's *Fire Protection Handbook*, 15th edition.[11]

Carpet Tests

Floor covering, specifically carpeting, has been an important contributing factor in a number of serious fires. In recent years, a number of tests and standards have been developed. Commercial carpeting now manufactured is required to meet a test that measures ignitability of carpeting from a small source such as a dropped match or cigarette. It is a screening test (Federal Specification FF-1-70, popularly known as the "pill test"), and seven samples out of eight samples must pass the test.

Wide expanses of carpeting, on walls as well as floors, are increasingly used as interior finish. Some materials may have extremely high flame spread characteristics. Fire departments should be well aware of the buildings where such problems may exist.

There are tests to rate carpeting for its flammability when attacked with a more severe fire than a cigarette: the UL Carpet Chamber test and NFPA 253, *Standard Method of Test for Critical Radiant Flux of Floor Covering Systems Using a Radiant Heat Energy Source*.

UL Carpet Chamber Test. Underwriters Laboratories now classifies flooring and floor covering materials in the carpet test chamber. The chamber is 8 feet long and 2 feet wide. The gas flame impinges downward on the carpeting.

Three levels of rating are assigned:
0-4   Overall flame propagation less than 48 inches. After-flaming inconsequential.
4-8   Overall flame propagation less than 96 inches. Possibly significant after-flaming.
8-25   Flame propagation over the entire 8 feet from 12 minutes to 5 minutes, 25 seconds.

If the flame spreads the full 8 feet in less than 5 minutes, 25 seconds, the material is too hazardous for listing.

The Radiant Flux Test. NFPA 253 is a test measuring the flammability of the carpeting. The result derived from the test is the Critical Radiant Flux (CRF) of the sample. This is the amount of external radiant heat energy (measured in watts per square centimeter) below which a flame front will cease to propagate.
The test uses the National Bureau of Standards Radiant Panel set at an angle. The carpeting sample is set flat in its normal position. The sample can be tested

The effects of radiant heat flux (radiation) can be seen in this completely destroyed house trailer. Combustible interior surfaces with close radiation and re-radiation spans can quickly intensify a fire.

with or without an underlayment. If a carpeting is flammable, the flame spread will probably increase, that is, the critical radiant flux will decrease if an underlayment is present. The underlayment acts as an insulator, keeping the heat from the carpeting from being dissipated to the floor.

For a comprehensive and easily understood discussion of this important subject, study *Flammability Testing for Carpeting*, by I. Benjamin and S. Davis of the National Bureau of Standards.[12]

The *Life Safety Code* now stipulates that for flooring systems, the minimum CRF permitted is 0.5 for health care facilities and 0.25 for other regulated facilities. The Critical Radiant Flux standard is being worked through the code adoption procedures of the model codes.

Industrial Risk Insurers (formerly FIA) recommends a CRF of .45 watts/cm^2 for unsprinklered areas and .22 watts/cm^2 in sprinklered areas.[13]

Critical radiant fluxes on carpeting involved in several serious fires have been recorded. Thus, when regulations are adopted specifying a minimum CRF, there can be a relationship to specific fire danger which is not so apparent in other standards.

Don't Be Mousetrapped

Developing adequate tests is very costly and time consuming. Tests must be repetitive, that is, tests performed in the same apparatus on the same material should produce the same results. Tests should be reproducible; others, using similar equipment and procedures, should observe the same results. Laboratories doing testing sometimes arrange "round robin" tests on samples cut from the same unit to help achieve uniformity.

Fire departments should be very wary of being induced by promoters to conduct "tests" which "prove" that a particular product is the answer to all problems. Testing is a difficult and exacting profession. Leave it to the experts.

All standards must be expressed in numbers to be legally enforceable. Early standards that spoke of "flammability no greater than wood or wallpaper or other commonly used materials" were struck down as vague and unenforceable.

Fire Combat

An estimation should be made of the potential for rapid flame spread over interior finishes. For instance, the many churches with exposed wood plank surfaces present the potential for extreme flame spread and, thus, a fully developed fire on arrival.

The fire can develop with unimaginable speed. In a Florida church fire, photographs showed only light smoke as units arrived. Within a minute, heavy fire was rolling out every window and door. A fire department that has not preplanned properly for this type of fire may well waste time using handlines that will be completely useless. Two or three heavy caliber streams positioned so as

(Top) A plank and beam church in Killarney, FL as it appeared to arriving fire fighters. Note the light smoke and fire in only one corner. (Bottom) Within a minute, flashover occurred and the entire planked ceiling was involved. Note the fire fighter carrying the "Memphis nozzle," a heavy stream appliance in the left foreground. The fire was controlled in a few minutes with three heavy hose streams. Photo credit Bob Cross, Orlando, FL Fire Department.

to sweep all surfaces can literally snuff out the main body of fire in a few minutes. Locations for the streams should be preplanned.

Tests have shown that burning carpeting alone will fill a corridor with fire. If carpeting is on fire as in a corridor, avoid the use of fog. The use of fog may drive the fire along the surface. If it gets around a corner in the corridor, the fire may get away. Direct a straight stream through the ball of fire to wet the carpeting on the opposite side. When this is done, revert to fog. The suggested tactic worked in a motel fire in Bethesda, MD. The carpeting was removed and sent to Underwriters Laboratories for testing. It had a flame propagation index of 23+.

If combustible tile ceilings have been left in when new ceilings were installed, make every effort to get the material removed. Point out the potential for a damaging fire from what is literally junk.

If the tile is left in, anticipate heavy smoke coming from concealed fire in the overhead void. Do not hesitate to make an aggressive assault on the suspended ceiling. Bring it all down to expose the combustible tile to fire streams.

Be aware that the tiles are probably burning freely on the unexposed surface. Determine if there are any ducts or other connections from the void above the tiles. In a hotel fire, a number of people died far above the fire floor. The fire was in an old combustible ceiling that had been left in when the new ceiling was installed. The smoke moved through old air ducts.

Fire burrows into combustible fiberboard and hangs on with amazing tenacity. "Extinguished" combustible tile should be torn out and removed to a safe outside location.

Expect rapid flame spread in plywood or other paneling interior finishes unless known to be adequately treated. Note that the installation almost always is on furring strips that create a void behind the paneling which is not hit by hose streams. If the paneling is glued, expect the glue to soften and release the paneling. The glue may or may not be flammable.

Some years ago, a government agency was building an extension on an existing university hospital. The excavations were in and winter was approaching. A temporary "snow roof" was erected to permit work to go on during the winter.

Buildings, such as this motel, should be carefully surveyed in a prefire plan. Back-to-back rooms often have a chase, or channel for pipes between them. Sound insulation with a high flame spread vapor seal in the chases presents a serious hazard.

The cheapest available material was selected for waterproofing. It proved to be a hemp-reinforced bituminous impregnated paper of the type used to protect finished lumber or overseas shipments. It has a very rapid flame spread. In one corner of the snow roof was a change room for employees. A stove for the burning of wood scraps was located here with the stack out through the roof. There was a potential for sending a sheet of fire up the face of a functioning hospital. People in the fire service must be aware of the possible hazards of situations like this and prepare for the worst.

It is difficult and sometimes impossible to determine the flame spread characteristics of materials in place. Low flame spread and high flame spread materials can have the same appearance. Field testing can be dangerous. In an eastern city, the fire department was conducting a campaign against flammable decorations. An inspector examined a display for a popular brand of blended whiskey. He decided to try it with his lighter. The display was indeed flammable. The fire went to a multiple alarm. The city paid a heavy judgment and two fire fighters were seriously injured.

If samples can be removed, a simple test may help determine the extent of the fire potential even if it is not adequate for any legal purposes.

References

[1]Fang, J., and Breese, J. N., *Fire Development in Residential Basement Rooms*, NBSIR 80 2120, Center for Fire Research, National Bureau of Standards, Washington, DC.

[2]"White Grotto Becomes Black Tomb," *Fire Journal*, Vol. 65, No. 3, May 1971, p. 91.

[3]Shaw, Gaylord, and Robert Gillette, "Urethane, a Deadly and Pervasive Peril," *Los Angeles Times*, Jan. 21, 1979.

[4]Lathrop, James K., "The Summerland Fire," *Fire Journal*, Vol. 69, No. 2, March 1975, p. 5. (This should be kept on hand. Many have not heard of this disaster involving clear plastic expanses.)

[5]Salvadi, Mario, *Why Buildings Stand Up*, W. W. Morton & Co., New York, 1980, p. 276.

[6]"Investigation Report on the Las Vegas Hilton Fire," *Fire Journal*, Vol. 76, No. 1, January 1982, p. 52. Fed by carpeting on the walls and ceilings of lobbies, fire spread up the outside of the hotel from the 8th to the 30th floor. Eight died.

[7]Juillerat, E., "The Hartford Hospital Fire," *NFPA Quarterly*, January, 1962.

[8]"Report of the MGM Grand Hotel Fire," Clark County Fire Department, Las Vegas, NV, 1981, p. v-16.

[9]Abbott, J. C., "Fire Involving Upholstery Materials," *Fire Journal*, Vol. 65, No. 4, July 1971, p. 88.

[10]McKinnon, G. P., ed., *Fire Protection Handbook*, 15th edition, National Fire Protection Association, Quincy, MA, pp. 5-49 to 5-50.

[11]See Reference 10, p. 5-51.

[12]Benjamin, I., and Davis, S., *Flammability Testing for Carpeting*, NBSIR 78 1436, Center for Fire Research, National Bureau of Standards, Washington, DC.

[13]"Carpeting," *The Sentinel*, 2nd quarter, 1982, p. 8, published by IRI.

CHAPTER 10

Smoke and Fire Containment

Smoke and Gases

There are a number of products of combustion. This chapter will deal with what the average citizen, and some researchers, call "smoke." In fact, one should distinguish between smoke, vapor, and fire gases. Smoke is solid particles or droplets of condensed vapor and is the usual visible product of combustion. Smoke probably provides the first alarm of most fires, can do more damage to property than the fire or even water, may reduce visibility to zero, and the irritant particles may cause retching to the point of immobility.

Under fire conditions, gases remain gases even when cooled to normal temperatures. Vapors revert to liquids or solids when cooled. Fire gases can cause injury or death when inhaled (or in some cases absorbed). Some investigators are examining the possibility that some fire gases can "paralyze" the ability to function so that the victim cannot escape.

Carbon monoxide (CO), a common fire gas, is flammable. Depending on the concentration in air, the CO may merely "light up" or may produce a deflagration or a detonation. (For a discussion of these terms see the NFPA's *Fire Protection Handbook*, 15th edition, p. 3-15.) The shock wave of a CO detonation has blown entire buildings apart. Carbon monoxide is not the only gas to be encountered under explosive conditions; for example, gases from burning polyurethane exploded and caused the collapse of a wall on the sound stage of the Samuel Goldwyn Studio in Hollywood in 1974.

Gas can accumulate in any enclosed area. Ventilation of the building does not foreclose the possibility of a CO explosion. The gas-air mixture can pocket in a fully or partially enclosed location. The sudden increase in the production of heavy black smoke is one presignal (unfortunately muddled these days by burning plastics). Another signal to veterans is the sucking of smoke back into the building (probably the origin of the term "backdraft"). Such explosions can occur at any time during the fire.

Many fire gases are toxic, and in sufficient dosage, lethal. Carbon monoxide is accepted as the principal toxic gas, but others are significant. The Cleveland (OH) Clinic, like many hospitals in 1929, used nitrocellulose X-ray film even though cellulose acetate (safety film) was available. A fire in film storage facilities sent clouds of deadly nitrous oxide gas (one breath can kill) up through the building and 125 doctors, nurses, and patients died. Some hospitals today use nitrocellulose "test tubes" for blood tests. There is very likely a substantial storage in the supply room. A fire in a "general supply room" may duplicate the Cleveland Clinic fire. *

Until the end of World War II, a nitrocellulose base film was used for commercial 35 mm motion pictures. Fires involving nitrocellulose film caused many disasters. There still are places where old, deteriorating (thus, specially dangerous) nitrate film is stored, such as libraries and museums and maybe even some hospitals. Such places demand intensive preplanning with emphasis on defensive operations.

Polyvinyl chloride (PVC) is a very effective electrical insulator and has made itself indispensable. When it burns, toxic gases are emitted.

It is the author's opinion that toxic gases from a burning combustible metal deck roof assembly were the chief cause of the loss of over 160 lives in the Beverly Hills Supper Club fire.

One gas or another might be more toxic in a given situation. Nothing generated by the fire is good to breathe. The message is clear: All personnel should wear self-contained breathing apparatus (SCBA) in potentially toxic atmospheres. * * Much more care is needed in deciding when it is safe to remove SCBA.

Smoke and gas have different physical effects. At one time, filter masks were popular in the fire service. They are very dangerous. The filter removed the solid particles allowing the odorless CO through. A chemical converted small amounts of CO to harmless carbon dioxide (CO_2), but in many cases the CO concentration was too much.

Smoke particles will plug up a screen. This is a problem in prisons. Screens are used on windows to prevent passage of contraband. The smoke from a mattress fire plugs the screen and the inmates die. (The same plugging effect, experienced with filter masks, should be expected in any location where the outgoing air is filtered, such as where toxic materials are handled.)

Smoke will "plate out" on surfaces especially if the electrical charge of the surface is opposite to that of the smoke. Gases simply ride with the air currents.

The removal of deposited smoke particles, particularly smoke from plastics, can be extremely costly. If the smoke is contaminated, the cost can be astronomical. In a nuclear weapons plant, a fire did about $1,000,000 worth of conventional damage. The cost of cleaning up the radioactive contamination was about 50 times that.

During a relatively small fire in a Binghamton, NY office building in 1982, a

---

*Inquire of your hospitals what they use for "disposable" tubes. Ask if they ever heard of the Cleveland Clinic. There is no "out-of-date" experience in the fire service except perhaps shoeing fire horses.

* *An Atomic Energy Commission Laboratory had a rule: "use a respirator when fighting a fire unless it is plutonium; then use a self-contained breathing apparatus." They changed the rule when it was pointed out that CO can kill you right now, while plutonium might kill you forty years from now.

transformer cracked, releasing the toxic products that are produced when the polychlorinated biphenyls (PCB) used in transformer coolant to make them nonflammable, are exposed to heat.

Clean-up costs were estimated at $11,000,000, with some estimates as high as $20,000,000. The cleanup was very slow and costly because of the precautions which had to be taken to protect the workers. Fire fighters should be aware of the extremely toxic nature of PCB and the fact that it can be released from any transformer in which it is used by fire or accident, or even in routine maintenance.[1]

It is only in recent years that the movement of smoke in a structure, as distinct from the extension of fire, has received serious consideration.

In fact, only recently has there been recognition of smoke and toxic gases as the most significant killers in fires. Not too many years ago, it was considered quite helpful to explain to the relatives of fire victims that "they did not suffer — they suffocated before they were burned!"

Confinement of Fire

When it was learned in such fires as the Parker Building in New York in 1908 that fire could extend up open stairways in "fireproof buildings," the concept of

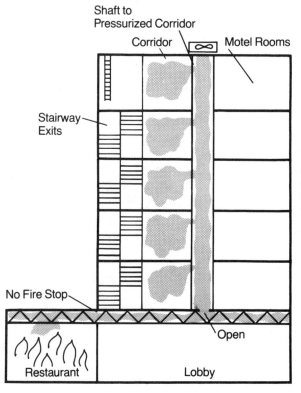

Smoke originating from a fire in one portion of a building will travel upward through vertical voids polluting the entire building. Deaths in many hotel fires have been the result of such smoke movement.

cutting the building up into compartments no larger than one floor was developed. Any connections between floors were to be cut off so that fire would not be transmitted.

The enclosure of stairways is not something which comes naturally. If there is no strong code, there is no enclosure. In some locations, a "monumental" stairway will be unenclosed.* The strange argument is that sufficient exits are provided as protected stairways. This completely begs the question of the open stairway as a transmitter of smoke and heat.

Elegant open stairways lend architectural charm, but provide easy passage of smoke and toxic gases to contaminate every portion of the building, endangering fire fighters and occupants.

Stair and corridor closure devices are of two types: self-closing and automatic.

A typical example of the self-closing type is the room or stairway door equipped with an automatic door check which closes the door after it has been opened. (Some self-closing doors are equipped with fusible links so that the door can stand open on the door check. If the link melts, the door will close.)

Automatic doors that close only in case of fire are equipped with fusible links. When the link melts, the overhead rolling door rolls down, the sliding fire door closes horizontally. Sliding fire doors may roll down a track by gravity or be pulled shut by a weight. In either case, they should be checked by lifting up on the weight which holds the door open. The door should close properly without any assistance.

Some overhead rolling fire doors in high bay buildings may be equipped with a fusible link high up in the fire wall. This link could trip with a flash of fire and drop the door behind fire fighters who have passed through the door. Never go

---

*The Grand Staircase of the Paris Opera House is the most famous monumental staircase in the world. It is truly impressive. It sometimes appears that it is the architect's dream to recreate as much of it as the client will buy.

This rolling fire door, recessed into the ceiling and designed for automatic operation, is doomed to fail. The pole (arrow) for manually pulling the door down hangs in the track recessed in the wall at left. If the fusible link releases, the door will jam and not close.

through an automatic fire door — vertical rolling or sliding — without blocking the door. In a New York department store fire, it took five fire fighters to get a sliding fire door open. Often the retainer is damaged and the door jams shut.

Dampers are required in ducts wherever they pass through a fire resistive barrier. These dampers normally are equipped with fusible links.

For a fuller discussion of protection of openings, see the discussion on Opening Protection in the NFPA's *Fire Protection Handbook*, 15th edition, p. 5-98.

All these concepts are inadequate to control smoke movement and may not even control spread of fire. It is a common practice to block open self-closing doors for the convenience of those who use them.*

The fact that stairway and corridor smoke and fire barrier doors are often blocked open is no bar to working for self-closing doors in apartment and hotel occupancies. Many fire reports note that the spread of fire and smoke in such occupancies was due to doors left open, not only in the unit of origin, but also in other units.

Non-fire resistive buildings and older fire resistive buildings were usually built without any consideration of the control of fire and smoke movement.

Forty-six lives were lost in Chicago's La Salle Hotel and 119 lives were lost in Atlanta's Winecoff Hotel in 1946 in fires that spread fire and toxic gases up open stairways. Municipal authorities were shocked into action. Across the country and Canada, hotels, office buildings, and hospitals were ordered to close up open stairways. Unfortunately, very often the stairway was not enclosed at the lobby level and sometimes not at the mezzanine and lower lobbies. All too often

---

*The author noted at the office of a prominent consumer advocate, whose offices occupy several floors, that the stairway doors were blocked open by piles of literature about hazards.

the effort was wasted as doors installed at great expense were blocked open. *

The fire department should bear down on the management of buildings to keep such doors closed. Signs should be provided, employees should be educated, and citations should be issued when necessary. There is no justification for the law requiring sizable expenditures to enclose a stairway and then taking no steps to see to it that the enclosure is maintained.

Doors provided in corridors may be fire doors or smoke barrier doors depending on what was required by the code. Smoke barrier doors should not be expected to serve as fire doors though one pair of such doors stopped the extension of a destructive fire from the combustible part to the fire resistive part of a hospital in the midwest.

Smoke Sensitive Trips

The development of the smoke detector has provided equipment which is sensitive to smoke and which can be arranged to eliminate the problem of blocked open doors. In one method, the doors are held open by electromagnets. The door latch system can be triggered automatically by any fire alarm system, or sprinkler water flow alarm, or manually. Any fire alarm cuts the current to the magnets and the doors release. Such a system is easy to test. A recommended procedure for a patrolled building is for the test switch to be thrown after the cleaners are through for the night. All doors should close. A guard making rounds reports any nonfunctioning door. In the morning the current is restored. As doors are pushed open, they engage the magnet and stand open thereafter. Some doors are too heavy for magnetic latches. They are held by mechanical latches that are released the same way. Two-leaf doors should be checked to see that the doors close in the designed sequence. The roller that accomplishes this is often damaged.

In other cases, the door control is individual. One or two detectors (one on each side) trip a door as smoke is detected. In some cases, the detector is integral with the door check. If the door is open and the detector senses smoke, the door is released. Such doors do not or at least should not have latches to hold them open. Fire fighters should carry wedges to block them open as necessary in fighting fires.

A variety of smoke detecting systems are used to manipulate air-handling systems in the event of fire. If the system depends on smoke reaching a detector in the ducts, the system must be operational at all times to be sure smoke is delivered to the detectors. The fire department should be in on the early planning of any smoke removal system. Fire department facilities to manipulate the system should be provided. These should be at a location not likely to be made untenable by the fire.

---

* An associate of the author was looking into hotel firesafety in the United Kingdom. He was told: "We put on a real push to enclose stairways and provide fire extinguishers. Now the extinguishers are used to block the stairway doors open."

Escalators

Escalators are widely used in department stores and public occupancies. A number of methods of "protecting" the openings with various types of sprinkler or spray systems have been adopted in various codes.

One method involves the use of water spray nozzles directed downward through the opening. The fire department should be aware of this type of system because the water damage may be considerable, and the blame for damage may be put on either the fire department or the fire protection sprinkler system.

In some systems, a line of sprinklers is located around the escalators. The sprinklers are shielded from one another (to prevent one sprinkler from freezing the other). While this might have some effect on a nearby fire, it would have no effect on smoke moving from a fire not in the immediate vicinity.

In short, the life safety of the occupants depends on the full sprinkler protection of the building and not partial protection at floor openings. This is another reinforcement for the concept expressed several times in this text: "If sprinklers provided for life safety are shut off, the occupancy should be closed."

Escalator openings, and atriums such as this one, permit smoke pollution of the entire structure. The sprinklers in this atrium may help to control fire spread but will not assist in smoke confinement.

Some escalator protection systems involve mechanical exhaust systems. These should be tested periodically. In other cases, motor-driven shutters are used to close the escalator opening. These also should be tested.

Venting

The movement of combustion products in a building is controlled by a number of interacting factors that are discussed in some detail in Chapter 11, High-Rise Construction. All these factors can be operative in any building, but in low-rise buildings, the effect of all but thermal energy and the wind are usually minor. If the air-conditioning system has been equipped to be used as a smoke removal system, its operation should be fully understood by the combat forces.

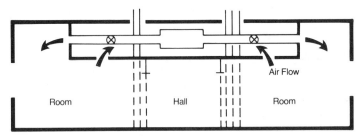

This typical motel-hotel air-conditioning system has a ducted supply and an unducted return interconnected from room to room and floor to floor. Smoke generated in one room will pollute all the others.

The simplified concept of ventilation, as presented in training texts, shows fire fighters opening the roof, heat and smoke rising by gravity through the opening, and the engine companies moving in on the fire accompanied by fresh outside air. This tactic was developed when buildings had sawn wood roof joists that provided some reserve strength even when burning as discussed in Chapter 2, Principles of Construction, and Chapter 3, Wood Construction. Many recent structures have trussed roofs for which the basic rule can be offered: "If ventilation is necessary, the roof is probably unsafe to work on." It would appear to be reasonable that fire departments should push for requirements for automatic vents on the roofs of buildings which present serious collapse potential in the event of a fire.[*]

A ventilation method that is safer than going on the trussed roof might be to open two holes opposite one another in the side of the roof or a mansard. The fire attack usually should be from windward, leaving the leeward opening for venting. Since the forest of lumber in a trussed roof might be burning on all surfaces, an attack from both openings might be effective if the building is clear of personnel.

If this problem is spotted on the building plans, and is discussed with the designer, perhaps two or more removable openings (or even window sashes) might be provided in the side of the roof void. Such openings might be obtained as a compromise in lieu of pushing for expensive automatic vents.

---

[*]Such a code requirement would not come easily. The author has been told more than once by fire officers who have dealt with code-making bodies that a common sentiment is: "It is not our function to worry about the safety of fire fighters. They must take care of themselves. We worry only about the occupants."

Shopping Malls

Shopping malls can present very severe smoke movement problems. The basic life safety provision in even the simplest mall is complete automatic sprinkler protection adequately maintained. Heavy smoke accumulations can build up from a transformer fire, a fire in an unsprinklered area (such as a store under construction), or in plastics burning so intensely that sprinkler discharge might be inadequate. (Note too, that some plastics can generate much more smoke when burning freely than ordinary carbonaceous materials, combustible metal deck roof may be protected from the sprinklers by the ceiling.)

Shopping malls are designed to create an infinite vista of goods that can't be resisted. The entire structure can become smoke polluted rapidly. The safety of all depends almost entirely on a functioning sprinkler system.

The mall is built to be open and the smoke will move freely throughout the structure. In many cases, there are no doors on stores, only security gates. When there are doors, personnel should be assigned on the prefire plan to close them to limit smoke damage.*

Roof vents may be provided, either manual or automatic. Those the author has examined do not appear to be nearly adequate in area, but they are better than no vents at all. If the vents are automatic, the mall management should be required to have them tested periodically. There also should be a backup manual method of operation of vents. If manual, the fire department obviously should know how to operate them. In one observed case, Christmas decorations were hung from the vent mechanism. The fire department should regard all such building fire equipment in the same light as its own equipment; insist on operational readiness and forbid tampering.

---

*It is curious that many fire departments which conscientiously try to limit water damage have no concept of the fact that smoke damage can be far greater. Perhaps it is because fire departments are often blamed for water damage but smoke damage is considered inevitable.

In this shopping mall, the two small smoke vents in the roof are probably inadequate. The skylights could be broken out, but careful coordination would be required to prevent injury below.

Exhaust Fans

Many times fire department exhaust fans are not properly used. Air is fluid like water and must be confined to be moved efficiently. The fan lowers the air pressure on the intake side. The atmospheric pressure pushes the nearest air into the fan stream. If the fan is placed in an open window, much of the air that is moving through the fan is coming from the outside entering between the window frame and the fan housing. In fact, some of the same smoke particles may be recycled repeatedly. The fan area around the opening should be closed up. It is useful to carry a roll of flexible reinforced sheet plastic, such as that used by some fire departments for salvage covers.

Salvage covers can also be used to close up the opening. Whatever can be done to restrict the backflow is useful. Fans should be located to take advantage of the wind, not oppose it. When a hole has been opened in a roof, or a skylight removed, a fan placed over the opening will greatly increase the efficiency of the ventilation. *

Large One-Story Sprinklered Warehouses

Fire suppression in large warehouses requires the utmost of prefire planning and cooperation between the fire department, mutual aid departments, the owners of the warehouse and the insurance company covering the risk. If the

---

*When removing a skylight, always turn it over so that another fire fighter with obscured vision is made aware that there is an open hole nearby in the roof.

Proper use of exhaust fans is essential for removing smoke and fire gases. The Kensington (MD) Fire Department uses salvage covers to seal off the door opening and get full efficiency out of the fans.

building is not sprinklered, or is inadequately sprinklered, a major, if not total loss, is a reasonable expectation. *

If the building is adequately sprinklered, major reliance must be placed on the sprinkler system. The system should be so arranged that the fire ground commander has access to the sprinkler pressure gages. If they are poorly located, a strong effort should be made to get gages located in an accessible location. If in a fire, the system is holding its pressure, or if the pressure is restored to what it should be by fire department pumping, keep it going! Although it will be churning the smoke and making visibility difficult, do not fall into the trap of shutting it down "to see what is happening." The fire may burst out and get beyond control in a very few moments.

Do not let "water damage" bother you. In a fire in a high bay building, it is inevitable that a large number of sprinklers will operate on any fire, and it is anticipated that a large area will be soaked to control the fire. If the sprinkler system is not holding the fire, the roof will be failing and it will be too dangerous to put fire fighters on the roof for manual venting.

For some years, an argument has raged as to whether sprinklered warehouses should be vented or left closed up allowing the sprinklers to extinguish the fire. There are competent people on both sides of the issue. The two organizations which specialize in insuring highly protected properties are in strong disagreement. One of them, the Factory Mutual System, argues for keeping the building closed up; the other, Industrial Risk Insurers, supports venting. A technical committee is working to attempt to resolve these divergent views. The problem is that theoretical evidence can be used to point either way, testing requires large-scale, expensive installations and equipment, and actual fire experience is very biased by the circumstances of a particular fire.

In any event, many factors enter into the design and construction of these huge warehouses, but it is unlikely that fire fighter safety is very high up on the

---

*In June 1982, a loss of over $100,000,000 occurred in a K Mart warehouse near Philadelphia. Those who are responsible for such warehouses should study the report of this fire when published.

list. The fire department may have been presented with an impossible problem. Be extremely conservative in taking risks.[2]

Some of the problems of high-stacked warehouses are also discussed in Chapter 7, Steel Construction.*

Regulation

The only measure of combustion products which has been used to regulate materials used in buildings is evaluating the degree of obscuration or loss of light due to smoke. There does not appear to be, at this time, any practical method of regulating materials based on toxic gases produced.

It was noted in Chapter 9, Flame Spread, that as a by-product of the ASTM E 84 Flame Spread Test (Steiner Tunnel Test), it is possible to provide a value for "smoke developed." The base points are the same as for flame spread: 0 for smoke emitted from asbestos cement board and 100 for smoke from red oak. The amount of light obscuration which occurs over the 10 minutes of the test is measured, compared to the base points of 0 to 100, and a value is assigned to the material. There are a number of problems with the test method, but it is used to a limited extent in a number of codes. Some cities specify limiting figures for smoke developed for finish materials in such areas as exit routes or in institutional occupancies.

The National Bureau of Standards has a smoke density chamber method of evaluating potential smoke density of building materials (ASTM E 662, NFPA 258). In this test, a sample of the material is burned. The value derived from the test is the Specific Optical Density, i.e., the concentration of smoke from a given sample of material collected in a chamber of a given volume, and examined through an optical path of a given length. The advantage of this method is that a code can take into account such factors as the size of the room, the area of the material, and the distance through which a person might have to see to escape. The NBS test is presently being used by the Port Authority of New York and New Jersey for certain applications and by some aircraft manufacturers.

Mis-education

Most experienced fire fighters are familiar with the heartbreaking cases in which it is apparent that victims who died in a fire had adequate warning to escape and failed to do so. There are of course many possible reasons. One that has not received the attention it deserves is the *mis*-education provided by TV and movies.

---

*The question of the interaction of sprinklers and venting is being studied by a committee representing a variety of interests. For information contact: Fire Venting Research Committee, c/o IRI Suite A2000, 175 West Jackson Blvd., Chicago, IL 60604.

The essence of TV and movies is to create the illusion of reality. Even those who have worked in those media have difficulty at times separating actuality from fantasy. Time after time the hero(ine) does the impossible in a fire, dancing through raging flames in the hold of an overturned ship, rushing into a blazing barn to save the horse.

It is simply not enough to tell people to get out if there is a fire. They should be told that what they saw on the screen is an impossible fake, created for entertainment. "If they showed what it is really like, you would think your tube was broken." It is bad enough when a life is lost inevitably in a fire but it is infinitely worse when it is realized that the person could have been out of the building, and worse yet, was out and went back in for no good reason.

Public fire education programs must stress the toxic nature of the fire environment. We are surrounded by huge quantities of toxic gas locked in the materials with which we live, requiring only a match for release.

References

[1] "Building Cleanup Costs Balloon," *Engineering News-Record*, May 13, 1982, p. 38.
[2] "Smoke and Heat Venting: The debate goes on." *The Sentinel*, first quarter, 1982, p. 7.

CHAPTER 11

High-Rise Construction

This chapter should be read in conjunction with Chapters 6, 7, 8, 9 and 10. Much of the material in those chapters is pertinent and is not repeated here except in passing references.

This chapter is not intended to present a complete discussion of all the problems of fires in high-rise buildings. The basic purpose is to direct attention to the fact that not all high-rise buildings are similar, that in addition to the height, the specifics of the construction of the building, and the era in which it was built, are very significant factors in planning to combat the high-rise fire.[1]

There are various definitions of high-rise buildings. The author participated in the International Conference on Fire Safety in High-Rise Buildings convened by the United States Government's General Services Administration at Airlie House, Warrenton, VA, in April 1971. At that conference, the generally accepted definition of a high-rise as a building beyond the reach of aerial ladder equipment was firmed up. The author protested that this definition was incomplete and that a mistake of two generations ago was being repeated. After New York's famous Triangle Shirtwaist Factory fire of 1911, laws were enacted providing that factory buildings over six stories in height (note the relationship to the aerial ladder) should be sprinklered. This provision of the law was changed after a fire in 1958 in the Monarch Undergarment building (a building less than six stories high in the same neighborhood) caused the loss of lives of over 40 garment workers.

The aerial ladder is a most limited tool and is no guarantor of life safety. The principal life safety question in any building is not its height but the time it takes occupants to reach a safe environment. Only a few feet from safety, more than 160 people died in the Beverly Hills Supper Club fire. Twelve people died in 1982 on the fourth floor of a newly opened high-rise hotel in Houston. The fire was confined to one room. In December 1980, 26 persons died on the third floor of the Stouffer's Inn in Westchester County, NY.

The definition is acceptable and valid, however, when it applies to fire department tactics, and is so understood in this chapter. Fire officers should bear in mind that buildings of any height can present any of the problems presented here. In short, this chapter is not to be skipped because there are no buildings of more than a few stories in your area.

It is a serious error to consider all high-rise buildings as a single problem. There are very fire-significant construction differences between one and another. The particular buildings in your area must be studied in detail so that all will be aware of the particular modes of failure to be expected.

High-rise buildings may not be what they seem. Tons of masonry were piled on New York's Flatiron Building to create the impression that the steel-framed structure was a masonry bearing wall building.

With a few specialized exceptions, such as the Vehicle Assembly Building at Kennedy Space Center, all high-rise buildings are designed to resist, to some degree, the effects of fire on the structural frame of the building and on the floors. Whether the design concepts are adequate to cope with all the possible effects of a fire is quite another matter.

Early fire resistive buildings were said to be "fireproof" and the designation persists in some codes. The more accurate term is fire resistive.

It is beneficial to understand the limitations of the term "fire resistive." In summary, fire resistance is intended to provide, within limits, resistance to collapse by structural members and floors, and resistance to the passage of fire through floors and horizontal barriers.

Fire resistance itself is not concerned with life safety, the control of the movement of toxic combustion products, or with the limitation of dollar loss.

Fire resistive buildings have evolved over the last century. The following classifications are broad and intended only for general guidance.

Early Fire Resistive Buildings, 1870-1930

Many of these buildings are still in use. The dates given are approximate. Some deficiencies were noted in fires and corrected in the local code of the city affected, while the same defects were continued in buildings erected in other cities. Again, do not be deceived by the "high-rise" title of this chapter. Courthouses, churches, libraries, college and school buildings were built to be fireproof and contain some or all of the defects cited here, and very possibly, some that are not noted here.

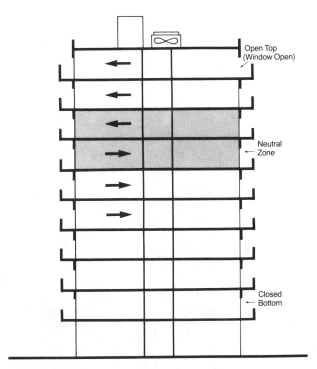

A graphic representation of conditions that tend to raise the neutral zone in a high-rise building.

Fire resistance was designed according to the opinion of the designer, usually without any real evidence to substantiate his opinion. Cast-iron columns were regarded as inherently firesafe and were often unprotected. Fire protection of columns, where provided, was of terra cotta tile.

In the Government Printing Office in Washington, the constructor surrounded steel columns with a brick box into which concrete was dumped from a height. We know now that this causes concrete to separate, and thus, render the fireproofing questionable.

In the Parker Building, another early fireproof building in New York, the electrical conduit was placed inside the tile. As it heated during a fire, it expanded and pulled the tile off the column. It was determined that a single line of columns was crucial to the stability of the building. It collapsed, and several fire fighters lost their lives.

In the 1970s, the author was asked his opinion of the Federal Building on Washington St. in New York City. It was built at the same time as the Parker Building. Buildings built at the same time often have the same defects. The General Services Administration made a structural analysis of the building and determined that it had the same basic construction as the Parker Building.

An article in the *New York Times* (Feb. 19, 1982), described how the old Federal Building will be remodeled into a mixed-use structure combining space for retail shops on the lower floors and more than 300 cooperative apartments on the upper floors. The article noted that an atrium would be installed.

This points up the fact that there is nothing "out of date" when studying buildings. The mistakes of the past live on for many generations.

The first fireproof floors were brick or hollow tile segmental (curved) arches sprung from wrought iron or steel beams. These arches were tied to resist outward thrust. The steel ties can be seen exposed across the ceilings of such buildings. If they are heated by fire, the consequences are unpredictable, but probably the floor section will fail.

This floor, shown from the underside, in Boston's famed Faneuil Hall, illustrates early "fireproof" construction. Brick segmental arches are supported by unprotected steel beams on cast-iron columns. A contents fire could cause failure of the beams and collapse of the structure.

The curved arches do not provide a walking surface. In some cases, piers of brick were built up to support wooden girders and a wooden floor structure. This creates a combustible void under the floor. This void is connected to other voids by holes in the cast-iron columns through which straps connect the steel floor girders. In the Parker Building, for instance, the leveling was provided by cinders which seriously overloaded the structure.

Segmental brick and tile arches were supplanted by flat terra cotta tile arches. In early construction, no protection was provided for the underside of the steel beams. Later, tile soffit blocks or "skew backs" were developed to provide protection to beams. These are often removed, and the steel is unprotected. Tile arch floors were delivered as packages manufactured to specifications. Often, the specifications for the floor units were in error and the floor was laid in an improvised manner. These floors pass water easily and a serious salvage problem should be anticipated.

Most importantly, these floors should not be cut through without taking precautions. The floor units are arches and cutting out a tile may drop the arch, such as was reported to have happened when a sprinkler riser was being installed in the old Post Office Building in Washington, DC. It is possible that the entire bay may collapse if the floor is disturbed. Even if personnel are protected against collapse, the collapse of a large floor section may cause a burst of fire upward.

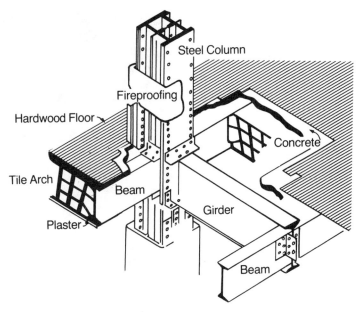

Detail of column to girder connection and floor construction of a store built in Chicago in 1892. Although fireproofing was added to the column, there was no real protection for the underside of the beam.

These structures were built without any appreciation of the fire effect of vertical openings. Ornate open stairways and light walls were a standard feature, and in most cases, still remain.

The Chicago Chapter of the American Institute of Architects starts its tour of Chicago skyscraper construction at the Rookery. This structure has an ornate iron stairway structure in the lobby leading to an open winding stairway. There is one other open stairway. Typical of Chicago and some other cities, but not universal, there is an outside steel fire escape. To get to the fire escape, a person breaks the glass in the corridor door of an office.[*]

Note that the glass doors typically used for offices provide no effective barrier to the extension of fire into the corridor.

The Southern Building in Washington is a typical early fireproof building. A florist shop occupied the first floor. A stairway (a so-called access or accommodation stairway that will be discussed later) had been cut through to the basement. A basement fire completely polluted the building via the unenclosed stairway so that hundreds escaped down the outside fire escapes. It was feared that the fire escapes would fall off, and fire department ladders were used to brace the platforms and take some of the weight.

The District Building, the city hall of the nation's capital, is a typical fireproof public building of its period with a wide open ornate stairway. The problem has

---

[*]On one floor, the entrance was through the glass door of the "Gee Whiz" Travel Agency. Going down an outside fire escape in a Chicago winter is undoubtedly a "gee whiz" way to escape.

To get to the outside fire escape of this building, it is necessary to break through the door of the Gee Whiz Travel Agency.

been discussed from time to time. The city council, concerned about the problem, moved its offices from the top floor to the first floor. In an article in the *Washington Star*, the building manager is quoted as saying that in his opinion, it would be difficult for a fire to spread from floor to floor. The fire service should consider being more aggressive in attacking opinions based on ignorance. The spread of fire and toxic gases is a technical problem. Technical problems require technical solutions developed by those with the requisite training and education.

In the same article, the *Star* quotes an architect who works in the building:

> "We have never had a fire drill in this building, and there is no need for one," said L S, an architect who works on the city council staff and whose views typified many in the building. "If a serious fire ever breaks out on one of the lower floors, there will be no way out for us."
> Then pausing, Miss S said, "Heavens, do not write that. If they brought this building up to code, it would mean the rape of the District Building. The long winding stairways would be ruined," she said.

Underlying this statement is the arrogance of "artistic" architects who presume to make life and death decisions for others. Put to a poll, it is very likely that the majority of those in the District Building might well vote for the "rape" of the stairways and a chance to go home to their families if a fire occurs.

Where outside fire escapes are provided, note that they may have been exposed to the weather for up to a century. They should be carefully inspected not only for the safety of occupants, but also for the safety of fire fighters.*

Live loads in older buildings may have increased over the years to a point where the building is in distress. Safes adequate to protect computer tapes from fire, for instance, are much heavier than other safes. In some cases, unprotected steel supports may have been inserted.

Fire loads may have grown to high levels over the years. This is particularly so in the case of public buildings. Records and files accumulate without any proper provision for storage. Many public buildings are domed. Usually there is an inner and outer dome. The space between makes a good void in which to store tons of paper. Basement rooms and attics are often similarly loaded with documents, many of questionable value. Conflicts of jurisdiction may prevent any effective fire prevention inspections but prefire planning is certainly within the duties of the fire department who will be expected to fight the fire. Prefire plans that disclose unmanageable situations should be brought to the attention of the responsible authorities, if only for self-defense against some politicians who will surely attempt to shift any blame if a disaster occurs.**

Almost all older buildings have been altered at some time. Walls are opened up, and doorways are closed up. The closure is often not as fire resistive as the masonry wall. Stairways are cut through and others cut off. High bay rooms may be made into two stories. Sometimes a wooden mezzanine is built in. In other cases, a steel bar-joist floor is installed to make an intermediate story.

Older buildings were built with high ceilings to provide summer ventilation. The ceiling may have been lowered one or more times. Each of these lowerings may have created a combustible void. Possibly, the latest alteration was done with relatively noncombustible construction, but this may conceal old combustible acoustical tile. (This subject is discussed in Chapter 9, Flame Spread.) In general, codes which may require new ceiling tiles to meet flame spread standards do not require the removal of combustible tile. This is a serious deficiency.

Standpipes in older buildings may be inadequate in size. It is possible that the standpipe has not been tested and would fail in a fire. It is not beyond the realm of the possible that the fire department would find that the standpipe outlets do not match the fire department's hose. The standpipe is a fire department tool as much as the pumper that supplies it. It is up to the fire department to assure that the owner is keeping it in proper condition. The city of Alexandria, VA has such a program.

Three fires of great significance occurred in New York City in the early 1900s. The Parker Building was a 20-story mercantile building considered to be the best type of fireproof building available. It was destroyed by a fire that started within the building and spread up the open stairways. This was of great concern to the

---

*There are many situations where a public agency or the owner attempts to block or succeeds in blocking the fire department from examining a problem. Wherever it can be shown that the problem relates to the safety of fire fighters, the fire department is on very good ground to justify its interest.

**A fire department, prevented from inspecting a building to preplan it, might consider announcing that for the safety of its personnel it will advance as far into the building for a fire as is permitted for the survey.

fire protection engineers because it demonstrated that the "best" buildings being built could be destroyed, despite the efforts of the nation's largest fire department. The Equitable Building was believed by the uninformed to be fireproof, but in fact, it was not. The fact that the fate of millions of dollars of negotiable securities (if burned, they were irretrievably lost) in the basement was undetermined for several days, due to tons of ice and debris, caused consternation in the financial community. In accordance with the golden rule,* then there was action. The National Bureau of Standards, Underwriters Laboratories, The National Board of Fire Underwriters, the Factory Insurance Association (now Industrial Risk Insurers) and Factory Mutual got together and developed the standard for fire resistive construction which exists today almost unchanged as ASTM E 119. (This standard is discussed fully in Chapter 6, Principles of Fire Resistance.)

The Triangle Shirtwaist fire, in which 145 died in a fire resistive building which is still in use, had tremendous political-sociological effects.** Of principal interest here is the requirement that factory buildings over six stories in height be sprinklered. The result was the construction of scores of high-rise sprinklered factory buildings in New York City. These provide a vast body of successful sprinkler experience in high-rise buildings that effectively gives the lie to the special pleaders who argue that sprinklers are unproven. (This subject is more fully developed later in the chapter.)

Buildings of 1920-1940

The high-rise buildings built after these developments and before World War II are excellent buildings. Those built after World War II, in the author's opinion, have retrogressed in fire protection features, in many cases, to a serious degree.

The pre-World War II buildings were universally of steel-framed construction. Floor construction and fireproofing of steel were of concrete or tile, both good heat sinks and slow to transmit heat to the floor above. The construction was heavy but no one gave this much concern because there was no alternative.

Floor areas were relatively small. This was dictated by the necessity for natural light and air. The ads for the RCA Building in Radio City proclaimed, "no desk any farther than 35 feet from a window." This limited both the fire load and the number of occupants.

Each floor was a well-segregated fire area. Wall construction was of wet masonry joined to the floor so that there was an inherent firestop at the floor line. Masonry in the spandrel area (the space between the top of one window

---

*He who has the gold makes the rules.
**Some disastrous fires pass into history without making a ripple. Others have a tremendous impact. The Triangle Shirtwaist fire was the springboard for the political careers of several advocates of far-reaching social legislation. Senator Robert Wagner, author of the Wagner Labor Relations Law; Alfred E. Smith, four-time governor of New York and Democratic presidential candidate in 1928; and Frances Perkins, Secretary of Labor in the Franklin Roosevelt administration, were three of the most prominent.

The Empire State Building probably represented the zenith in high-rise fire resistive construction.

and the bottom of another), was adequate to restrict outside extension. However, the author recalls such an extension in a government building in Washington, DC. The cause was an added combustible acoustical tile ceiling which provided for heavy fire out the window.

All vertical shafts were enclosed in solid masonry with openings protected with proper closures.

Fire department standpipes of adequate capacity were provided. These were wet and immediately pressurized by gravity from a tank (or tanks) in the building.

Fire tower stairways, the finest escape device available, were provided. Such a stairway can be compared to an enclosed tower located away from the building which is reached by a bridge open to the weather, so that smoke cannot pollute the tower. In fact, the same objective is accomplished, but the tower is built within the perimeter of the building. They were readily accepted as an alternative to outside fire escapes because such escapes on the side or rear of a building require that the land below the escape be taken from the lot, reducing the size of the building.

The fire loads were low. The typical office was quite spartan, though executive suites and eating clubs often were paneled with huge quantities of wood.

Windows could be opened. This provided local ventilation and relief from smoke migrating from the fire.

The windows leaked, literally, like sieves. Heat was cheap. The author observed fires in such buildings. They usually presented very little problem to the fire department. Some serious fires occurred in sprinklered garment lofts* where accumulations of cuttings under the wide cutting tables could burn unaffected by water from the sprinklers. The occupants of the affected loft would evacuate, but thousands of others would continue to work. In his boyhood, a New York fire officer worked in such a loft. He left during a fire to watch it from the street. He was docked for the lost time.

Such buildings were almost immune to any serious fire attack except from exposures. In 1973, a six-story brick and wood-joisted building was being demolished in downtown Indianapolis.[2] Only the floors remained. A violent fire erupted and ignited surrounding fire resistive buildings. In Chapter 8, Concrete Construction, the damage to the fire resistive Mid Hudson warehouse in Jersey City from an exposure fire is discussed.

One caveat must be noted. Many of these buildings were "modernized" by the installation of combustible acoustical tile.[3] This tile, whether installed on furring strips or glued up, presents the serious possibility of a general spread of fire through the building. It should be removed. It cannot be rehabilitated satisfactorily by painting it with an intumescent coating (see Chapter 9, Flame Spread.) It is most certainly asking for disaster to leave it in place and then install a code-approved ceiling below it.

Modern High-Rise Buildings

After World War II, a number of significant developments occurred. The development of fluorescent light and air conditioning removed any limits to floor area. Thus, building populations could be enormously increased.

There was a definite push to lighten and thus cheapen buildings. The Empire State Building weighs 23 pounds per cubic foot. A typical modern high-rise weighs 8 pounds per cubic foot.

Because of the development of better reinforcing steel and new techniques, reinforced concrete became a serious competitor to steel. No longer could the steel industry be indifferent to the weight penalty caused by the fireproofing of steel. This necessity is a perceived competitive disadvantage to steel. Fireproofing of a steel building is a separate item in the cost estimates while the cost of fireproofing concrete is hidden in the cost of the concrete and is not separately identified.

The spray-on materials and the floor and ceiling assemblies developed to provide lighter construction are discussed in Chapter 7, Steel Construction. One of the results of reducing the concrete in the building is the loss of a valuable heat sink. Every Btu that is absorbed into the concrete is just one less available to keep the fire going.

This retained heat can be very distressing to fire fighters, and personnel may

---

*In New York City, a "loft" is a building in which space is rented out to factories.

The 19th-century Fort Worth, TX Court House shown here under construction, and still in use, is an early fire resistive structure. Many such buildings still exist. They can all be assumed to have many of the firesafety defects discovered by fire loss experience. Geneology and Local History Department, Fort Worth Public Library.

require frequent rotation. On balance, though, fire fighters are better off because the fire has been robbed of heat.

There has been an unimaginable increase in the requirements for electrical service and communications. Insulations are flammable, and the products of decomposition and combustion are toxic. Communications systems breach the fire and smoke barriers from floor to floor and from vertical shafts to horizontal voids.

The steel-truss floor and ceiling assembly provides a useful void. It carries utilities and communications. It can be one side of the air-conditioning system.

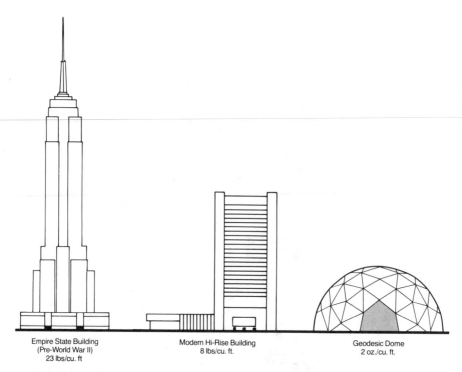

Empire State Building
(Pre-World War II)
23 lbs/cu. ft

Modern Hi-Rise Building
8 lbs/cu. ft.

Geodesic Dome
2 oz./cu. ft.

The comparative weights of three different types of construction. Elimination of deadweight may decrease inherent fire protection of structures.

Slab floor concrete buildings do not have an inherent void, but voids are useful. By the use of dropped ceilings, voids are created and then connected via utility shafts. In a modern office building, possibly 25 percent of the floor volume is in the ceiling void.

Gypsum board rather than masonry is often used to enclose elevator and other shafts. In the One New York Plaza fire, these enclosures were displaced leaving the shafts unprotected.

In earlier buildings, there was but little concern about using the exterior of the building for fire towers and even for rest rooms. Promoters then discovered the value of exterior offices, particularly corner offices. All outside space is sold. Utilities are relegated to an interior "core" structure. Stairways are located in the core, thereby eliminating the principle of remote exits.

Often the stairways are "scissor" stairways, two stairways in the same shaft. Often there is no fireproofing on the underside of the stairways so heat or fire in one stairway makes the other untenable. If there is a standpipe, the outlets will be on alternate stairways. Call the stairways A and B. The outlet will be on the odd numbered floors in stairway A and on the even numbered floors in stairway B.

Elevators are grouped together in the core; thus, it is possible that all will be affected by the same problems. Not all high-rise buildings are of core construction,

Massive steel trusses, both vertical and horizontal, frame the Federal Reserve Bank of Boston. Trusses provide for lighter weight construction and useful, but often dangerous, voids.

using the term structurally. Some particularly high high-rise structures are of tube construction, that is, the wind load is taken principally on the exterior. This does not eliminate the use of a central core for utilities and services.

In an attempt to provide the equivalent of a smokeproof fire tower in the interior core, a vent shaft is provided adjacent to the stairway. This shaft can become a chimney. Heat and fire reaching it are directed upward. In the 299 Park Avenue fire in New York in 1980,[4] aluminum guard rails on such a shaft melted from the heat, exposing fire fighters to a potential fatal fall.

Modern design emphasizes an uncluttered exterior. Prefabricated panels, glass, aluminum, and other curtain wall developments make it unlikely that there is a real barrier to the vertical extension of fire between the edge of the floor and the skin of the building. A glass wall system is described in detail in an engineering publication, but the article is utterly silent as to how this will stop a fire.

Fire does not spread only upward through inadequately protected gaps between the curtain wall and the floors; it can also spread downward. In December 1981, fire gutted the 23-story Barao de Maua Building in Rio de Janeiro, Brazil. Burning electrical wires in the 15th floor combustible tile ceiling void started the fire. It spread downward and upward through a 6-inch gap between the floor slab and the aluminum and glass curtain wall.

Fire spread upward because of inadequate perimeter protection in the One New York Plaza fire in New York City.

The enclosure of balconies by tenants or condominium owners may increase the potential for exterior vertical extension.

Core construction makes operation of handlines from a safe location, i.e., the stairway, utterly inadequate. Prepare to use monitor nozzles to knock down fire and reach distant points if the standpipe system can supply enough water.

High-rise fire plans may envision cutting through floors to reach the fire. Note from Chapter 8, Concrete Construction, that it is not safe to cut through post-tensioned concrete floors.

Because of changes in office decor, and the elimination of interior partitions, in many cases heavy fire loads in huge unbroken areas will be found. Conversely, compartmented floor areas, with mazelike corridors, require long hose stretches and are extremely confusing when smoke-polluted.

Each building is a problem all its own. A high-rise with concrete floors and masonry panel walls built on each floor presents no problem of fire extending up the outside behind the wall panels. Another concrete-floored building, with precast concrete or imitation concrete panel walls, may have a serious problem of perimeter extension.

When floor slabs project like those shown here, the problem of vertical extension of fire at the perimeter is negligible.

Just as is the case in older type fireproof buildings, do not be misdirected by the "high-rise" title of this chapter. Many buildings up to six stories in height have the same construction defects as high-rises.

The occupancy changes the problem. Offices, hotels, apartments, homes for the elderly, factories and showrooms are all different. Some buildings have mixed occupancies.*

The construction may not be consistent throughout the building. The author has observed several concrete buildings on which top floors or penthouses of lightweight steel are erected. The penthouse of one high-rise in Polk County, FL, has a lightweight wood truss and plywood roof.

---

*A sad example of inflexible thinking is found in a high-rise apartment building. An architect occupies several apartments as offices with the customary substantial fire load with a high potential rate of heat release. The fire department was overseeing the retrofitting of smoke detectors. No detectors were installed in the architect's office; "It's not an apartment." The hazard that a fire in the offices presents to the tenants was just ignored in the literal application of the law.

Air conditioning requires fixed windows to control loss of treated air. Many buildings have no openable windows. Breaking glass is extremely hazardous because glass pieces can plane for several blocks. If windows are covered with a sun-screening plastic, the glass tends to hold together when broken and can be pulled into the building. More recently, windows which can be opened are provided and keys are located in strategic but inconspicuous locations.

When Boston's John Hancock Tower was first built, mysterious defects caused glass panes to fall out. They were replaced with plywood panels until the tower looked almost like a wooden structure. When the flame spread hazard was realized later, the plywood panels were fire-treated.

The open light well is back. Today, it is called by an ancient Roman name, "the atrium." This huge void presents tremendous difficulty in the control of smoke, as was experienced in a fire in 1977.[5] In the Citibank Building in New York, the atrium is glass enclosed to prevent smoke migration. A separate line of sprinklers protects the glass.

Building codes have recently come to grips with the atrium problem. Typical requirements are sprinklers throughout the building, smoke exhaust systems with standby power, and separation from floors by a fire barrier, such as the sprinkler protected glass wall cited above. The number of floors that can be open to the atrium without a fire barrier is limited.

Parking garages present a number of problems. The garage area may extend out beyond the building. The covering slab may be designed only to carry the weight of cars. The approach of heavier fire apparatus may be restricted or totally denied. One shopping mall in the midwest consists of many buildings built on a lightweight slab over the garage. No horizontal standpipe was installed. Fire apparatus cannot drive on the slab. Impossibly long hand hose line stretches are required.

The parking garage may be partially or totally above grade and open, at least nominally, to the atmosphere. In one high-rise, this led to permission to use ABS (acrylonitrile-butadiene-styrene) conduit (a very heavy smoke producer) on the ceiling of the garage. A rubbish fire involved the plastic conduit and wiring. The high-rise building then, fortunately, under construction, was heavily smoke-polluted. All garage areas under buildings, open to the atmosphere notwithstanding, should be sprinklered.

The huge populations of high-rise buildings may make evacuation an almost unmanageable problem.[6] Where will the evacuees go, particularly if the weather is bad? The lobby is safe and all sorts of interesting things are going on. People will be anxiously, perhaps even hysterically, searching for co-workers. A strong police presence may be required to provide the working space needed.

The need for evacuation or rescue should never be overlooked. Do not be deceived by the "nine to five" aspect of office buildings. A building that is guarded may have sign-in and sign-out sheets, but many buildings may have no night security and cannot be assumed to be empty. Such persons as cleaners, computer operators, and attorneys researching cases may be working at any hours, often alone, and unknown to anyone.

Fire fighters may be denied access to stairways by the sheer numbers of people descending. In residential occupancies, it may be possible to train the occupants to keep to the wall side, leaving the rail side for fire fighters.

All exits should provide a clear path to the outside. In one high-rise building, the exits opened to a rear yard. The yard was then planted and enclosed with an unbroken 10-foot high masonry wall.

In another case, the exit situation in a group of three interconnected buildings leased by the government was very poor. Several alterations were made to improve the situation. The added exits lead through a small garage area to the open street. The occupying agency decided to use the garage for the storage of cardboard boxes for files, a huge smoke-producing fire load. They asked the building department for advice. The advice was to enclose the garage in 2-hour fire resistive walls. By this "cookbook" advice, a serious situation was created. The pressure generated by a fire would be contained. Smoke and gases would be forced into the exitways through the space under the exit doors. Though the author was without responsibility, he had many friends in the agency and he prevailed on the management to install a dry-pipe sprinkler system in the area. His recommendation to the fire department was to break through the gypsum board of the fire resistant wall assembly, tie a rope around the studs, pull the wall out, and make a massive attack on the fire unless the sprinklers had it under control.

The Problem of Elevators

Elevators are a real problem[7]. The author spent a number of years on a federal interagency committee concerned with building fire problems. The committee took on the problem of trying to determine how elevators could be made safe in the event of fire. The only real solution appeared to be the construction of high-rise buildings in at least two sections completely cut off from one another on every floor with respect to void spaces, utilities, and any other connections, except for properly protected openings where corridors passed through the dividing wall. Persons threatened by a fire in one area could move laterally into the next area and either remain there or proceed leisurely to the street using the elevators in the unaffected section. This is particularly suitable to the movement of the disabled, and this concept has been adopted in some hospitals.

The problem of the elevators is not solved just by having them return to the first floor and providing for fire department operation. This may only expose fire fighters to the hazards from which the occupants have been removed. Unless there is proper planning and training, fire fighters may get into serious trouble. Chief O'Hagan's[8] book gives some excellent advice. Some points are: limit the number of personnel in any elevator; all personnel should have SCBA slung and carry extra cylinders; all units should have forcible entry tools.

Perhaps another addition to high-rise equipment should be an elevator pack consisting of spare air bottles and forcible entry tools which is placed and left in the elevator used by the fire department.

Two Chicago, IL fire fighters fell 16 stories to their deaths during a high-rise office building fire.

The fire initially involved an elevator that had been called to the 25th floor of the 38-story building. With doors open, the elevator burned for some time and then dropped to the 9th floor. By now, the fire had spread into the 25th-floor lobby area.

Members of two responding companies took an elevator to the 24th floor. Because the stairwell door was locked, they returned to the elevator which inadvertently went to the 25th floor. Wearing SCBA, they attempted to make their way through heavy smoke and heat to a safety area. A 56-year-old fire fighter apparently lost his way and fell through the open doors where the burning unit had been. A 23-year-old fire fighter, who went back to find him, also fell through the open door. Their bodies were found on the roof of the fire-damaged elevator on the ninth floor.

The many problems of elevators in fires are beyond the scope of this book. However, all experience points to the valid possibility that at some fires, all elevators may be totally unavailable to the fire department. Plans should embrace this contingency. *

Accommodation or Access Stairs

Many tenants occupy more than one floor in the building. They find it convenient to have an "accommodation" stairway or access stairway installed. This is usually done as an alteration and is rarely enclosed. The result is that two or more floors become one fire area, completely negating the concept of floor integrity.

Duplex and triplex apartments, two- or three-story "houses," can be found in high-rise apartments. Often, there are no exits from the upper levels.

The elegant restaurant located at the top of the structure may be a multi-level structure with an open stairway connecting the levels.

All of these violations of good practice increase the fire area and can multiply the fire problem. In some codes, and among some designers, there is little concern about open stairways, provided the "required exits" are properly enclosed.

A Canadian fire department responded to an automatic fire alarm from the 28th floor of a 43-story building. They took the elevator to the 27th floor (which they assumed was the floor below the fire). The door opened on the fire. Fortunately, they were able to get the car door closed and descend. One firm occupied the two floors. There was an access stairway. The smoke detector on the 28th floor detected and alarmed the fire. Smoke detectors that are far removed from the fire may be the first to operate. The "floor below the fire floor"

*One suggestion from an experienced high-rise fire commander who is a physical fitness enthusiast: "If you have to walk, carry your boots. It is less stressful to have them around your neck than on your feet."

A Christmas tree fire may send smoke and fire gases up this accommodation stairway, causing automatic alarm indications to report a false location of the fire.

operating base was adequate for the slender pre-World War II towers in which it was developed, but is no longer safe. References 9, 10, and 11 are pertinent.

Forcible Entry

Because of today's high crime climate, security is a serious consideration in all occupancies. Heavy doors with multiple locks may delay entry for some time. The fire department should be aware of how fire separation was accomplished. For instance, if the building is of reinforced masonry construction, the corridor walls may be of reinforced masonry. In other cases, the fire resistance may be achieved with gypsum wallboard. It may be much simpler to go through the

Exit doors may lock occupants and fire fighters in stairways. All units, not just ladder companies, should carry forcible entry tools.

wallboard and ignore the lock. This may also be possible from apartment to apartment and into the upper levels of duplex and triplex apartments.

Smoke Movement in High-Rise Buildings

For many years, the only factor considered by the fire department as affecting smoke* movement in a building was the thermal energy of the fire. This is discussed in Chapter 10, Smoke and Fire Containment. In high-rise buildings, there are a number of additional factors. These factors can be combined in an infinite number of ways which are ever changing. It appears to be impossible to obtain and process information rapidly enough at the fire to make accurate decisions as to action to be taken if, indeed, the action indicated would be feasible. In the author's opinion, an objective examination of the disaster potential inherent in the combination of huge quantities of toxic gases locked up in solid fuels, which can be released literally with a match, and the thousands of people trapped in the typical enclosed structure, leads inescapably to one conclusion. There must be absolute assurance that the amount of toxic gases that can be released will be severely limited, not dependent on "control" of unlimited amounts of toxic gas. This can be accomplished by limiting the quantity of fuel present. This does not appear to be very productive, though at least one city, Boston, is attacking the problem of flammable fuels, and the quasi-governmental Port Authority of New York and New Jersey has applied some restrictions after the disastrous fire in the furnishings of the BOAC terminal at Kennedy Airport.

The other alternative is to accept the fact that buildings will have combustible contents, which really represent solidified toxic gases, and provide a system which will limit the amount of gas produced. The only such system available or on the horizon is automatic sprinklers. Fire department reaction time** is not the fabled 1½ minutes but more nearly 20 minutes or more. Burn time between ignition and alarm can be any length. The relatively short reaction time of the sprinklers, plus the fact that they can automatically transmit the alarm, makes them infinitely superior to any other system. If the sprinkler system water is tied to the domestic water so that the toilets fail before the fire protection water is unavailable, the system is almost foolproof.

This is not to say that other systems are not necessary to supplement the sprinkler protection, but sprinklers are the core of firesafety for the occupants of high-rise buildings.

It is very possible that we will see the world's worst fire disaster in an unsprinklered high-rise office building. Study the reports of some of the high-rise fires which have occurred. Add one or more of these very reasonable contingencies:

---

*Smoke is used in this chapter as shorthand for fire-generated toxic and explosive gases and liquid droplets and solid particles.

**Time between receipt of alarm and water on the fire.

- The fire forces committed heavily to another fire.

- Heavy snow conditions.

- Gale wind conditions.

- Building fully occupied.

- Total elevator power failure.

The list is endless and the danger of catastrophe is real.

Unfortunately, building codes, in the author's opinion, are political not technical documents.

Therefore, we must assume that in an unsprinklered building,* there will be huge amounts of smoke which may be delivered in devastating quantities to persons far removed from the fire. We must be aware of the mechanisms which may be involved in the distribution.

The factors which can be significant in smoke movement in any building, but which are particularly significant in high-rise buildings, are discussed below.

Thermal Energy

The movement of smoke by the thermal energy of the fire is covered in Chapter 10, Smoke and Fire Containment. Thermal energy is, of course, also important in high-rise fires. In 1961, the famous Times Tower (where the midnight ball falls on New Year's Eve), was under renovation. The only tenant was a magazine recycling business in the second subbasement, 70 feet below the street. During the fire, two truckmen had the elevator operator take them to the top floor. They left their SCBA in the lobby. They were found dead on the top floor. The elevator operator was found dead in his car at the third floor. The thermal energy of the fire can move toxic gases great distances.

The huge fire load in the MGM Grand Hotel Casino in Las Vegas created a tremendous thermal updraft which overwhelmed the high-rise structure. In an article in the *Bulletin of the Society of Fire Protection Engineers*, David Breen estimated the burning rate of the fuel at 3 tons per minute.[12]

Atmospheric Conditions

When the atmospheric temperature is constantly decreasing as height increases, the condition is called "lapse." Under lapse conditions, smoke will

---

*In the definition of an unsprinklered building, the author includes a building in which the sprinklers are turned off.

move up and away from the fire. If there is a layer of air warmer than the air below, this layer is called the inversion layer. It acts as a roof to rising smoke. A high-rise building may actually penetrate the inversion layer. This would cause substantial differences in the smoke conditions above and below the inversion layer.

Wind

Wind is very powerful and influences smoke movement. Take note of the tremendous bracing required in the high-rise to overcome the tendency of the wind to overturn the building. The wind exerts a pressure on the windward side of the building and a suction on the leeward. If the windows are out and the fire is on the leeward side of the building, the suppression may be "a piece of cake." Given the same fire, the windows out, and the fire on the windward side of the building, it may be impossible to get onto the fire floor.

Closely spaced high-rise buildings can create a canyon effect, an increase in velocity as the wind squeezes through a narrow opening. The wind can shift direction many times during the fire. Ground level observations are not valid. At upper levels, the wind can blow harder and from a different direction.

Consider a high-rise fire in an apartment house with large glass areas. Smoke is showing from an upper floor on arrival. By the time the units reach the fire floor, the windows have failed, the wind is blowing in and the occupants face an inferno. Wind direction and apparent velocity should be noted on arrival and changes reported to attack units by outside personnel. No matter how simple the fire looks, entrance to the fire floor from the stairway should always be cautious.

Wind blowing against the building seems to split about two-thirds of the way up the building. The upper portion flows up and over the roof. The lower portion flows downward forming a vortex next to the building. It increases in velocity as it flows downward. The effect of this wind on fire pouring out windows appears to be unevaluated in the design of high-rise structures.

The effect of wind on the structure is becoming better understood. As previously mentioned, modern high-rise buildings weigh about 8 or 9 pounds per cubic foot. The result is that the top stories of some such buildings sway noticeably in the wind. To counter this, some buildings have "tuned mass" dampers installed high up in the building. Heavy weights sliding on films of oil are adjusted by computer to counter the wind-induced oscillations.

Stack Effect

Stack effect is the term used to describe the movement of air inside a relatively tightly sealed building due to the difference in temperature (Δt) * between the air

---

*The Greek letter delta, shown as Δ, is the engineering symbol for difference.

inside the building and outside the building. Stack effect is most significant in cold climates in the winter time because of the great difference between the inside and outside temperatures. Stack effect also occurs, with opposite airflow direction, when the outside temperature is greater than the inside temperature. This condition is, however, much less significant, because the amount of stack effect is proportional to the difference between the two temperatures. There is a much greater difference between the inside temperature and outside temperature during the winter than during the summer, and the winter differential lasts for longer periods of time. In a cold climate, a Δt about 70°F might exist for weeks on end, while in the heat of summer, the Δt might be only about 15°F, and as much as that only in the afternoon of days on which the building is occupied.

Stack effect is not caused by the fire. The products of combustion ride the stack effect currents.

For the moment, assume winter conditions. Later, "cold smoke" at a summer fire in a refrigeration plant will be discussed. Under winter conditions, stack effect causes a movement of air from the floors into the vertical shafts, stairways, elevators, etc., in the lower portion of the building. The greatest flow will be at the first floor, with the flow gradually decreasing as the height of the floor above ground increases. At a point about one-third to one-half the height, the flow is reduced to zero; this is called the neutral zone. Above the neutral zone, the flow reverses and is out to the floors from the shafts. The pressure increases floor by floor and is greatest at the top.

The complete proof of this phenomenon is difficult to understand. For our purposes it is necessary only to recollect that the external pressure is the pressure of the outside atmosphere, and it decreases as the altitude increases. In the opposite way, the pressure in the building stacks increases as the altitude increases due to the heating of the air in the building. Thus, the greatest outside pressure corresponds to the least inside pressure, at the first floor, causing the greatest flow into the stacks. The pressure differential decreases floor by floor, as the altitude increases. When there is no pressure differential, there is no airflow. Above the neutral zone, the differential commences again but in the opposite direction.

Take the elevator to the top of a tall building on a cold day. Walk down the stairways, and feel the intensity and direction of the air under the door at each floor. Air-conditioning mechanics use an anemometer (wind speed gage) that counts the revolutions of a propeller placed in the air stream. By relating the number of revolutions during a measured time to the size of a vent opening the airflow can be determined.

The formula for stack effect is as follows:

$$P_c = 0.52 \, ph \left[\frac{1}{T_o} - \frac{1}{T_i} \right]$$

The pressure differential between the top and the bottom of the building, P_c, expressed in "inches of water," $= 0.52$, a constant multiplied by p, the absolute atmospheric pressure (at sea level 14.9 psi, use 15) multiplied by h, the height of the building in feet, multiplied by a term consisting of the reciprocal of the temperature outside (T_o) in degrees F absolute (fahrenheit thermometer reading plus 460°) minus the reciprocal of the temperature inside (T_i) in degrees F absolute.

This formula is well known and is used by engineers concerned with heating, venting and air-conditioning (HVAC) systems.

Inches of water pressure are used for the measurement of differences in air pressure because very small differences can be measured without long decimals as would be necessary if inches of mercury were used. The same pressure (one atmosphere) that will raise a column of mercury 29.92 inches will raise a column of water 406.8 inches.

The flow of air is always in the direction of the lower pressure. As the difference increases, the speed increases in proportion to the square roots of the two pressures. The following table will be helpful:

Pressure	Draft speed	Flow under door opening 39 inches × ½ inch
Inches of water	Miles per hour	Cubic feet per minute
0.1	14.4	103
0.2	20.4	146
0.5	32.2	230
1.0	45.0	326
2.0	64.5	461

If the stack effect formula given above is applied to a 400-foot-high building, on a day when the Δt is 75°F, we find that the ΔP is 1.0 inch of water. Since the flow of air at the top of the building is outward from the vertical openings to the floors under stack effect conditions, there would be a high airflow through any openings under the stairway door. A ½-inch clearance is permitted. The flow might be less than the maximum flow of 326 cfm given above due to differences in exterior wall leakage, but there would be substantial pressure against the exit door to a fire stairway. The pressure might make it difficult for an ordinary person to open the door.

If a building was composed of floors completely cut off from one another, the stack effect would exist separately on each floor. Air would flow in at openings at the bottom of each floor and out the top. There would be no overall stack effect since there would be no connection between the floors.

The buildings in which fires are fought do have interconnected vertical openings, stairways, elevator shafts, and pipe openings, so the stack effect must be considered when fighting a fire.

Many variables can affect the stack effect; openings and their locations are important.

The interrelationship of all factors is shown by the formula:

$$\frac{h_2}{h_1} = \frac{A_1^2}{A_2^2}\frac{T_i}{T_o}$$

- A_1 is the total area of openings above neutral zone.* In the formula it is squared.

- A_2 is the total area of all the openings below the neutral zone. In the formula it is squared.

- h_1 is the difference in feet from the lower opening to the neutral zone.

- h_2 is the distance in feet from the upper opening to the neutral zone.

- T_i, as before, is the inside temperature.

- T_o, as before, is the outside temperature.

The following conditions tend to raise the neutral zone:

- Increased openings above the neutral zone.

- Decreased openings below the neutral zone.

The following conditions tend to lower the neutral zone:

- Decreased openings above the neutral zone.

- Increased openings below the neutral zone.

In general, it would be best to raise the neutral zone and thus, prevent pollution, or clear the smoke out of the greatest number of floors. If it were possible to mass enough fans at the top openings, the neutral zone could be raised substantially. This is probably impossible but it is ridiculous to set a fan in the stair doorway attempting to blow smoke from the corridor when stack effect is directing the airflow into the floor from the stairway.

It is a common practice to hook open the lobby doors to make it easy to handle equipment. This may lower the neutral zone several floors and increase the number of floors polluted by a fire below the neutral zone.

When stack effect is a significant factor due to weather conditions, all should be reminded of the situation just as reminders are issued when the burn index is

---

*Note the use of the word zone, rather than plane. It is probably a better word because there is not a sharp line of demarcation but rather a zone in which the direction of airflow is zero or vacillating.

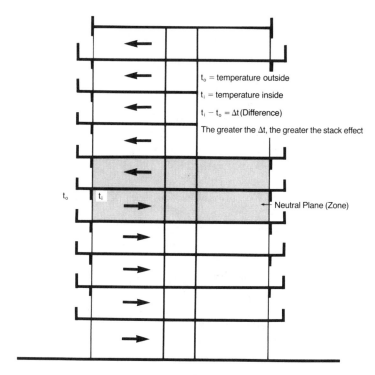

t_o = temperature outside
t_i = temperature inside
$t_i - t_o = \Delta t$ (Difference)

The greater the Δt, the greater the stack effect

Neutral Plane (Zone)

The stack effect on a high-rise building in winter conditions.

high. All should be aware that the first alarm might be from a person or a detector many floors removed from the scene of the actual fire. Fire ground commanders should not commit all forces until the location of the fire is determined. Alarm operators should be trained to take particular note of callers who say *fire* or *flames* when multiple calls about smoke are being received and pass this to the fire ground commander and not assume that the combat forces will find the fire location immediately upon arrival.

A fire that vividly demonstrated winter stack effect occurred in an office building under construction. The fire was in a pile of rubbish on the 9th floor. Fumes from degraded and burning plastics went to the 31st floor by stack effect. The workmen started down the stairways. They were so affected by the fumes that they went into the 25th floor for refuge. They were in such distress that they started to break windows despite the hazard to those below.

In another case, a winter fire on a loading dock was first reported from almost the top floor of a 40-story building.

In a sealed air-conditioned building in the summertime, the stack effect is reversed. The flow is downward. The situation is rarely as extreme as the winter situation because the Δt is less, but it still can be serious. In a hospital in Puerto Rico, in the summer, smoke pollution occurred several floors below the fire floor due to summer stack effect.

At a very serious summer high-rise fire in New York, five floors below the fire floor were polluted by smoke so as to be unusable for staging the fire. Air bottles were seriously depleted before personnel reached the fire floor.

Fire fighters experienced in fires in refrigerator plants have long been acquainted with "cold smoke" which falls downward. Units under the author's command assisted the Norfolk, VA Fire Department in a refrigerator plant fire. The fire was in fungus-treated construction lumber in a cold box. The boxes around the box on fire were at zero temperature. The box where the fire was located was at 30°F. By rigging salvage covers and setting up fans as we had preplanned for a similar building on the Naval Base, we were able to pump the smoke into the elevator shaft where it fell down the shaft to the street. At that time, the term "stack effect" was unknown to fire fighters, but this was a perfect example of summer stack effect conditions.

Stack effect might lead fire departments to change some operations. Consider a winter fire in a high-rise building. The fire is below the neutral plane and accessible from the outside of the building. Using conventional operations, the fire department would attack the fire from the interior, thereby freeing all the smoke to ride the available stack effect currents and pollute the upper floors. An alternate attack might be to protect the interior exposure with charged lines but not to open the fire floor to the stairways. The attack is made from the outside. Most of the smoke is vented out the windows by thermal pressure, fire department fans, and, with a little luck, by the wind blowing in the right direction. The fire loss would probably be the same, but the smoke damage and apprehension throughout the building would be substantially reduced.

Stack effect is but one of the factors which can affect smoke movement. In a given situation, it may be dominant, as in a relatively small rubbish fire in the basement which causes acute distress to tenants on upper floors. In other cases, it may be subordinate or inconsequential.

In a particular fire, all the factors affecting smoke movement may be operative, assisting or cancelling one another. The situation changes from minute to minute in indeterminate ways. Each is discussed separately out of necessity. How they combine in a specific situation remains a mystery.

Air Conditioning

Whether or not air conditioning is a significant factor in smoke movement depends on the type of system.[13] Individual room units drawing air from the outside, as are seen in many recently built motels, will have no effect on the general situation.

On the other hand, if the individual units draw air from the corridor, as was the case in the MGM Grand Hotel fire in Las Vegas,[14,15] polluted air from the corridor will be drawn into the rooms. It appears that the opening windows at this fire increased the flow from the corridors.

Single-floor systems are seen in many smaller office buildings. In a typical system, there is a fan unit where the air is tempered on each floor. Air flows from the corridor to the fan unit, is tempered, and distributed through ducts which spill into the ceiling void. The air comes into the offices through openings in the ceiling tile. It returns to the fan unit via the corridors. Smoke from any fire on the floor will be distributed throughout the floor. If the building code has permitted the dangerous situation of the floor and ceiling assembly "protecting" the columns in the plenum space, then smoke can move along the reentrant space of the column to the voids on other floors. Telephone service is provided by cabinets located one above the other.[16] Holes in the floors connect the cabinets. Openings from the cabinet to the ceiling void provide another path for smoke distribution. In short, the fact that the building has single-floor separate systems will not prevent smoke migration throughout the building.

When the entire building is air conditioned by one or more building systems, the problems become incredibly complicated. Each system is different, depending on many factors.

In the 1950s, for instance, when total building systems were first coming into use, the treated air was regarded as a precious commodity. Only the exhaust from toilet rooms was passed to the outside; all other air was tempered and returned to the occupants. Regarding one building, the author was assured that in the case of fire, the fans could be put on "full exhaust" which was taken to mean that all air reaching the fan would be exhausted. In fact, at any given pass, only 14 percent of the air could be exhausted so that most of the smoke would be returned to the building. As technology progressed and energy costs rose, it was learned that there are many days when the outside air is just lovely and could be pumped into the building to the advantage of the occupants and the electric bill. More modern systems often have a true full exhaust capability.

This is but one example of the questions that can arise. The fire department must develop officers who are familiar with all the nuances of HVAC systems and able to hold their own with designers and installers. It is simply not adequate to accept somebody else's idea of what will be an adequate functioning of the system when there is a fire in the building.

This is particularly true when the system is supposed to be designed to function in a smoke removal mode.[17] Will the fire department operate it? Will the building engineer operate it? In one high-rise, the operating station for the engineer who allegedly will operate the system in an emergency is within the main plenum of the system. The instructions from the union are for him to leave immediately in the event of a fire. The control room for the air-conditioning system must be invulnerable to the effects of the fire with which the system is allegedly designed to cope.

There are those who have insisted that high-rise fires can be controlled simply by manipulating the air supply. If such a system is proposed as the principle defense, the proponents should be required to provide an evaluation of the efficiency of the system on a credible fire, with the windows out and a 35-mile-per-hour-wind blowing into the building.

One such advocate argues, "Let the fire burn, but let it burn clean," presumably so that there will be no products of combustion. This is a gross fallacy. There is no such thing as a "clean-burning" hostile fire. Even if the fire itself is burning freely, above, below, and on all sides of the fire, materials will be in various stages of decomposition and degradation and generating toxic and explosive gases. Another advocate distributed material that referred to recommendations for automatic sprinklers as "paranoid."

Smoke removal is an extremely complicated task.[18]. A well-designed and properly maintained smoke removal system can certainly supplement the prime defense, automatic sprinklers, but it is certainly no cheap substitute for adequate protection.

The author looked at a smoke removal system in a five-story sprinklered office building with a huge atrium. The system is triggered by detectors in the air ducts, by the operation of a sprinkler, or by a manual fire alarm. The normal air conditioning is a ducted supply with an unducted return through the ceiling void. In the smoke removal mode, the supply is cut off to the fire floor and all return air from the floor from which the alarm was received is spilled outside the building. Air from other floors is drawn to the fire floor through the atrium. It is possible that a person looking across the atrium might see a fire on another floor and pull the nearby alarm. The effect of this contingency is unevaluated.

One tenant installed an insulation barrier in the void to control sound from a typing pool. This completely negated the system. The void must remain unobstructed, and all ceiling tile units must be in place for proper operation.

In larger buildings, the smoke removal system can be incredibly complex and undoubtedly will be unique. The fire department must be aware of the installation and be completely involved from the inception. One of the possible deficiencies is that it may be necessary that the system remain running, even if not refrigerating or heating, so that smoke will reach the detectors when the building is closed.[19] A future energy-conscious manager may turn it off over weekends.

Shortly after New York's Citibank building opened, there was a serious fire in a store in the atrium. To clear the smoke, the supply fans were shut down, doors were opened, and the exhaust fans started. The smoke did not clear even though the exhaust fans were running at full speed. The fan system was interlocked. If the exhaust fans received a signal from the supply fans that they were not passing air, the variable pitch blades automatically adjusted to no airflow, even though the fans kept running. After this experience, the system was modified to make it possible to bypass the automatic controls.

This demonstrates the importance of studying how the system will react to every imaginable set of circumstance.

All the necessary information on the HVAC system should be developed as part of the fire department's retrievable information system. It is dangerous to rely on the building staff. When the fire occurs on a weekend, you may find a temporary replacement on duty who knows nothing specific about the fire operations mode. Similarly, the fire ground commander may be from another area and may never have seen the fire building.

Pressurized Stairways

One or more of the stairways in the building may be equipped to pressurize when fire occurs. Fans are installed to pump outside air into the stairway so that the pressure differential will keep the stairways free of smoke. This requires a dependable source of power and intakes located so that they will not pick up smoke from the fire. Occupants must be trained to use the proper stairway and should be drilled since the fans can make enough noise to be frightening to people already upset. Since the temperature should be close to the outside temperature, there should be no stack effect. The overpressure may make opening doors difficult and mechanical assistance may be required. The system does nothing to clear corridors. The fire department must not use this stairway to attack the fire as opening the fire floor door will admit a heavy airflow to the fire causing it to accelerate. The author was told that this happened to a west coast fire department.

A 21-story high-rise in Liverpool, England was built with only one exit stairway because the stairway was pressurized. The fire started in rubbish in a passenger elevator. The smoke from the fire vented out at the roof level in proximity to the intakes for the pressurization fans. The entire stairway was polluted. Fortunately, there were only a few occupants in the building.

This demonstrates the necessity of careful attention to the location of intake fans for pressurized stairways and for ventilation systems generally.[20]

In any event, any equipment designed to function in case of fire is the equivalent of fire department equipment. The fire department should approve its installation, be totally familiar with its operation, and supervise its testing and maintenance by the owner or his contractor.

The logical basis for this argument, which would be opposed by some designers and building officials, is the fire chief's unquestioned authority over the conduct of fire fighting operations. Of course, being logical is no guarantee that the concept will be upheld either administratively or legally, but it is the position with which the fire department should start. Over 100 years ago, when automatic sprinklers were first developed, the International Association of Fire Chiefs voted to have nothing to do with them. The decision was reversed at the next annual meeting. Those who adopt the former position with respect to installed equipment are courting disaster.

Fire Load and Rate of Heat Release

Consideration of the fire problem of a high-rise building must go beyond the building code and other legal constraints to the question of contents. In general, the construction of the building itself is noncombustible.

In his book, *From Bauhaus to Our House*, Tom Wolfe describes the problem, albeit in another context:[21]

Every great law firm in New York moves without a sputter of protest into a glass-box office building with concrete slab floors and seven-foot-ten-inch-high concrete slab ceilings and plasterboard walls and pygmy corridors — and then hires a decorator and gives him a budget of hundreds of thousands of dollars to turn these mean cubes and grids into a horizontal fantasy of a Restoration townhouse. I have seen the carpenters and cabinetmakers and search-and-acquire girls hauling in more cornices, covings, pilasters, carved moldings, and recessed domes, more linenfold paneling, more (fireless) fireplaces with festoons of fruit carved in mahogany on the mantels, more chandeliers, sconces, girandoles, chestnut leather sofas, and chiming clocks than Wren, Inigo Jones, the brothers Adam, Lord Burlington, and the Dilettanti, working in concert, could have dreamed of.

The fire load of the nonstructural elements, such as interior finishes and utility insulation, and the contents of the building are at the heart of the problem. They are essentially unregulated, though Boston has some regulation of the flammability of furnishings. Even when there is some regulation as for the flame spread of ceilings and wall finishes, alterations made after the certificate of occupancy is issued may introduce serious hazards.

Furniture and Finishes

The nation was astonished to see on television a huge column of fire rising up the side of the Las Vegas Hilton Hotel not long after the fire at the MGM Grand Hotel.[22] Eight lives were lost, and 300 persons were injured. An arsonist was convicted of setting the fire and this seems to satisfy the national predilection for determining the "cause of the *fire*" rather than the cause of the *disaster*.

The cause of the disaster was the use of flammable carpeting on the ceilings and walls of the elevator lobbies, the provision of fuel in the form of combustible furniture, and the omission of any dependable system of control in the event a credible incident (an arson fire) occurred.

The Casino of the MGM Grand Hotel was 68,000 square feet in area, equivalent to about one and a half football fields. The fire load is not reported in terms of Btu per square foot but the figure of 12 tons of combustible adhesive holding up the plastic tiles is indicative. Las Vegas hotels exist primarily for gambling so the Casino is open to the hotel lobby.

Automatic protection did not extend throughout the hotel complex. The management apparently was determined to do only what they were legally required to do and possibly placed great faith in the fact that the Casino was always guarded. It is possible that the management fell prey to a common fallacy which is often stated by salesmen for automatic fire detection devices and sometimes by fire departments in a worthy endeavor to get automatic detectors installed. "All fires start small. If the fire department is called immediately, the fire will be put out with slight loss." This is very wrong. There are many situations in which, if a person had an open phone to the fire department in one hand, and a match in the other, and called as the match was struck, a disaster would result.

Suppose the argument had been put to the MGM bottom-line-minded management this way: "If there is a fire and if we get there in time and put it out, the smoke and water and general mess will make it necessary to close the Casino for at least a day. The gamblers will go elsewhere. A day's profits will be lost. Chances are a day's profits would pay for the sprinkler system."

The fire load problem is well summarized in this extract from an article in the April 1975 *Bulletin* of the Society of Fire Protection Engineers alerting to the hazards of plastic polymers used in furniture cushioning, and padding:

> The rapidity of burning produces extraordinary amounts of gas and smoke. Smoke from this burning plastic is generally different from the familiar small, spherical, gray particles of burning cellulose (wood, paper, and cotton) and often consists of long, black, carbon-heavy strings of large size. Data on toxicity of the smoke/gas from this burning are not yet available. The rate and volume of smoke/gas produced by this type fire usually present a greater than normal potential of life loss, injury, and property damage.
>
> The speed of flame advance and the rate of development of dangerous conditions are greater than in similar spaces involving more traditional uses of furniture based on wood, wool, and cotton. The rate of fire development can create a condition that may tax or overpower traditional fire defenses. Defenses of the past, both passive and active (evacuation, alarm, ventilation, and manual fire control), have not been designed, by either engineer or code, to anticipate this hazard.

This is primarily a furnishings problem and is significant in all types of occupancies.

Heavy fire loads may be found in special locations such as club rooms and restaurants. Wood paneling and imitation wood beams and heavy loads of plastic are common.

Storerooms for office supplies and telephone rooms are other high fire load areas. Not only is the wiring insulation combustible, but a typical telephone room is the supply room for the installer with boxes of equipment usually packed in foamed plastic.

Rubbish

The handling of rubbish is another fire load problem. As rubbish is gathered up, it is concentrated in volume. The condition of the material provides a high rate of heat release. The concentration takes place in halls, elevators and basements or lobbies, all particularly bad locations. Collecting rubbish on elevators tempts the potential vandal and provides the circumstances where otherwise inconsequential fire can pollute an entire building and damage or destroy the elevator system.

Many apartment houses have huge compactor units for rubbish located in the basement. Rubbish is delivered through chutes. It is real "fun" to pour a flaming liquid down the shaft. In one apartment house, the aluminum electrical conduit passed through the compactor room. A "simple" rubbish fire destroyed the electrical system and forced all the tenants out of their homes for an extended period. The rubbish handling system is important. It should be properly planned with the anticipation that there *will* be fires.

A 4- by 15-foot, free-burning rubbish fire, on a first-floor stair lobby of an 18-story apartment hotel in New York, quickly became an inferno that almost cost the life of a fire fighter and spread 185 feet vertically and 100 feet horizontally in less than 10 minutes. The "rubbish" was found to contain cans which probably had held flammable liquids.

Rubbish is usually accumulated in the service lobby on each floor. Fire departments should be wary of using the service elevator routinely to approach all fires.

Partial Occupancy and Alteration

A serious hazard is created when the building is altered or rehabilitated while being occupied. In a university hospital being altered, an untreated plywood wall was erected to separate the construction area from the maternity ward and nursery. The construction area had the typical fire hazards of any construction job. There are no real standards for construction work in occupied buildings. The fire department should be alert for any preliminary signs that this hazardous condition is about to take place. Newspapers, including the real estate pages, should be read with this possibility in mind. All clues should be forwarded through channels to the proper authority. A pre-construction conference may eliminate many of the problems and all concerned will be on notice that the fire department intends to give the operation very close attention.

Hotels are "done over" periodically. This usually involves replacing all the furniture, particularly beds. The new furniture is stored wherever space is available, including the basement. Sometimes a floor is placed out of service and used for furniture storage. Mattresses removed from rooms may be stored in halls, awaiting disposal. Two serious fires occurred in Maryland suburban motels from such mattresses, inviting targets to the hotel arsonist.

Watch for the sign "pardon our dust, we're remodeling." It may be the clue to a potential disaster.

One of the most costly fires in the nation's history, the General Motors Transmission plant at Livonia, MI, in 1954, was caused by alterations being performed in an operating plant.

It is a most dangerous situation to have a building partially occupied and partially being finished, but this will be the case because of economic pressure. Fire protection systems are not complete; doors may not yet be installed on stairways and elevators. Temporary heating using LPG may be used in some areas and all the hazards of a building under construction exist. The strictest special precautions should be demanded if the building department permits partial occupancy.

The author examined one high-rise building where the fire protection was the pride of the management. We looked at a floor under construction. There were a number of wood construction shanties erected on the floor and a large accumulation of combustibles; the doors were not on the elevator shaft (they go in last to avoid damage). There were many possible fire causes, and this floor was not yet sprinklered. The piping was in, but plugged off. The heads go in after the ceiling. There were occupied floors above this floor. The sprinklers should go in first in such a situation.

A similar situation happens in shopping malls. The promoter wants to minimize his investment. A sprinkler main is installed. Each tenant installs his own valve and sprinkler system. As a result, the mall is occupied, and a most hazardous area, a store being installed, is unsprinklered until just before it is opened.

Automatic Sprinklers

When the World Trade Center was being designed in New York City, staff members of the Port Authority of New York and New Jersey were very confident that fire problems would be minimal since they would have a well-trained building staff. The Center has had several serious fires. After a third-alarm fire which extended to several floors, a number of changes were made. Sprinklers were installed in high fire load areas and lobby areas with additional capacity to cover future high fire load developments. A number of other improvements, which should have been installed when the gargantuan structures were built, were retrofitted.

After the Triangle Shirtwaist fire, New York law required sprinklers in factories over six stories. This led to a mass migration of the garment industry from the area east and south of Greenwich Village to new high-rise, fully-sprinklered loft buildings in midtown. The record of sprinkler protection of these buildings is impressive.

Sprinkler systems in such high-rise buildings are fed from a gravity tank on the roof. A smaller pressure tank is necessary to develop adequate pressure on upper floors. In most cases, only the pressure tank water is expended on the fires.

Sprinkler Experience in High-Rise Buildings[23] by Robert Powers, superintendent (retired) of the New York Board of Fire Underwriters, is a most useful document and certainly gives the lie to those who allege that "sprinklers are untried in high-rise buildings." The New York Board of Fire Underwriters supports the New York Fire Patrol, the last of what once were numerous "Salvage Corps" maintained by insurance companies. The Fire Patrol is in a position to accumulate extremely valid statistics of sprinkler effectiveness. Generally, the performance of automatic sprinklers is probably better than recorded since many small one-sprinkler fires probably fall within insurance deductibles and are not reported.

The report shows that in 254 fires in office buildings, the performance was "satisfactory" in 98.8 percent of the cases. Of 1,371 fires in non-office buildings, 98.4 percent were satisfactory. In high-rise buildings (other than office buildings), 1,394 fires were recorded in a 10-year period. Only 23 fires were not controlled by sprinklers. In 21 of the 23 cases not controlled by sprinklers, closed valves caused the loss.

Powers also mentions that the overwhelming number of fires occurred in "rubbish," again pointing up the seriousness of this problem, which is no doubt "beneath the notice" of an architect designing his own monument.

Powers also notes the fact that water supply reliability is achieved by taking the domestic supply from part of the sprinkler supply. When the domestic supply fails, corrective action is immediate.

In short, the consideration of available fire loads in high-rise buildings, the fire department reaction time (realistically as much as 20 minutes minimum), and the numbers of people exposed to toxic gases as packed within the contents, needing only a match to be released, forces the inescapable conclusion that full automatic sprinkler protection is vital to the safety of occupants of high-rise structures.

Some of the worst buildings are in areas where the fire department, at best, is of modest strength. The 920 Park Avenue fire in New York was essentially confined to the 7,500-square-foot compartmented area specified in the code as an alternate to sprinklers.[24] Yet, the task took more than 500 fire fighters of whom over 100 were casualties. It took more than 300 fire fighters to suppress the Occidental Building fire in Los Angeles. How many fire departments can even approach these numbers?

What gives the developer of a high-rise the right to erect a structure which places such a burden on the public purse? The argument against sprinklers is usually an economic one, buttressed at times with arguments from some construction materials suppliers and devotees of managing fires by "controlling smoke."

The developer wishes to place thousands of people on a limited plot of ground for his profit. They refuse to stand on one another's shoulders, so he must supply floors. As soon as you get above the first floor, *everything costs more*. To be blunt, it costs the builder a fortune to provide facilities for a person to answer a call of nature on the uppermost floors. Why doesn't the builder require all patrons to use the facilities on the lower floor? It costs more to move furniture onto the top floor, install carpeting, send a maintenance man to handle a complaint, etc.

The basic argument for sprinkler protection is one of equity. The builder is creating the problem for his own profit. It is up to him to provide the solution.

This parallels what is done in the case of parking facilities, sewage facilities, and other amenities.

The fire chief is often all alone in an argument before the city council. Enormous social, political, and financial pressures may be placed on the councilmen.* The fire chief is arguing the case on behalf of people who probably do not even know they are going to be in the high-rise, and who in many cases, would be oblivious to the hazard. It is hard to tell what argument will work. In today's management climate, "options" is a buzz word. One option might be to have the promoter purchase $250,000 indemnity (not liability) insurance for each fatality up to some enormous amount. If the probability of a serious fire is as low as he argues, the insurance should be low in cost. The contract might also be sweetened by a hold harmless clause freeing the city of liability. Another option might be to request an appropriation for a large number of body bags and a standby contract for refrigerator trucks to store bodies. The chief must pick his targets; one might respond to one stimulus, another to another stimulus, and another might be a hopeless case.

If fire detection or smoke control is being offered as an alternate to sprinklers, lay out a realistic scenario of time and events, and request from the council enough funds to hire the necessary additional personnel to control the fire manually at the same time the "alternatives" are being discussed. Fire detection is not fire protection. It is simply another way of crying, "Fire!" The question is: What happens after the fire is detected? Compare the time sequence in Bureau of Standards publication *Fire Development in Residential Basement Rooms*[25] with a realistic response time by the fire department. This report is discussed in more detail in Chapter 9, Flame Spread.

When the improvement of existing buildings is being discussed, "guard house lawyers" and many licensed attorneys immediately dismiss any effort to change the fire resistance rating of an existing building as being unconstitutional. The United States Fire Administration asked Professor Vincent Brannigan, J.D.,** to research the subject. He found no appellate court case which accepted a constitutional barrier to the enforcement of a new code to an existing building.[26]

It is well worth repeating that when sprinklers are installed for life safety, they must be in service when the building is occupied. If evacuation of the building is not feasible, then special precautions should be taken during the out-of-service period. Many other code provisions are waived when sprinklers are installed. One sprinkler company has published an attractive publication showing the code benefits to be obtained from installing sprinklers. If all the eggs are placed in the sprinkler basket, the basket requires close watching.

---

*The Washington, DC area subway is a combined effort of the District and the Maryland and Virginia suburbs. When fire officers were protesting the use of huge quantities of plastics and other defects, one board member suggested to another that, "We each go back to our jurisdiction and teach these fire people who they work for." The manager of the project was a former major general. One fire marshal was rebuked by his superior, "You don't talk to a major general like that." "He's not a major general in *my* army," was the apt reply.

** The author's son. The complete article should be made available to the attorneys who represent the fire department.

Fire Suppression

There are good texts available on the problems of fighting high-rise fires.[27,28] Here only some observations will be presented rather than an attempt at a complete dissertation.*

Each fire resistive building is different. The fire department should make a systematic evaluation of each building, evaluating the fire origin potential and the potential for the spread of smoke and fire. The cooperation of the building management should be actively sought, but lack of cooperation is no reason not to make the evaluation.

All the information the fire department gathers that would be useful at the fire scene should be reduced to a single record and retrieval system which should be immediately available to the fire ground commander at the scene of the fire. In simple situations, plastic protected pages in a loose-leaf book may be adequate. Where there are a number of buildings, one of the many information storage and retrieval systems available should be used. It is possible that the storage function would be maintained at a central computer, and at the scene there would be units with display and hard copy printout capability.

There are innumerable items of information which would be necessary for the fire ground commander to have at the scene. The following text is not provided as a check list but is merely suggestive of some information relative to the building which might be included in the information system.

At first glance, the objection might be raised: "Some of these items are fire prevention duties, why is the fire ground commander concerned?" It is an error to assign responsibility to anybody but the combat arm responsible for any system which is supposed to function in any way to assist in suppressing a fire. It is a tool of the fire ground commander just as much as hose or pumpers or aerial platforms. An automatic vent system may be the equivalent of another truck company. An exterior water spray system may make an engine company available for other duty. The building fire pumps are as important as the apparatus pumps.

Even though a sophisticated retrieval system is not available, the information should be gathered systematically. The day is past when we can depend on information known only to "Captain Joe, the old-timer," who won't be around when he's needed. The information must be institutionalized. The building may be there for a century. It may outlive several generations of the fire department personnel.

---

*Note that many fires in high-rise buildings occur at the lower levels where ordinary tactics will suffice (except for the problem of smoke pollution of upper floors). One Sunday night, the entire second floor of a New York high-rise was ablaze. A well-placed deck pipe knocked the fire out before inside units could force the door.

Building Inventory Items

Structural

Frame

> ***Cast iron*** What is the estimated value of "fireproofing'? Are there connections from column voids to floor voids?
>
> ***Steel*** What type of fireproofing? Are columns unprotected in plenum space?
>
> ***Concrete*** Are there visible rods and cracks? Any fire experience in *this building?*
>
> ***Reinforced masonry*** Are corridor walls reinforced? (Breaching would be difficult.)

Floors

> ***Bar-joist floor and ceiling assemblies*** What is the quality of maintenance? Early collapse is possible. Floor above may be hot.
>
> ***Concrete*** Are they of solid reinforced construction (good insulator and heat sink) or composite decks (Q-floors or cored slabs — poor insulator)? Floor above may be hot. Posttensioned: Don't drill or cut.
>
> ***Tile arches*** Dangerous to cut — entire panel may collapse.
>
> ***Overloaded floors*** Potential for multistory collapse.

Containment

Floors

> ***Floor and ceiling assemblies*** Ceiling tile failure may permit at least partial collapse and open fire and smoke passage.
>
> ***Unprotected columns in plenum*** Smoke and fire may pass to voids above, via reentrant space.
>
> ***Firestopping*** Is there effective firestopping in void spaces?

Masonry floors   Utility openings may pass fire.

Accommodation or access stairways.   These create huge fire areas.

Perimeter Integrity

Joints   Are masonry wall to masonry floor joints solid (good inherent firestops)?

Panel walls   How are exterior panel walls joined to floor? What is their value as firestops?

Glass walls   Is vertical extension of fire probable?

Exterior architectural panels   Voids between panels and walls may negate firestops.

Contents   High rate-of-heat release furnishings near windows or combustible tile ceilings increase potential for exterior extension.

Lower levels   What is the potential for exterior hose streams to block extension of fire?

Open floor areas   What is the potential for floor flooding ("improvised sprinkler system")?

Horizontal Containment

Compartmentation

Integrity of compartmentation   Are there utility openings or under-floor openings such as for computer cables? Do barriers extend to underside of the floor above? What about integrity of fire doors?

Gypsum board compartmentation   Consider penetration of relatively lightweight gypsum partitions as a substitute for forcible door entry.

Stairways   Are there deficiencies of stair enclosures, particularly on lower floors? Are there proper operating door closures on all floors? Are stairways locked to deny entry to floors? Do all units carry forcible entry tools?

Elevators Gypsum board enclosures may be destroyed in fire.

Radio Interference

The steel frame or steel in the concrete frame may interfere greatly with radio communications. Systematic tests should be conducted to determine poor radio communications areas and alternative locations designated. The New York City Fire Department has developed equipment to cope with this problem and may be contacted for updated information.

References

[1]Gerard, John C., "An Incident Command System for High-Rise Fires," *International Fire Chief,* Vol. 47, No. 1, January 1981, p. 17, published for the IAFC, Washington, DC. A picture of the command staff required to control a high-rise fire manually. A help in demonstrating why building codes in smaller cities must be even more stringent than in larger cities.

[2]Sharry, John A., "Group Fire, Indianapolis, Indiana," *Fire Journal*, Vol. 68, No. 4, July 1974, p. 13.

[3]"Dallas Fire Kills Two Fire Fighters," *Fire Journal*, Vol. 71, No. 4., July 1977, p. 108. Fire which started in an apartment flashed along glued-up combustible tile in a corridor.

[4]Bell, James R., "137 Injured in New York High-Rise Building Fire," *Fire Journal*, Vol. 75, No. 2, March 1981, p. 38.

[5]Lathrop, James K., "Atrium Fire Proves Difficult to Ventilate," *Fire Journal*, Vol. 73, No. 1, January 1979, p. 30. A fire in a small office did $300,000 smoke damage because smoke removal system failed to operate properly.

[6]Fisher, Charles A., "Lessons in High-Rise Rescue," *Fire Service Today*, Vol. 49, No. 3, March 1982, p. 24. Lessons from a fire in which eight persons died. Recommended for departments which have never had a high-rise fire.

[7]Kravontka, Stanley, J., P.E., "Elevator Use During Fires in Megastructures," *Technology Report 76-1*, Society of Fire Protection Engineers.

[8]O'Hagan, John, *High-Rise Fire and Life Safety*, Dun Donnelly Publishing Co., New York, 1977.

[9]"'Small' Fire in High-Rise Building," *Fire Command!* Vol. 41, No. 7, July 1974, p. 16.

[10]Watrous, Laurence D., "High-Rise Fire in New Orleans," *Fire Journal*, Vol. 67, No. 3, May 1973, p. 6. Fire in a 15th floor wood-panelled room blew out windows and corridor doors. Five women jumped to death; two fire fighters died in a rescue attempt. Inadequate hose streams contributed to problem.

[11]Sharry, John A., "An Atrium Fire," *Fire Journal*, Vol. 67, No. 6, November 1973, p. 39. A fire in a second-floor nightclub completely charged the atrium of the Hyatt Regency O'Hare Hotel. Automatic smoke removal did not work because the detectors were disconnected. Elevators could not be controlled.

[12]Breen, David, P.E., "The MGM Fire and the Spread of Flames," *Bulletin* of the Society of Fire Protection Engineers, January 1981.

[13]"High-Rise Office Building, Los Angeles, CA," *Fire Journal*, Vol. 69, No. 6, November 1975, p. 87. Flammable paint thinner was being used while building was occupied. A light switch ignited vapors. Air-conditioning system spread smoke through two connected floors.

[14]"Report of the MGM Grand Hotel Fire," Clark County Fire Department, Las Vegas, NV, 1981.

[15]"Fire at the MGM Grand," *Fire Journal*, Vol. 76, No. 1, January 1982, p. 19 and "Human Behavior in the MGM Grand Hotel Fire," *Fire Journal*, Vol. 76, No. 2, March 1982, p. 37.

[16]Lathrop, James K., "World Trade Center Fire," *Fire Journal*, Vol. 69, No. 4, July 1975, p. 19. Fire spread to several floors through unprotected floor openings connecting telephone cabinets. Smoke removal system did not operate because HVAC system was shut down for the night; functioned well after manual turn-on.

[17]See Ref. 16.

[18]Boyd, Howard, "Smoke, Atrium, and Stairways," *Fire Journal*, Vol. 68, No. 1, January 1974, p. 9. Smoke control in completely sprinklered Hyatt Regency Hotel with the characteristic atrium.

[19]See Reference 16.

[20]"Pressurization Failed to Keep Smoke from Tower Staircases," *Fire*, April 1981.

[21]Wolfe, Tom, *From Bauhaus to Our House*, Farrar Strauss Giroux, New York, 1977.

[22]"Investigation Report on the Las Vegas Hilton Hotel Fire," *Fire Journal*, Vol. 76, No. 1, January 1982, p. 52. Fed by carpeting on the walls and ceilings of lobbies, fire spread up the outside of the hotel from the 8th to the 30th floor. Eight died.

[23]Powers, Robert, P.E., "Sprinkler Experience in High-Rise Buildings," *Technology Report 79-1*, Society of Fire Protection Engineers, 1979.

[24]Mulrine, Joseph, "Compartmentation vs Sprinklers," *Fire Engineering*, Vol. 133, No. 12, December 1980.

[25]Fang, J. B. and Breese, J. N. *Fire Development in Residential Basement Rooms*, NBSIR 80 2120, Center for Fire Research, National Bureau of Standards, Washington, DC.

[26]Brannigan, Vincent, J.D., "Record of Appellate Courts on Retrospective Fire Codes," *Fire Journal*, Vol. 75, No. 6, November 1981, p. 62.

[27]See Ref. 8.

[28]Mendes, Robert F., *Fighting High-Rise Fires*, National Fire Protection Association, Boston, MA, 1975.

Additional Readings

Best, Richard, "High-Rise Apartment Fire in Chicago Leaves One Dead," *Fire Journal*, Vol. 69, No. 5, September 1975, p. 38. Chicago Fire Department uses master streams on high fire load of furniture fire in 17th-floor apartment. Fire confined to unit of origin by self-closing door despite unsprinklered tenant storage area.

Birr, Timothy, "Landmark Hotel Closed for Fire Code Violations," *Fire House*, June 1981. The Eugene, OR Fire Department is successful in its battle to close a hotel which failed to comply with retrospective fire code provisions requiring sprinklers.

Lathrop, James K., "Two Die in High-Rise Senior Citizens' Home," *Fire Journal*, Vol. 69, No. 5, September 1975, p. 60. Intense fire in Albany, NY. Gypsum board over polystyrene disintegrated. One-quarter-inch plywood concealing a drain pipe provided a "pinhole" in the fire resistive containment, allowing extension to the next apartment.

Lathrop, James K., "Building's Design, 300 Fire Fighters Save Los Angeles High-Rise Office Building," *Fire Journal*, Vol. 71, No. 5, September 1977, p. 34. Building had perimeter protection of wet masonry. Heat detectors detected the fire just before it broke out the windows and extended to the floor above. A guard went to the alarm floor to investigate. Building personnel should be instructed in the hazard of going to the fire floor.

Lathrop, James K., "Nineteenth-Floor Dormitory Fire Kills One Student," *Fire Journal*, Vol. 70, No. 2, March 1976, p. 77.

Paul, G. H., Clougherty, E.W., and Lathrop, J. K., "Federal Reserve Bank," *Fire Journal*, Vol. 71, No. 4, July 1977, p. 33. An excellent source of information on what can go wrong when foamed plastics are being installed. Exterior panels are aluminum with foamed insulation. There is no statement of any test to validate the system. The Boston Fire Department required a number of corrective actions.

Sharry, John A., "Gas Explosion, New York, New York," *Fire Journal*, Vol. 68, No. 6, November 1974, p. 28. The article tells of a gas explosion which ripped the curtain wall off a 25-story building.

Sharry, John A., "South America Burning," *Fire Journal*, Vol. 68, No. 4, July 1974, p. 23. Recounts three serious high-rise fires: the Joelma Building, São Paulo, Brazil, in which 179 lives were lost; Avianca Building, Bogota, Columbia; Caixa Economica Building, Rio de Janeiro, Brazil. Exterior extension, open stairways and combustible interior finishes were principal factors.

Sharry, John A., "A High-Rise Hotel Fire, Virginia Beach, Virginia," *Fire Journal*, Vol. 69, No. 1, January 1975, p. 20. A useful article when instructing hotel personnel. Assistant manager died trying to fight the fire. With standpipe out of service, fire department used ladder pipe as auxiliary standpipe.

Willey, A. Elwood, "Baptist Towers Housing for the Elderly," *Fire Journal*, Vol. 67, No. 5, March 1973. In Atlanta, GA, ten died in an 11-story fire resistive home for the elderly. The door to the fire apartment was left open (no self-closer). The windows were out. High flame spread carpeting extended fire into the corridor.

Bimonthly Fire Record, *Fire Journal*, Vol. 67, No. 3, May 1973, p. 59. A Christmas tree fire in a Dallas high-rise did $340,000 in damage. Pipe chases were enclosed only to the ceiling and voids above were pierced for air conditioning. The building lacked integrity. A force of 85 fire fighters prevented a more serious outcome.

"Arson in Seattle Hotel," *Fire Journal*, Vol. 69, No. 6, November 1975, p. 5. Fire started on an open stairway mezzanine and smoke polluted the entire stairway.

CUT-AWAY DRAWING
SHOWING SECOND FLOOR
SUPPORT SYSTEM.

COLUMN

DUCT OPENING

DUCT OPENING

VANCE
'73

E   D-NOT SHOWN   C   B-NOT SHOWN   A

APPENDIX A

The Hotel Vendome Fire

The following account of the Hotel Vendome fire is presented with permission from "Without Warning — A Report on the Hotel Vendome Fire, June 17, 1972," written by District Fire Chief John P. Vahey and published by The Boston Sparks Association.

On June 17, 1972, Boston's 100-year-old Hotel Vendome succumbed to fire and a collapse that cost the lives of nine fire fighters and left another nine injured. The cause of the fire which occurred during extensive renovations is not known, but the cause of the collapse is a different matter. A board of inquiry determined that it came from a penetration through one of the interior bearing walls.

The fact that the penetration existed is an example of the accidental, or in this case, probably deliberate, effort to circumvent building codes and requirements to save money.

During the time the building was under renovation, the scope of the work changed at least three times. Permits were sought for some of these changes. One application stated, "No structural changes," although the plans called for both vertical and horizontal penetrations for duct work. There was no certificate signed by a registered architect or engineer stating that the work would not involve structural members, although it would.

About this time, the owner suspended the services of the architect and his structural engineer for budgetary reasons. A building inspector found that the work being done exceeded the scope of the permits and issued a notice of violation. Though court action was recommended, there is no indication that it was ever taken.

Although the contractors were required to file daily reports, "Among the reports missing . . . are those for . . . August 6, 1971, to September 24, 1971. This is the same period . . . the architect and structural engineer were off the job; and it is the same period in which it is believed the duct openings were made in the basement walls."

Sometime in the early history of the hotel, two first floor walls were removed to provide a larger open area. The loads were picked up by wrought iron beams as shown in the accompanying sketch. The interior masonry-bearing wall was supported on two 15-inch beams supported in the center by a seven-inch cast iron column. The column was based on a 12-inch square, two-inch thick metal plate centered above the basement bearing wall. The bearing stress on the wall was about 200,000 pounds, "about seven to eight times the allowable stress for a masonry wall of good grade brick with a good lime cement mortar."

It was under this column that a 12- by 12-inch penetration was made for an air conditioning duct.

"There is nothing to indicate that the architect or structural engineer offered any comment to the owner or the contractor relative to this part of the work during their absence from the job.

"This is not consistent with the complete reliance placed upon the architects and engineers by the Boston Building Department; and in this case, it left the owner and builder free to do things their own way.

"The difficulty illustrated here is that it appears to be possible for an owner to hire well established architectural and engineering firms, only to dismiss them or limit their roles after the building permits have been issued, while the Building Department goes on believing that the public interest is being protected by these professionals."

On the fateful day, two workmen on the first floor noticed smoke drifting down from upper floors. One went to the fourth floor where he noticed smoke and noise overhead. He called to the other to ring in an alarm.

That alarm was sounded at 2:35 p.m. The next hour and a half saw 20 engines, six ladders, two aerial towers, and one rescue vehicle deployed at the fire. At 5:28, with overhaul going on and some companies preparing to leave, the building collapsed with tragic results.

The board of inquiry found, "The loss of support at the base of the circular cast-iron column was sufficient to trigger the collapse of the entire building starting at the second floor and proceeding upward. It is apparent that the cutting of the opening in the twelve-inch bearing wall directly below the base of the column weakened the wall to an extent that any additional weight put on the upper floors, such as fire fighters and their equipment moving about was enough to initiate the collapse. While there was some water on the floors, it appears that it was draining out quickly through construction holes cut by fire fighters and added very little weight to the floors."

The cause of the fire remains unknown.

The Cocoa Beach Collapse

The following account of the Cocoa Beach collapse is presented with permission from *Fire Engineering.*

Although no specific cause has yet been defined, the collapse of the Harbor Cay Condominium in Cocoa Beach, FL, on March 27, 1981, typifies the possible disaster that can occur during building construction and the fire department's need for a predetermined plan.

The building was being constructed of precast reinforced concrete slabs stacked on concrete columns. The roof concrete had just been poured when the structure collapsed into twisted metal and broken concrete. All five floors and the roof had pancaked and still-fluid concrete covered the wreckage.

The rescue operation was difficult. Approximately 60 men were working on the job, but there was no quick way of determining how many were buried. At times, sound-sensing devices brought in from Patrick Air Force Base were used to locate the sounds of life. After 60 hours of rescue operations, the final toll was 11 dead and 23 injured.

When Fire Chief Bob Walker arrived, he recognized the need for an organized rescue effort. A triage station, manned by department EMTs and county paramedics was set up. The whole area was roped off and traffic patterns established to expedite movement of ambulance and rescue equipment. City and county disaster plans went into effect.

An inch-by-inch search plan was initiated. Obvious fatalities were marked and left in place. Victims who could be seen or heard were removed and treated. Heavy equipment was called in; crane riggers, torch men, cutters, and others helped to remove debris until it was determined that all casualties were found.

The county's civil defense communications van provided the multi-channel capabilities needed for coordination with police, Air Force, and Coast Guard units controlling air, sea, and highway traffic.

Rescue workers were put on four-hour shifts and hard hats and gloves were required equipment. Within three hours of the collapse, all of the survivors had been removed and triaged. Five fatalities had also been removed and it was known that at least five more were in the rubble. Because it was impossible to get an accurate list of all who had been working, it was decided to continue the search until, literally, every stone had been turned.

The collapse occurred at 2:45 p.m. on Friday. The search and rescue operation continued until 2:00 a.m. on Monday to make sure that all casualties had been found. Five county fire districts and 16 municipal fire departments had assisted in the rescue.

Chief Walker is preparing for the next disaster. "I feel our resource list was not thorough enough," he says. "Fire officials need to realize that they will have to reach out beyond the resources of fire departments when a disaster occurs."

ABOUT THE NFPA

The National Fire Protection Association (NFPA) is a nonprofit organization committed to making both the home and the workplace more firesafe. Its 32,000 members cooperate to promote scientific research into the development and updating of firesafety standards, and to support educational programs which heighten firesafety awareness. In its role as a publisher, it produces informative and practical publications by recognized authorities and leaders in the fire protection community on matters of vital interest to all who are concerned with preservation of life and property from fire. This book, Building Construction for the Fire Service, Second Edition, *by Francis L. Brannigan, is presented as part of that publishing mission.*

The Association has no enforcement power, and governmental or institutional compliance with its standards is entirely voluntary. Nevertheless, the Association and its principles are internationally acclaimed, and its standards have for many years been implemented at all levels of government. NFPA's National Electrical Code® *for example, is the most widely accepted safety code in existence. The Association's* Life Safety Code®, *adopted for use by some two-thirds of the states, ensures that the construction of high-rise buildings, nursing homes, hosptials, theaters, and other large facilities is governed by life safety principles.*

Today, millions of Americans and their property are protected by the NFPA: an organization of people and principles.

SUBJECT INDEX